ENGINEERING
MECHANICS

An Introduction to Statics, Dynamics
and Strength of Materials

ENGINEERING
MECHANICS

An Introduction to Statics, Dynamics and Strength of Materials

VAL IVANOFF

This text is designed to meet the requirements of the following modules from the TAFE Engineering Technician and Engineering Associate Curriculum:

STATICS (EA859)
INTRODUCTORY DYNAMICS (EA772)
INTRODUCTORY STRENGTH OF MATERIALS (EA804)

The McGraw-Hill Companies, Inc.

Beijing Bogotà Boston Burr Ridge IL Caracas
Dubuque IA Lisbon London Madison WI
Madrid Mexico City Milan Montreal New Delhi
New York San Francisco Santiago Seoul
Singapore St Louis Sydney Taipei Toronto

WCB/McGraw-Hill

A Division of The McGraw-Hill Companies

Reprinted 1997, 1999, 2000, 2002, 2003, 2004, 2006 (twice), 2007, 2008
Text © 1996 by McGraw-Hill Book Company Australia Pty Ltd.
Illustrations and design © 1996 McGraw-Hill Book Company Australia Pty Limited
Additional owners of copyright are named in on-page credits

National Library of Australia Cataloguing-in-Publication data:

Ivanoff, Val.

Engineering mechanics.

ISBN 13: 978 0 07 101003 0

1. Mechanical engineering. I. Ivanoff, Val. Mechanical engineering science. II. Title. III. Title: Mechanical engineering science.

621

Published in Australia by
McGraw-Hill Australia Pty Limited
Level 2, 82 Waterloo Road, North Ryde, NSW 2113

Acquisitions Editor: Nicola Cowdroy
Production Editors: Karin Riederer and Jill Read
Designer: Kim Webber

Typeset in Australia by Monoset Typesetters, Brisbane
Printed in Australia by Griffin Press

My dear young friends:

It is not enough that you should understand about applied science in order that your work may increase man's blessings. Concern for man himself and his fate must always form the chief interest of all technical endeavours, concern for the great unsolved problems of the organisation of labour and the distribution of goods—in order that the creations of our mind shall be a blessing and not a curse to mankind.

Never forget this in the midst of your diagrams and equations.

Albert Einstein

From an address before students of the California Institute of Technology, 1938

Contents

Preface

This book is an introductory text in applied engineering mechanics for technical college students. It covers topics in statics, dynamics and mechanics of solid materials, with an emphasis on fundamental mathematical and graphical analysis of practical engineering problems, at a level suitable for the engineering technician courses.

It is also intended that this book serve as a source of basic revision and reference material for engineering technicians at work and when undertaking further studies. In order to satisfy this aim, the book contains a summary of the metric system of units, a list of standard formulae and symbols, some tables of useful data, and a glossary of technical terms cross-referenced to the main text.

My first book published in 1984 under the title *Mechanical Engineering Science* had an excellent reception, particularly in New South Wales. Since then, the development of the new National TAFE Engineering Technician and Engineering Associate Curriculum provided a strong impetus for a new book, focused more narrowly on topics in the field of applied engineering mechanics, and more closely aligned with the structure of the new Australia-wide modular courses.

The new book is largely based on my considerable teaching experience, the insights gained from using the original book as a prescribed textbook, and many years of involvement with the development of engineering curriculums in technical education. I am confident that the book will serve well as a basic text for the three new modules of the national technican-level curriculum in Statics, Dynamics and Strength of Materials.

The main objective I had in mind during the preparation of this text was to present the fundamental principles of applied engineering mechanics in a clear and unpretentious manner, intelligible to the beginner. The many worked examples in the text were selected from as broad a range of practical engineering applications as possible. While some degree of oversimplification was inevitable, care was exercised to keep all data and calculated answers within realistic limits. I make no particular claim to excellence or originality, and if practical usefulness and simplicity of approach prove to be the most redeeming features of this book, my aims will be quite satisfied.

My special thanks are due to my wife Nonna, who prompted me to proceed with the work and was always extremely patient, helpful and supportive. I was also encouraged by my sons Victor, Serge and Herman, who never seemed to lose their faith in my abilities.

But ultimately, it was my love of teaching, combined with the needs and attitudes of my college students, past and present, which provided the necessary incentive for me to embark on this project and to complete it.

Finally, it would be surprising if there were no errors in this book. I hope that they are few and relatively insignificant in nature. Should any errors, omissions or misprints be found, I would be grateful if they were brought to my attention.

V. Ivanoff

THE FOUNDATIONS

PART 1   **An introduction**

AN INTRODUCTION

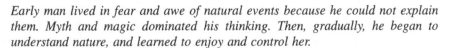

Early man lived in fear and awe of natural events because he could not explain them. Myth and magic dominated his thinking. Then, gradually, he began to understand nature, and learned to enjoy and control her.

Henry Margenau

CHAPTER 1

Mechanical engineering science

It is difficult to imagine our world without all the fascinating machines which over the centuries helped to shape and transform society into what it is today through the development of modern industry and technology. Machines are mechanical devices that augment or replace human effort for the accomplishment of physical tasks. The number and complexity of machines are increasing all the time, and each new machine is a combination of the knowledge of the scientist, the ingenuity of its designer and the skill of its builder.

This is a book about a branch of engineering science called *applied engineering mechanics*. It is written for the technical college student, and its primary objective is to provide a study guide that will help the student learn the fundamental concepts, units and laws of engineering mechanics, in a context of applied examples, with only basic mathematical skills. In addition, it is the author's wish that the book will encourage students to become interested in the history of people whose ideas created the technology that permeates and affects so many aspects of our lives.

1.1 *ENGINEERING*

Engineering is the art and science of applying the knowledge of the physical world to the conversion of the resources of nature, in order to improve or control our environment.

It is said that in the ancient world there were only two kinds of engineering: military and civilian. The former was concerned with the building of such engines of war as assault towers, catapults and floating bridges, while the latter with roadworks, water supply and sewerage systems.

Imhotep, who built the famous Egyptian pyramid near Memphis in about 2500 BC, was the first engineer known to history by name. The construction methods he used must have combined the best of the practical engineering skills of his time. His successors in Greece, Rome, medieval Europe and other regions of the world developed many sophisticated techniques of metallurgy, construction and hydraulics, which helped to establish and maintain advanced civilisations, responsible for a variety of creative forms of art, literature and government.

The methods used in those days were based principally on trial and error, supported by the engineer's knowledge of arithmetic, geometry and draughtsmanship, and of the materials used. A reflection of this knowledge survived not only in large ancient structures, such as the Roman aqueducts or Persian road systems, some of which endure to this day, but also in many ancient manuscripts, intended by their writers to preserve the accumulated knowledge for the education of future generations of engineers. One such work was written by a Roman engineer and architect, Marcus Vitruvius, in the 1st century AD, and contained ten volumes covering city planning, building materials, construction methods, mensuration, clocks, military machines and hydraulics.

The gradual growth of specialised engineering knowledge and the formulation of scientific theories relevant to engineering practice necessitated the establishment of systematic engineering education. The first of the engineering schools, the National School of Bridges and Highways, was founded in France in 1747, at about the same time as the term *civil engineer*, separate from *military engineer*, came into general use. The British Institution of Civil Engineers, incorporated by royal charter in 1828, became the world's first engineering society.

The birth of the second branch of modern engineering, mechanical engineering, was largely the result of the Industrial Revolution in England and Scotland in the 18th century, with the invention of the steam engine and the development of machines for the textile industry, coalmining and later for transportation. The formal recognition of mechanical engineering, dealing with machinery of all types, came with the founding in 1847 of the Institution of Mechanical Engineers in England.

Modern developments in the knowledge of electricity, chemistry, electronics and nuclear physics gave the impetus for further growth and branching of engineering into separate specialised fields.

1.2 MECHANICAL ENGINEERING

Mechanical engineering is a branch of engineering concerned with the design, construction and use of machines for the production of goods and power, and for environmental control. The areas of mechanical engineering comprise the development of machine tools, materials-handling equipment, furnace and boiler technology, automotive and power-generating plant, pneumatic and hydraulic systems, refrigeration and air-conditioning.

Mechanical engineers are particularly concerned with harnessing natural sources of **energy**, such as the chemical energy of fuels, and its conversion into heat and work. In doing so they make use of **forces** and **motion**. They employ two kinds of materials: solids and fluids. Solid materials form stationary and moving parts of mechanisms used to transmit or change forces and motion. Fluids, i.e. liquids and gases, are used as working agents to transform force into pressure and to facilitate conversion of energy from one form to another.

The functions of the mechanical engineer include research, development, design, construction, production, operation and management of mechanical plant and associated services, all of which require some degree of knowledge and understanding of the physical world and the ability to adapt this knowledge to practical problem solving. The types of occupational requirements in industry vary greatly in the degree of emphasis placed on the applications of science. Between the research and development engineer, who is largely involved in the application of scientific methods in search of new ideas, and the tradesperson who is using manual skills in the construction, operation and maintenance of machines, there are engineers, draughtspersons and technicians, whose work brings the knowledge of some aspects of engineering science to bear on practical problems.

1.3 MECHANICAL ENGINEERING SCIENCE

Mechanical engineering science is based principally on mathematics and physics and their extension into mechanics of solids, fluid mechanics and thermodynamics. Its purpose is to explain and predict physical phenomena and to serve as a basis for engineering analysis and design.

Mechanics of solids, which is the oldest of the physical sciences, is usually subdivided into statics, dynamics, and strength of materials. *Statics* deals with bodies at rest and enables the engineer to predict forces in the solid members of a machine or structure. *Dynamics*, on the other hand, deals with bodies in motion and considers relations between the forces involved and the motion. *Strength of materials* takes into account the behaviour of solid materials under stress.

Fluid mechanics is concerned with liquids and gases at rest or in motion and is important for the design of tanks, pumps, turbines and pipework systems. Fluid mechanics deals with the relations between pressure and the mechanical properties of fluids, such as density and viscosity. It can also be subdivided into *fluid statics*, which is the study of pressure measurement, fluid forces on submerged surfaces and buoyancy, and *fluid dynamics*, which deals with fluid flow and fluid machinery.

Thermodynamics is the science of temperature and heat, and is particularly concerned with the theory of heat engines, i.e. mechanical devices for the conversion of thermal energy into mechanical work.

It should be noted that the division of mechanical engineering science into its component parts is purely a matter of logical convenience and, to some extent, historical tradition. Its applications, particularly in the area of energy systems design, usually involve a combination of factors and principles derived from different parts of engineering science. The automotive engine and the steam power plant are two good examples.

1.4 THE HISTORICAL FOUNDATIONS

The foundations of mechanical engineering science rest firmly on two principles: mathematical reasoning and experimental observation. The former provided the tools and the latter the material from which the great body of specialised, systematised and verified knowledge was developed.

As far as is known, the practical use of arithmetic for the measurement of land areas was first made by the Egyptians, without any mathematical proof that the ideas they used were correct. The true development of mathematical deduction, which is the cornerstone of all science and technology, started with Pythagoras in the 6th century BC. Two centuries later, Euclid's *Elements* was written, in which he introduced such basic concepts as point, line, plane and angle, and related these to physical space. Mathematics had reached its climax in ancient Greece with Archimedes in the 3rd century BC. He did a great amount of work with the sphere and cylinder, and was able to calculate the value of π with remarkable accuracy, sufficient even today for most practical purposes.

In the 17th century AD, the century of genius, wonderful discoveries followed one another in rapid succession. John Napier of Scotland published his discovery of logarithms. René Descartes, the French philosopher, invented analytical geometry, in which a point can be represented by its distances from two perpendicular axes. Sir Isaac Newton and the German mathematician Gottfried Leibniz, working independently of each other, developed the methods of differential and integral calculus, which were destined to become an invaluable tool for engineering science, as well as the mainspring of further developments in higher mathematics, which go beyond the scope of this book.

The names of Archimedes and Newton are also associated with the second fundamental principle of science—that of objective experimental observation. The well-known story—perhaps it is legend—of how Archimedes studied the effects of buoyancy and arrived at what is stilled called Archimedes' principle, while in the bathtub, illustrates his ability to observe, to understand what is observed, and to use the observation to discover new ideas. In the words of Newton:

> The best and safest way of doing scientific work seems to be, first to inquire diligently into the properties of things, and of establishing these properties by experiment, and then to proceed slowly to theories for the explanation of them.

Another early exponent of experimental science was Galileo, who lived in Italy in the 16th century, and is best known for his defence of the Copernican concept of the solar system. He is rightly regarded as the founder of modern experimental scientific method, who had set up the basis for Newton's developments. In an age when equipment had not yet been developed to make very accurate measurements, Galileo conducted experiments to confirm his reasoning on motion, acceleration and gravity, particularly in relation to the problems associated with falling bodies. He also anticipated the idea of inertia, which was later used and refined by Newton in his laws of motion.

Each area of engineering science is based on the work of many great scientists throughout the centuries. Archimedes, Galileo, Hooke and Newton in solid mechanics; Leonardo da Vinci, Torricelli, Pascal and Bernoulli in fluid mechanics; and Boyle, Rumford, Carnot and Joule in thermodynamics, are only some of them.

1.5 *FROM SCIENCE TO APPLICATIONS*

The progress of the machine from its early primitive forms to the modern marvels of the technological age is the result of brilliant inventions followed by the patient work of development, adaptation and improvement. The relation between scientific knowledge and the ability to apply it to practical purposes has varied from person to person, and from one historical period to another. Some excelled in the depth of their theoretical perception, while others succeeded in building and perfecting useful machines. In general, most scientists have been aware of the practical possibilities of their scientific knowledge—some were also engineers. On the other hand, most great inventors have had a good understanding of engineering science, on which their inventions were based.

Fig. 1.1   *Leonardo da Vinci, 1452–1519*

Leonardo da Vinci, born in Italy in 1452, is considered by many historians to have been the greatest genius of all time—a great artist, scientist and the greatest inventor of his age. Throughout his life, Leonardo was an inventive builder, for whom an interest in pure science merged increasingly with an interest in applied mechanics. His model book on the elementary theory of mechanics contains thousands of beautifully illustrated pages describing his observations and outlining inventions of all kinds. He was particularly interested in problems of frictional resistance, and described various combinations of machine elements, such as screw threads, gears and hydraulic jacks, designed to overcome friction and transmit or modify forces. His mechanical inventions are interesting and varied: a machine gun, a military tank, a submarine, an anemometer, a pump, a flying machine. Many mechanisms invented by Leonardo are in use in similar forms today. He anticipated variable-speed drives, roller bearings and screw-cutting machines. As a scientist–inventor, he was so far ahead of his time that many of his inventions, all feasible, had to wait for centuries before they could be realised. The importance of Leonardo's contribution to engineering science lay in his ability to apply the principles of mechanics to the invention of practical machines.

Fig. 1.2   *James Watt, 1736–1819*

Another name which marks a turning point in the successful combination of science and engineering is that of James Watt, born in Scotland in 1736. Watt was not the original inventor of the steam engine. His predecessors included Hero of Alexandria in the 1st century AD, Thomas Savery (1698) and Thomas Newcomen (1705). However, Hero's novelty was not much more than a scientific toy, and Savery's and Newcomen's engines were very inefficient. Once while repairing a Newcomen engine, Watt was impressed with its waste of steam and became interested in steam engines. In 1765 he invented the separate condenser for steam engines, and later developed the more efficient double-acting engine, in which the piston both pushed and pulled. He also adapted the steam engine for rotary motion. In 1769, James Watt took out his famous patent for 'A New Invented Method of Lessening the Consumption of Steam and Fuel in Fire Engines'. This event ushered in the age of practical efficient technology and introduced the social change known as the Industrial Revolution.

From the time of James Watt, mechanical engineering has developed in response to the demand for increased efficiency, accuracy and complexity. Production machinery has been developed for highly automated manufacturing processes; thermal efficiency and output of power-generating plants throughout the world has steadily increased, making possible other advances in industrial development; steam and internal-combustion engines completely revolutionised transport; and mechanical refrigeration has been applied to food preservation and to comfort air-conditioning.

There have also been some undesirable side effects of the products of mechanical engineering, such as noise, pollution of air and water, and the rapid depletion of the natural sources of energy, particularly oil.

In the future, the demand for mechanical engineering skills will continue, with an emphasis on conservation and efficient utilisation of scarce material resources and on maintaining a satisfactory environment.

1.6 *MACHINES AND MACHINE COMPONENTS*

This is not a book about machines. It is a book about mechanical forces at work in the world of machines and machine components. In general, it will be assumed that the student has an elementary knowledge of how mechanical things work, or that he or she is learning about machines and machine components concurrently in another subject of the engineering course. However, it would be useful to briefly define and summarise some of the more common mechanical devices before proceeding to the study of fundamental principles of applied engineering mechanics.

Machines

A **machine** can be defined as a mechanical device for overcoming a resistance at one point by the application of a force at some other point. It is usual to call the output resistance force the **load**, and the input force the **effort**. Some basic mechanical devices, known as **simple machines,** have been in use in some form for thousands of years, since the dawn of recorded history. These are:

inclined plane   a plane surface inclined at a small angle to the horizontal, used to lift a heavy mass up the plane by applying an effort along the plane

lever   a bar of rigid material pivoted at a point called the **fulcrum,** used to move a load applied at some part of the bar by means of an effort applied at another part

pulley   a wheel with a grooved rim, or a combination of such wheels mounted in a block, for a cord or chain to run over for changing the direction or magnitude of a force

screw   a cylinder with a spiral groove around its outer surface, called the **thread,** used for conveying motion or bringing pressure to bear

wheel and axle   a cylindrical axle on which a wheel is fastened concentrically, the difference between their respective diameters supplying the leverage

Mechanisms

One of the distinctive characteristics of a machine is that its parts are interconnected and constrained to move only in a particular predetermined way relative to each other. The way in which the parts are interconnected and guided is called the **mechanism** of the machine. The most common **mechanisms** include:

cams   eccentric projections on revolving shafts, shaped so as to give some desired linear motion to a follower, which is usually returned by a spring

friction drives   in which one wheel causes rotation of a second wheel with which it is pressed into contact, by means of friction force at the point of contact

gears   operating in pairs, to transmit rotational motion by means of successively engaging projections called **teeth**. Gearboxes containing multiple pairs, or trains, are often used to obtain speed ratios that cannot be obtained with a single pair of gears

linkages   assemblies of solid members, or links, connected to each other by hinges or by sliding joints, for the transmission of motion in a machine. The piston, connecting rod and crankshaft, for example, constitute the mechanism of a reciprocating pump or engine

wrapping connectors   such as belt, rope or chain drives, used for transmitting rotational motion over some distances

Machine components

In addition to what are usually referred to as mechanisms and simple machines, as described above, there are various machine components that students of engineering will learn about in the course of their studies. The following is a brief summary of the more common machine components:

bearings   connectors that support a rotating shaft relative to the stationary parts of a machine

couplings   devices for connecting the ends of two adjacent rotating shafts, conveying a drive from one to the other

flywheels   heavy wheels attached to rotating shafts for the purpose of storing up energy or moderating fluctuations in the speed of a machine

shafts   cylindrical bars, rotating and transferring rotational motion

springs   elastic members, usually of bent or coiled metal, used for a variety of purposes in many mechanical devices

1.7   *ABOUT THIS BOOK*

Mechanical engineering science is essentially a mathematical science, based on a few fundamental principles and laws. The primary objective of this book is to enable the student to understand these fundamental principles and laws. In order to assist the student to achieve this goal, a number of worked examples have been provided throughout the text. In addition, each chapter contains problems which should be used by the student to reinforce and test understanding of the material covered. Most of the examples and problems are of a practical nature and should be relevant to the student's interests. However, care has been exercised to ensure that engineering jargon has not obscured the basic concepts presented. As far as possible the problems are graded.

Since this book is intended as an introductory text in engineering mechanics, new concepts have been presented in elementary terms and, as far as practicable, one at a time. The emphasis throughout is on learning fundamental principles. For this reason, applications and methods which do not have an immediate and direct relationship to a concept to be learned have been avoided. It has also been assumed that in the majority of cases the student will pursue the study of engineering mechanics beyond the scope of this text. This book can therefore be regarded as a basic introduction to subject matter that is covered in greater depth in books such as *Applied Mechanics and Strength of Materials* by R. Kinsky.

The material presented in this book requires from students a reasonable degree of competence in elementary mathematics. This includes competence in the four basic arithmetical operations: addition, subtraction, multiplication and division, with the aid of an electronic calculator. Students must be competent in manipulating numbers presented in either scientific or engineering exponential notation. The geometry and trigonometry of triangles are also required. Mensuration of areas and volumes of basic geometrical shapes must be understood. Above all, transposition of terms in an equation and the solution of equations must be handled confidently and correctly. An elementary understanding of vectors would be helpful, but not essential. Knowledge of differential and integral calculus is not required.

Graphical methods are used in the book where appropriate for vector addition of forces. They are one of the analytical tools of the engineer, and should not be regarded as inferior to mathematical methods. One should recognise that some engineering problems are better solved mathematically, while others lend themselves better to graphical solutions. This means that a small drawing board, a pair of set squares, a rule, a protractor and a compass are required for solving some of the problems.

In engineering, the possible accuracy of answers is limited by the accuracy of the original data contained in the statement of a problem. Unless stated otherwise, this must be assumed to be known with a degree of accuracy comparable with that of ordinary engineering measurements, which is seldom greater than three or four significant figures. Students must resist the temptation of false accuracy when using electronic calculators with eight or more digits displayed. Usually three or four significant figures are quite sufficient for an answer to an engineering problem. As a general rule, answers given to three significant figures have been preferred in this book. In the case of graphical solutions, two significant figures are adequate.

On the other hand, crude approximations of a number of successive intermediate steps can often lead, through accumulation of errors, to a very inaccurate final answer. In the majority of worked examples in this text, the answers have been calculated with the aid of a pocket calculator and not approximated until the final answer has been obtained. However, the explanations often show intermediate steps with only approximate intermediate answers. Therefore, students may occasionally observe a slight discrepancy if they use these approximate intermediate values to check the final results.

Where standard computed or natural constants such as π (3.141 59. . .) or g (9.806 65) are required, it is usually quite sufficient to simplify their values to three or four significant figures. However, if the appropriate function key is available on the calculator, it may be used without affecting the essential accuracy of the results from the engineering point of view. In this book the value of the gravitational constant is taken to be $g = 9.81$, and calculator function keys are used for all mathematical constants and trigonometric functions without approximation.

Unlike problems in pure mathematics, engineering problems are about structures and machines which always have some practical purposes and limitations. In problem solving, students must strive to appreciate the engineering significance of each problem by drawing on their own experience and by relating engineering science to other subjects in their engineering course. Considerable importance should be attached to a proper understanding of the problem and to the logical, clear and precise recording of its solution. A neat diagram showing only the essential information, such as forces acting on a body, is always very helpful in analysing any engineering problem.

Finally, a word of advice for the student from an old Chinese proverb:

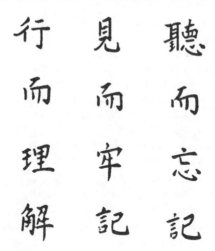

which translated means:

I hear and I forget,
I see and I remember,
I do and I understand.

which translated again means:

Do not just read this book.
Work through the problems and you will learn.

CHAPTER 2

Mathematical tools

Engineering mechanics is a mathematical science which is based on remarkably few fundamental principles and laws, and requires only a basic minimum of observable data. It is about solving applied engineering problems, particularly those that relate directly to engineering design, by means of rigorous analytical techniques.

A collection of mathematical and graphical problem-solving methods and techniques has sometimes been described as the design engineer's 'toolbox'. The basic 'tools' this box contains come from different branches of elementary mathematics, and can be broadly categorised into the following four areas:

1. **arithmetic**—numerical computations with the aid of electronic calculators and computers
2. **algebra**—analytical techniques employing formulae and algebraic equations
3. **geometry**—mensuration and graphical methods based on properties of geometrical shapes
4. **trigonometry**—application of trigonometric functions to problem solving involving angles

It is obvious that in a book written about statics, dynamics, and mechanics of materials it is not possible to give a detailed account of various mathematical rules and operations, which you, as a student of engineering mechanics, are already expected to be familiar with. However, this chapter offers a brief overview of the prerequisite mathematical skills needed for the solution of problems in applied engineering mechanics covered elsewhere in this book, along with some problems to help you ascertain if some revision of your mathematical knowledge is necessary.

Only those mathematical tools which are applicable in some degree to the subject matter and to the level of this book are summarised and discussed in this chapter. More specialised or advanced mathematical topics, such as logarithms and calculus, are not mentioned.

Expected learning outcomes

After careful study of this chapter, students should be well aware of the essential mathematical knowledge and skills required for satisfactory learning of applied engineering mechanics at the engineering technician level.

As a result, they should be able to decide for themselves if revision of some prerequisite mathematical knowledge and skills is necessary.

2.1 ARITHMETIC

Arithmetic is the study of computations involving numbers and basic operations such as addition, subtraction, multiplication and division, and their application to the solution of problems. It is in performing various arithmetical operations, i.e. in manipulating numbers and quantities expressed in numerical form, that you will find your electronic calculator particularly useful.

Quantities and numbers

In engineering, we always deal with objects and phenomena which are measurable in terms of their extent, amount, position, duration, temperature, or the like. These measurements are always quantifiable, i.e. they can be expressed, compared and manipulated using numbers. Quantities that are exactly alike in certain respects are said to be **equal** in those respects, and their measured magnitudes are assigned the same numerical values. Comparative relationships, such as **larger than** or **less than**, can also be expressed in precise numerical terms as a difference or as a ratio of two numbers.

The starting point of all mathematics is the process of counting. The whole numbers used in counting objects, i.e. 1, 2, 3 etc., are called **natural numbers**, or **positive counting numbers**.

Zero is a number which stands for the absence of any quantity, or for the origin of any kind of measurement, such as the point from which all divisions on a scale rule are graduated. Any number that is less than zero is referred to as a **negative number**. Any number with a positive or negative sign, indicating that it is measured in a certain direction from the origin along a line, is sometimes referred to as a **directed number**.

Collectively, all positive and negative whole numbers together with zero are called **integers**:

$$\ldots -4, -3, -2, -1, 0, +1, +2, +3, \ldots$$

Note that the plus sign indicating positive direction is often omitted, and is automatically implied if not shown. On the other hand, the minus sign must always be shown in front of the number if negative direction is to be indicated.

When parts of a whole have to be expressed we use **fractions**. The fraction 'two-thirds' is written as $\frac{2}{3}$, where the top number is the numerator, and the bottom number is the denominator.

A **decimal fraction** is one with the denominator being a power of ten, such as:

$$\frac{3}{10}, \ \frac{5}{100}, \ \frac{27}{1000}$$

When a number is expressed in decimal notation, fractional parts are indicated by the position of a decimal point as follows:

$$0.3, \ 0.05, \ 0.027$$

A **mixed decimal** is one consisting of an integer and a decimal fraction, separated by a decimal point, e.g. 5.73, read as 'five point seven three'. The majority of engineering calculations are usually carried out with the aid of electronic calculators and computers, using decimal notation.

Elementary arithmetical operations

1. **Addition** is the process of increasing one number by uniting it with another number to give a third number called the **sum**.

2. **Subtraction** is the inverse operation to addition. It is the process equivalent to removing one quantity from another, or calculating the **difference** between two numbers.
3. **Multiplication** is a process which can be regarded as repeated addition, for example:

$$3 \times 5 = 5 + 5 + 5$$
$$= 15$$

The result of multiplying two numbers is called the **product** of these numbers. Therefore 15 is the product of 3 and 5.

Certain rules apply to multiplication of directed numbers:
(a) The product of two numbers with *like* signs is *positive*.
(b) The product of two numbers with *opposite* signs is *negative*.

$$(+3) \times (+2) = +6$$
$$(-3) \times (-2) = +6$$
$$(+3) \times (-2) = -6$$
$$(-3) \times (+2) = -6$$

Fractions are multiplied by multiplying the numerators and denominators separately, for example:

$$\frac{3}{5} \times \frac{2}{7} = \frac{3 \times 2}{5 \times 7}$$
$$= \frac{6}{35}$$

4. **Division** is the inverse operation to multiplication. The result, called the **quotient** of two numbers, is determined by the numerical operation associated with the process of splitting one quantity into a number of equal parts.

Dividing a fraction by another fraction is equivalent to multiplying the first fraction by the reciprocal of the divisor, i.e. the second fraction is inverted and the two fractions are then multiplied. For example:

$$\frac{2}{5} \div \frac{3}{7} = \frac{2}{5} \times \frac{7}{3}$$
$$= \frac{2 \times 7}{5 \times 3}$$
$$= \frac{14}{15}$$

The rules concerning division of positive and negative numbers are the same as for multiplication.
5. **Raising any number to a given power** is equivalent to repeated multiplication of the number by itself. The numerical symbol shown as a superscript placed next to the base number is called the **power index** or **exponent**. It signifies the number of times the base number is to be used in repeated multiplication. For instance, the fourth power of three is given by:

$$3^4 = 3 \times 3 \times 3 \times 3$$
$$= 81$$

If the exponent is negative, then the expression is the reciprocal of the number with a positive value of the exponent, as follows:

$$3^{-2} = \frac{1}{3^2}$$

$$= \frac{1}{3 \times 3}$$

$$= \frac{1}{9}$$

Any number (except zero) with an exponent of zero is always equal to 1:

$$3^0 = 1$$

Any number to the power 2 is said to be **squared**, and any number to the power 3 is said to be **cubed**. For example, three squared is equal to nine, and two cubed is eight.

6. **Extraction of roots** is the inverse operation to raising a number to a power. It is indicated by the sign $\sqrt{}$ (the radical sign). A small numeral placed within the sign shows the type of root, e.g. $\sqrt[4]{}$ is the fourth root. If there is no small numeral shown, the root is a square root.

Extraction of roots is the process of finding the number which when raised to a given power produces the given number, e.g. 3 is the fourth root of 81, since $3^4 = 81$. It is important to note that -3 is also a fourth root of 81, since $(-3)^4 = 81$.

In many engineering calculations, negative roots are often discarded if they do not appear to represent the physical reality behind the calculation; e.g. it is not possible to have a real shaft with a negative diameter, even if the calculations tell you so.

However, in some situations, negative roots provide perfectly meaningful alternative answers to problems which can have more than one solution; e.g. the velocity of a projectile fired upwards and observed at a certain height above the ground may be positive on the way up and negative on the way down. In any case, negative roots should never be dismissed without some consideration being given to their possible relevance to the problem at hand.

Order of arithmetical operations

Arithmetical operations must always be performed in this strictly ordered sequence:

1. expressions in brackets
2. powers and roots
3. multiplication and division
4. addition and subtraction

Brackets, also called **parentheses**, indicate the grouping of some operations together to override the normal order of arithmetical operations. When brackets are used, the operations shown inside the brackets must be performed first so as to reduce the expression within brackets to a single numerical value before proceeding with the normal sequence of remaining computations. For example:

$$5 + 2 \times (7 - 3) = 5 + 2 \times 4$$
$$= 5 + 8$$
$$= 13$$

When a radical, i.e. a root sign, contains an expression consisting of more than one number, the entire expression under the radical sign must first be reduced to a single numerical value, as if it were contained in brackets, before the root can be extracted. For example:

$$\sqrt{25 - 9} = \sqrt{(25 - 9)}$$
$$= \sqrt{16}$$
$$= 4$$

Likewise, when division is indicated in the form of a composite fraction, with expressions consisting of more than one number above and below its dividing line, both the numerator and the denominator of such a fraction must be reduced to single numbers, as if they were enclosed in brackets, before division can be carried out. For example:

$$\frac{9 + 6}{12 - 7} = \frac{(9 + 6)}{(12 - 7)}$$
$$= \frac{15}{5}$$
$$= 3$$

Therefore, the radical sign and the dividing line of a composite fraction must always be regarded as grouping symbols, implying the presence of brackets without the necessity of actually showing them.

Exponential notation and significant figures

When very large numbers or very small fractions are involved, ordinary decimal notation becomes somewhat inconvenient to use. In this case, numbers can be written in what is known as **scientific exponential notation**, also called **standard exponential notation**. This is a method of writing a number as a product of a number between 1 and 10 and a power of 10. For example:

80 257.63 in scientific notation is $8.025\,763 \times 10^4$
0.000 074 613 59 in scientific notation is $7.461\,359 \times 10^{-5}$

In order to achieve this transformation, the decimal point had to be moved a number of places either to the left or to the right to obtain a number between 1 and 10. It can be seen that the power of ten is related to the direction and number of decimal point moves.

In engineering practice, preference is given to a modified form of exponential notation, known as the **engineering notation**, where only the powers of ten that are multiples of 3 are allowed (10^{-6}, 10^{-3}, 10^3, 10^6 etc.), and the position of the decimal point is adjusted so that the number in front is somewhere between 1 and 1000. For example:

80 257.63 in engineering notation is $80.257\,63 \times 10^3$
0.000 074 613 59 in engineering notation is $74.613\,59 \times 10^{-6}$

The reason for this preference lies in the ability to convert the powers of 10 from the engineering notation directly into decimal prefixes as prescribed within the metric system of units. (The use of decimal prefixes will be explained in some detail in Chapter 3.)

Since measurements of physical quantities are always inherently inexact, one should avoid giving the impression of false accuracy. This often occurs when numerical values are presented with too many significant figures, without realistic regard to the practical limits of their degree of precision.

A **significant figure** in a number is defined as a figure that may be considered reliable as a result of measurements and subsequent mathematical computations. In recording measured data, only the first of the doubtful digits is retained, and it is considered to be a significant figure. The position of the decimal point has nothing to do with how many significant figures there are in a number.

When rounding off numerical values to a specified number of significant figures, the last figure retained should be increased by 1 if the first figure dropped is 5 or greater. For example, if the above-mentioned numbers were to represent data of physical measurements which were deemed to be accurate to only three significant figures, the appropriate alternative ways of expressing these numbers without giving the impression of false accuracy would be:

$$80\,300 \quad \text{and} \quad 0.000\,074\,6 \quad \text{in ordinary decimal notation}$$
$$8.03 \times 10^4 \quad \text{and} \quad 7.46 \times 10^{-5} \quad \text{in scientific exponential notation}$$
$$80.3 \times 10^3 \quad \text{and} \quad 74.6 \times 10^{-6} \quad \text{in engineering exponential notation}$$

Most scientific calculators allow data to be entered in the exponential notation form using the EXP key, and some have special function keys that allow direct conversion from the ordinary mode of decimal notation into scientific and/or engineering exponential notation, and vice versa.

2.2 *ALGEBRA*

Algebra is a branch of mathematics which has been described as the art of abstract mathematical reasoning. In a more restricted sense, algebra is a generalisation and extension of arithmetic in that it summarises and represents, by means of symbolic notation, various results and patterns of relations that exist between mathematical operations involving numbers.

Arithmetic becomes algebra as soon as general rules are developed whose validity holds for any arbitrary choice of specific numbers. To use a very simple illustration, the fact that $20 added to $30 gives the same result as $30 added to $20 is arithmetic of specific numbers. When translated into a general rule using the symbolic language of algebra, this becomes $a + b = b + a$, which is valid for any two numbers a and b. This signifies the fact that the order of addition does not affect the final result, and is known as the **commutative law of addition**. However, if you think that all this is too obvious anyhow, think of another arithmetical operation. For example, is division commutative?

From the practical point of view, algebra is probably the most all-embracing and powerful of all mathematical tools. Typical problems in elementary algebra involve setting up and evaluating algebraic expressions, transposing formulae, and solving linear and quadratic equations.

Evaluating algebraic expressions

The most direct and obvious link between arithmetic and algebra is found in the process of numerical evaluation of algebraic expressions.

An **algebraic expression** is a mathematical statement obtained by combining letters that represent arbitrary numbers with symbols for various arithmetical operations, which may contain powers or roots. For example, $p^2 - 2pq + q^2$ is an algebraic expression.

When specific numerical values are assigned to each of the letters in an algebraic expression, the entire expression assumes a fixed numerical value. The process of evaluating the expression, i.e. finding its specific numerical value, consists of two steps: substituting the given values into the expression and then calculating the result. For instance, if $p = 5$ and $q = 3$ in the expression $p^2 - 2pq + q^2$, then the expression is evaluated as follows:

$$p^2 - 2pq + q^2 = 5^2 - 2 \times 5 \times 3 + 3^2$$
$$= 25 - 30 + 9$$
$$= 4$$

Transposing formulae

A **formula** is any identity, general rule, or law of mathematics or physics, stated in the form of an algebraic equation.

When a formula is presented with a single letter on the left-hand side of the equality sign, the single letter is said to be the **subject** of the formula. The process of changing the subject by rearranging the formula is called **transposition**. For example, the formula linking temperatures on the old Fahrenheit scale, which is still in widespread use in the United States of America, to the Celsius scale, which is used in Australia, is:

$$F = 1.8C + 32$$

This formula can be transposed to make C the subject:

$$C = \frac{F - 32}{1.8}$$

Solving equations

An important part of algebra is concerned with the study of methods of solving various types of equations.

A **conditional equation** is one which is true only for certain values of the variables involved. The **root** of an equation is a number that, when substituted for the variable in a given equation, satisfies the equation, i.e. makes both sides demonstrably equal. For example, $2x + 3 = 11$ is a simple linear equation which is conditional upon the value of x. The root of this equation is $x = 4$.

Common types of equations include the following:

1. **Linear equations** are equations which contain the unknowns in the first degree. A single linear equation with one unknown, such as the one mentioned above, is solved by a process similar to the transposition of formulae. It involves a series of steps necessary to isolate the unknown variable on one side of the equation, resulting in the transfer of all known numerical terms to the other side.

2. **Quadratic equations** are equations which contain the unknown in the second degree. A quadratic equation is usually solved by first converting it into the form

$ax^2 + bx + c = 0$, where a and b are numerical coefficients and c is a numerical constant, and then evaluating the roots by applying the quadratic equation formula:

$$x = \frac{-b \pm \sqrt{b^2 - 4ac}}{2a}$$

A quadratic equation usually has two distinct roots resulting from the positive and negative signs before the radical. For example, the roots of the quadratic equation $2x^2 + 3x - 14 = 0$ are found as follows:

$$x = \frac{-3 \pm \sqrt{3^2 - 4 \times 2 \times (-14)}}{2 \times 2}$$

$$= \frac{-3 \pm 11}{4}$$

Hence the two roots are $x_1 = 2$ and $x_2 = -3.5$.

In some cases, the expression under the radical sign in the quadratic equation formula may be equal to zero. The two roots are then equal to each other and are said to be coincident.

3. **Simultaneous equations** are two or more equations that apply simultaneously to several variables. The solution of simultaneous equations involves finding values of the variables that satisfy all given equations. For example, the equations:

$$x + 2y = 5$$
$$2x + 3y = 8$$

can be satisfied simultaneously by only one pair of values for x and y, namely $x = 1$ and $y = 2$.

(a) **Substitution**. The most common method of solving simultaneous linear equations is the method of substitution. The first equation is put in the form:

$$x = 5 - 2y$$

and then the expression on the right, i.e. $5 - 2y$, is substituted for x into the second equation:

$$2(5 - 2y) + 3y = 8$$

Solving this as a single equation in terms of y gives $y = 2$. This value of y is substituted back into the first equation, which yields $x = 1$.

(b) **Elimination**. A somewhat different process for solving simultaneous equations is called the method of elimination. There are various alternative means of elimination. For example, the same two equations can be solved by multiplying the first equation by 2 in order to match the coefficients of x in both equations:

$$2x + 4y = 10$$

Subtracting the left-hand and right-hand sides of the second equation from this modified form of the first gives:

$$2x + 4y = 10$$
$$-(2x + 3y) = -8$$
$$\overline{y = 2}$$

The value of x is then found by substituting $y = 2$ into either one of the original equations and then solving it for x.

2.3 *GEOMETRY*

Geometry is the branch of mathematics concerned with the properties of points, lines, angles, plane figures, and solids in space.

From the practical engineering viewpoint, the importance of geometry is twofold. First, it gives us the description of properties and the necessary formulae for calculating dimensions, areas and volumes of different geometrical shapes used in engineering. Second, it provides the theoretical foundation for many graphical methods of problem solving used in structural and mechanical engineering and in related disciplines.

Geometrical shapes and solids

It is necessary for every student of engineering to be able to recognise the names and to be familiar with the essential properties of the more common geometrical shapes and solids. These include the polygon, triangle, parallelogram, rectangle, square, hexagon, circle, cuboid, cube, cylinder and sphere.

1. A **polygon** is a closed plane figure having three or more straight sides. **Regular polygons** have all their sides and all their angles equal.
2. A **triangle** is a three-sided polygon. The internal angles of any triangle always add up to 180°. An **equilateral triangle** has three equal sides and therefore three equal angles, each 60°. A right-angled triangle has one of its interior angles equal to 90°.
3. A **parallelogram** is a four-sided polygon with its opposite sides equal and parallel. The sum of all the internal angles is 360°. Opposite internal angles in any parallelogram are always equal. If all four angles of a parallelogram are right angles, then the parallelogram is a rectangle.
4. A **rectangle** is a four-sided polygon with each of the four angles equal to 90°. Opposite sides of a rectangle are equal and parallel, and the four angles obviously add up to 360°. If all four sides of a rectangle are equal, the rectangle is a square.
5. A **square** is a four-sided polygon with all sides equal and each of the four angles equal to 90°. A square is a regular polygon.
6. A **hexagon** is a six-sided polygon. A **regular hexagon** has equal sides and each of its interior angles is equal to 120°.
7. A **circle** is a closed plane curve which has all its points located at a fixed distance from the centre. The fixed distance is called the **radius**. A straight line joining two points on the circle and passing through its centre is called the **diameter**. The distance around the circle is referred to as the **circumference**.

The ratio of the circumference of a circle to its diameter is a constant for all circles, designated by the Greek letter π, and equal to $3.141\,592\,653\,5\ldots$ (with infinitely non-repeating decimals!). An approximate value of 3.14 (to three significant figures) was in common use, sufficiently accurate for most practical purposes, until the introduction of scientific calculators, which always have a special key for this ratio.

8. A **cuboid**, generally recognised as a 'matchbox' shape, is a geometrical solid with six rectangular faces, the opposite faces being identical in shape and size. The cuboid is also known as the **right rectangular prism**. If all six faces of a cuboid are square, the cuboid is a cube.

9. A **cube** is a regular geometrical solid which has six identical square faces.

10. A **cylinder** is a geometrical solid formed between two identical plane circular bases and bounded by a tube-like curved surface parallel to its longitudinal axis of symmetry. A cylinder has a uniform circular cross-section.

11. A **sphere** is a geometrical solid bounded by a closed curved surface that has all its points located at a fixed distance from the centre. The fixed distance is called the **radius**. A straight line joining two points on the surface of a sphere and passing through its centre is called the **diameter**.

As a student you are expected to know how to calculate the areas enclosed within the plane figures described above (with the exception of the irregular polygon), as well as the surface areas and volumes of the solids. The formulae required for these purposes can be found in Appendix B at the end of this book.

Pythagoras' theorem

This extremely useful theorem states that in any right-angled triangle, the square of the hypotenuse (the longest side which is opposite the right angle) is equal to the sum of the squares of the other two sides:

$$h^2 = a^2 + b^2$$

For example, the following triple of numbers corresponds to the lengths of the sides of a particular right-angled triangle known as the **Egyptian triangle** because of its use in the ancient world by the 'rope-stretchers' for setting out right angles on construction sites:

$$5, 4 \text{ and } 3 \quad (\textit{Note: } 5^2 = 4^2 + 3^2)$$

The first general demonstration of this theorem is attributed to the Greek mathematician and philosopher Pythagoras, who lived in the 6th century BC, although the practical use of specific sets of such numbers was known to the Babylonians a good thousand years or so ahead of him.

In its original formulation, Pythagoras' theorem is applicable only to right-angled triangles. However, it can easily be adapted and is frequently used to find unknown lengths in all kinds of geometrical configurations. Looked at in this way, the theorem forms a lasting and indispensable cornerstone of mathematical science and its many practical applications.

You will come across one of the engineering applications of Pythagoras' theorem when discussing the addition and resolution of forces in Chapter 4.

Graphical scales

The key to any form of graphical analysis or presentation is the ability to designate various measurable quantities by directed line segments (arrows) whose lengths represent magnitudes, or by marks on graduated reference lines, and to select and use suitable scale factors for this purpose.

The **scale factor** can be defined as the amount represented by the unit division on a linear scale; e.g. a scale factor of 50 kg/mm means that each millimetre of the scale length represents 50 kilograms.

A scale factor must be chosen to suit the size of the paper used and the estimated range of the variable. Avoid a scale that may require awkward relationships between the quantity being represented and the graduations on the line drawn or on the scale rule used to measure it. The best scale ratios to choose for easy plotting and reading are 1, 2 and 5, or one of these numbers multiplied or divided by 10, 100, 1000 etc.

To select a suitable scale factor, make a trial calculation:

$$\text{Scale factor} = \frac{\text{range of variable}}{\text{line length available}}$$

For example, if it is necessary to represent a range of car speeds between 0 and 130 km/h by lines not longer than 70 mm, a suitable scale would be determined as follows:

$$\text{Estimated scale factor} = \frac{130 \text{ km/h}}{70 \text{ mm}}$$

$$= 1.86 \frac{\text{km/h}}{\text{mm}}$$

Therefore, a convenient scale factor to choose would be 2 km/h per millimetre. With the aid of the 1:2 scale on your scale rule, the values can be plotted and read without any further calculations.

2.4 *TRIGONOMETRY*

Trigonometry is a branch of mathematics whose primary concern is with the measurement of triangles. Unknown angles or side lengths can be calculated by using trigonometric functions such as sine, cosine and tangent. Trigonometry is of immense practical value in engineering, surveying and related fields.

Trigonometric ratios

The difference between geometry and trigonometry lies in the ability of the latter to numerically link linear dimensions to angular measure within triangles, something that even Pythagoras' theorem is unable to accomplish.

To achieve this we must be familiar with the meaning of at least three trigonometric functions, sometimes called trigonometric ratios: sine, cosine and tangent. For an acute angle, these ratios are defined as the ratios of the sides in a right-angled triangle that contains the angle:

1. The **sine** of an angle is the ratio $\dfrac{\text{opposite side}}{\text{hypotenuse}}$.

2. The **cosine** of an angle is the ratio $\dfrac{\text{adjacent side}}{\text{hypotenuse}}$.

3. The **tangent** of an angle is the ratio $\dfrac{\text{opposite side}}{\text{adjacent side}}$.

Although they were originally and conveniently defined in relation to acute angles within right-angled triangles, the usefulness and interpretation of trigonometric ratios is much broader than that, and can be extended to apply to triangles of any shape, to circles, and to angles greater than 90°. It is therefore preferable to regard the trigonometric ratios as properties of angles rather than of triangles.

For any given angle, the values of its trigonometric ratios are fixed and can be obtained with a single keystroke using appropriate trigonometric function keys on a calculator. Use your calculator to verify these to three significant figures:

$$\sin 27° = 0.453\,99 \ldots$$
$$\cos 27° = 0.891\,00 \ldots$$
$$\tan 27° = 0.509\,52 \ldots$$

Inverse trigonometric ratios

If it is possible to determine the value of a specified trigonometric ratio with the aid of a calculator, it should also be possible to perform the inverse operation, i.e. given the value of the function, determine the angle to which it belongs.

The **inverse trigonometric functions** are usually written as $\sin^{-1}$, $\cos^{-1}$ and $\tan^{-1}$, as follows:

If $\sin 43° = 0.682$, then $\sin^{-1} 0.682 = 43°$.
If $\cos 36° = 0.809$, then $\cos^{-1} 0.809 = 36°$.
If $\tan 53° = 1.327$, then $\tan^{-1} 1.327 = 53°$.

Check these statements using the inverse trigonometric function keys on your calculator, probably in conjunction with the alternative-function key labelled SHIFT, INV or 2ndF.

Solution of right-angled triangles

The **solution of triangles** is the process of calculating the unknown sides and angles of a triangle when sufficient data are available to specify the triangle. The method of solution depends on the type of triangle and on the known parameters.

Right-angled triangles are solved by a combination of Pythagoras' theorem and trigonometric functions, plus the fact that the sum of the two acute angles in such a triangle is 90°.

If two sides are given, the third side is found by Pythagoras' theorem, and one of the acute angles is found by using an appropriate trigonometric ratio of two of the sides. The other acute angle is found by subtraction from 90°.

If one side and one acute angle are given, the other acute angle is found by subtraction from 90°. Then a second side can be found by a trigonometric function involving the known side. The third side is found by Pythagoras' theorem.

Solution of oblique triangles

An **oblique triangle** is a triangle that does not contain a right angle. Pythagoras' theorem does not apply. Therefore, when solving oblique triangles it is necessary to employ two additional relationships known as the sine rule and the cosine rule.

1. The **sine rule** states that within any one triangle, the ratio of a side length to the sine of its opposite angle is the same for all three sides. This can be written as a formula:

$$\frac{a}{\sin A} = \frac{b}{\sin B} = \frac{c}{\sin C}$$

where a is the length of the side opposite to angle A, b is opposite to angle B, and c is opposite to angle C.

The sine rule is used when only one side and two angles are given. The third angle is found by subtraction from $180°$ and the two unknown sides can then be found from the sine rule ratios.

2. The **cosine rule** is given by the following formula:

$$c^2 = a^2 + b^2 - 2ab \cos C$$

where C is the angle opposite side c, i.e. the angle included between sides a and b.

The cosine rule is used to find the third side, c, when only two sides, a and b, and the included angle, C, are known. The other angles can then be determined by the sine rule.

The cosine rule can also be used to determine an unknown angle in a triangle, when only the three side lengths are given. For this case, the formula can be transposed to become:

$$\cos C = \frac{a^2 + b^2 - c^2}{2ab}$$

and then the inverse cosine function key is used to determine angle C.

2.5   REVIEW

The following review problems are not intended to teach you all you need to know about mathematics. They are just a spot check on some essential mathematical skills; you can do them as a self-test. Use an electronic calculator where necessary.

If you can work out the correct answers to these problems, you will have no major difficulties with the mathematical level of this book.

Everyone is entitled to one or two mistakes! However, if you do not get at least ten of these problems right, and you do not feel comfortable with the calculations required here, then now is the time to start revising and improving your mathematical skills.

Problems

2.1 Evaluate the following expression:

$$\frac{0.231\sqrt[3]{64} - 22 \times 18.5}{\sqrt{81 \times 10^{-6}} + \sqrt{0.05^2 - 0.03^2}}$$

2.2 Given that:

$$y = \frac{Fa(3L^2 - 4a^2)}{24EI}$$

what is the value of y when $F = 30\,000$, $L = 12\,000$, $a = 2500$, $E = 200\,000$ and $I = 554 \times 10^6$?

2.3 Given that:

$$y = \frac{Fa^2(3L - a)}{6EI}$$

make F the subject of the formula.

2.4 Solve the following equation:

$$\frac{3x - 5}{x - 2} = 5$$

2.5 Solve the following equation:

$$x^2 + 2x - 35 = 0$$

2.6 Solve the following simultaneous equations:

$$2x + y = 12$$
$$7x - 3y = 3$$

2.7 Calculate the volume of a cylinder with diameter 3 m and height 4 m.

2.8 What is the length of the hypotenuse in a right-angled triangle if the lengths of the other two sides are 35 mm and 84 mm?

2.9 Determine, accurate to three significant figures:
(a) $\sin 27°$ (b) $\cos 70°$ (c) $\tan 58°$

2.10 Given that:

$$\theta = \tan^{-1}\left(\frac{L}{\pi D}\right)$$

calculate θ for $L = 10$ mm and $D = 35$ mm.

2.11 In a triangle, the side opposite to a 35° angle is 2 m long. In the same triangle, what is the length of the side opposite to a 47° angle?

2.12 Two sides of a triangle are 5 m and 3 m long, and the angle between them is 61°. What is the length of the third side?

CHAPTER 3

Fundamental concepts and units

Engineering is concerned with the world of physical objects and their properties. In engineering analysis and design we are continually confronted with questions such as how long? how much? how fast? how hot? how strong? and many others. Answers to questions of this kind require an understanding of the physical nature of the objects or phenomena concerned. They also require a common basis of measurement, involving a standardised system of units. This chapter introduces the science of measurement and the few fundamental concepts of physical science on which the modern international system of units is based.

Expected learning outcomes

After carefully studying the material presented in this chapter, working through all numerical examples, and successfully completing all practice problems, students should be able to:

1. describe the essential features of the metric system of units as practised in Australia;

2. state the names and symbols, and correctly use, the SI and related units of length, mass, time, plane angle, area and volume;

3. recognise and apply standard decimal prefixes in the range between micro and giga.

3.1 PHYSICAL QUANTITIES AND DIMENSIONS

Our experience of the world around us reveals qualities of different kinds, some simple and obvious, others complex and difficult to describe or define. Some of the qualities are thought to belong to the objective reality of the physical world, while others are subjective judgments found 'in the eye of the beholder'. Size, speed, colour, beauty and justice represent in some ways certain aspects of a particular object or event. However, only some of these categories can be regarded as physical quantities.

An important characteristic of a physical quantity is its ability to be described quantitatively in terms of precise physical measurements. In other words, a **physical quantity** can be defined as a measurable attribute of a physical object, substance or process. When you drive a car, the speed of the car is a physical quantity, because it has to do with a physical object, the car, and it can be measured. On the other hand, if you are a passenger in a car driven at 150 kilometres per hour, the feeling of exhilaration or of fear is not a physical quantity, because it is subjective and does not allow description in terms of physical measurements.

Through experience, the scientist and the engineer have learned that certain physical quantities are **fundamental**, in a sense that they can be used to define or describe all other physical quantities and relations. A limited number of such fundamental physical quantities have been selected arbitrarily to form the fundamental **dimensions** on which our system of measurement is based. The fundamental dimensions with which we are concerned in this book are **length, mass, time** and, to a limited extent, **temperature**.

Somewhat related to the concept of length, and supplementary to it, is the geometrical concept of **plane angle**. For our purposes, it is possible to consider angular measure as an independent fundamental quantity. However, unlike length, plane angle is a dimensionless quantity.

In addition to the above-mentioned fundamental quantities, it is also convenient to include electric current, amount of chemical substance, luminous intensity of light, and solid angle as independent fundamental quantities. However, these are outside the scope of this book.

Fundamental dimensions can be combined in numerous ways to form **derived dimensions**. For example, *two* linear measurements of a rectangular shape can be combined to express an essentially different physical quantity called **area**. Similarly, *three* measurements combine to express **volume**. Throughout the book, we will introduce derived dimensions of various physical quantities, such as force, energy, pressure, density and acceleration.

All equations involving physical quantities must be **dimensionally homogeneous**, i.e. they must be balanced dimensionally as well as numerically. For example, the volume of a rectangular prism is equal to the product of its three sides, for which the formula is:

$$V = a \times b \times c$$

When the dimensions of a particular prism are given in a problem such as 'What is the volume of a room 4 m long, 2.5 m wide and 2.8 m high?', the solution is:

$$V = 4 \text{ m} \times 2.5 \text{ m} \times 2.8 \text{ m}$$
$$= 28 \text{ m}^3$$

Note that regardless of the magnitudes or units used, dimensional homogeneity of this equation must always be true and can be shown as:

$$l \times l \times l = l^3$$

where l represents a single linear dimension, while l^3 stands for the derived dimension of volume, or 'cubic measure'.

3.2   *MEASUREMENT AND SYSTEMS OF UNITS*

All physical quantities, no matter how complex, are measurable using instruments and techniques appropriate to the quantity to be measured. Some physical quantities can be measured with relative ease and a high degree of precision. Others can only be measured indirectly, by measuring other physical quantities associated with the object or process under investigation and then computing the magnitude of the required quantity as a function of these measurements. Some measurements are made with precise instruments, while others are only crude approximations.

The science of engineering measurement is called **metrology** and deals with the essential requirements of uniformity of units and standards, as well as the instruments and techniques used for practical engineering measurements.

Uniformity of measurement requires reliable units and standards, as well as accuracy in collecting and recording of results. A **unit** is an agreed-on part of a physical quantity, defined by reference to some arbitrary material standard or to a natural phenomenon. In essence, the process of measurement involves making a comparison between the quantity being measured and a standard unit appropriate for the particular physical quantity. The result of measurement is, therefore, a statement of magnitude as a number of standard units, or a fraction of a standard unit. For a surface measuring 2 metres long by 0.7 metre wide, we can say that in comparison with the standard unit, the metre, its length is equal to two units and its width is equal to seven-tenths of the unit.

The units devised and used in earlier times were closely associated with the practical trades, such as carpentry and surveying, and with commercial transactions. They were based on common, but ill-defined and unreliable, standards such as the digit, a finger's breadth, the foot (which was later redefined as 'one-third of the Iron Yard of our Lord the King'), the wine-jar and the barley corn. These and similar units were inaccurate and unrelated to each other, and could hardly provide a sound basis for methodical and precise measurement of physical quantities.

Throughout history, a number of systems of measurement emerged, competing with each other and gradually replacing each other. Two types of systems can be distinguished: an evolutionary system, which grows haphazardly out of usage, and a planned system. The British system, which survived with some local variations in the USA, is an example of an evolutionary system; the metric system, which originated in the Netherlands and France, is an example of a planned system.

The latest worldwide move is towards a universal, simple and internationally accepted system of units known as the **International System of Units**, with the abbreviation 'SI' (from Système International d'Unités). In Great Britain, all units of the British system were redefined in terms of the metric system by an Act of Parliament in 1963, with a national changeover to SI beginning two years later. The progress in the United States has been slower, but quite discernible within its scientific community. In the meantime, the traditionally metric countries of Europe have also been adjusting to the new International System, which is essentially an expansion of the metric system, incorporating scientific and technological developments of the 20th century.

In Australia, conversion to SI units began with the Metric Conversion Bill, introduced in the Commonwealth Parliament in 1970, with the aim of progressively introducing the use of metric measurement as the sole system of measurement of physical quantities in this country. The inherent advantages of the metric system, particularly its decimal nature and coherent relationships between units, as well as the fact that all major countries were already using metric or were in the process of changing, suggested that it was desirable and practical for Australia to make the change.

The *Metric Conversion Act 1970* provided for a Metric Conversion Board to be established to coordinate the conversion program, and to compile and disseminate appropriate information and advice. The Act also defined the metric system to be used in Australia as measurement in terms of basic and derived SI units, with certain additions of special units declared as part of Australia's metric system (see Appendix A).

With its task of guiding the initial steps in the process of the introduction of SI units into Australia largely completed, the Metric Conversion Board was eventually abolished and the role of maintaining the integrity of standards of measurement in this country passed to other organisations, such as the National Standards Commission, the Commonwealth Scientific and Industrial Research Organisation (CSIRO), the Standards Association of Australia, and the National Association of Testing Authorities (NATA).

3.3 *THE INTERNATIONAL SYSTEM OF UNITS (SI)*

A decimal system of units was originally proposed by the Flemish mathematician, engineer and public servant Simon Stevin in a small pamphlet *La Thiende* (translated *The Tenth*), published in 1585, in which he declared that the universal introduction of decimal coinage and measures would be only a matter of time. Other European scientists also discussed the desirability of a new rational and uniform system of units to replace the multitude of local inconsistent weights and measures, which made scientific communication difficult. The establishment of the metric system is often associated with the social and political upheavals of the French Revolution, being one of its most significant and lasting results.

In 1790, the French National Assembly requested the French Academy of Sciences to establish a set of units suitable for international use. The French formally adopted the metric system of measurement in 1840. By that time the Netherlands, Belgium, Luxembourg and Greece had already done so. Internationally, the Metre Convention was signed in 1875, leading to the establishment of the International Bureau of Weights and Measures. The United States, along with fourteen other nations, was a party to the original treaty, Great Britain became a signatory in 1884, and Australia joined formally in 1947.

The original metric system was based on the metre, gram and second as base units, but in 1873 the centimetre replaced the metre to define the CGS (centimetre–gram–second) system. Later, the MKS (metre–kilogram–second) system was introduced. The International System was developed from the MKS system and adopted by the General Conference on Weights and Measures in 1960 as 'a practical system of units of measurement suitable for adoption by all signatories of the Metre Convention'.

The International System of Units comprises a set of **base units** and **derived units**, corresponding to the fundamental and derived dimensions as discussed previously. One of the main advantages of the system is that for each physical quantity there is only one SI unit, with its decimal multiples and submultiples, without any odd multiplying factors to be remembered. Furthermore, the system is coherent, i.e. one in which the product or quotient of any two unit quantities in the system is the unit of the resultant quantity.[*]

Altogether, there are seven SI base units. Of these, we are only concerned with the units of length, mass, time and temperature. In addition, there are two SI supplementary units for angular measurement, but we shall only use the radian as a measure of plane angle.

---

[*] The full significance of this characteristic will become apparent when derived units of force, work and stress are discussed:

$$1 \text{ N} = 1 \text{ kg} \times 1 \text{ m/s}^2 \text{ (see Ch. 13)}$$
$$1 \text{ J} = 1 \text{ N} \times 1 \text{ m (see Ch. 16)}$$
$$1 \text{ Pa} = 1 \text{ N/m}^2 \text{ (see Ch. 25)}$$

Table 3.1   *SI base and supplementary units*

Physical quantity	Name of unit	Symbol
length	metre	m
mass	kilogram	kg
time	second	s
temperature	kelvin	K
plane angle	radian	rad

It should be noted that as a general rule, names of units (e.g. metre, kelvin) and prefixes (e.g. kilo) when written in full are written in lower case letters.* Unit symbols are also written in lower case letters (e.g. m, kg) except the symbols for units named after people (e.g. K). Unit symbols do not change in the plural and a full stop should not be used after a symbol, except at the end of a sentence. However, when spelled out, unit names take a plural 's', e.g. 3 kilograms *but* 3 kg.

In order to express decimal multiples and submultiples of SI units, the system provides a number of decimal prefixes which, when attached to a particular unit, indicate the relationship with the parent unit. Preference is given to the use of prefixes related to the parent unit by a factor of 1000.†

Table 3.2   *SI prefixes*

Prefix	Symbol	Value
giga	G	10^9
mega	M	10^6
kilo	k	10^3
milli	m	10^{-3}
micro	μ	10^{-6}

When stating the value of any measurement, the appropriate unit should be chosen so that the numerical value of the statement is between 0.1 and 1000. For example, the distance between Sydney and Canberra is stated as 309 km, not 309 000 m, and shaft diameter as 55 mm, not 0.055 m. However, some engineering disciplines have their own special rules, e.g. all dimensions on mechanical engineering drawings are shown in millimetres, regardless of the size of the component shown.

3.4   *FUNDAMENTAL DIMENSIONS AND UNITS*

It has already been stated that a limited number of physical quantities can be selected as fundamental dimensions on which a coherent and comprehensive system of units can be built. Let us now consider length, angle, mass, time and temperature, as such independent basic concepts, and their associated units.

---

* The sole exception is 'degree Celsius'.
† This rule does not preclude the use of intermediate prefixes such as centi (= 10^{-2}) for non-engineering applications, e.g. dressmaking.

The concept and units of length

Definition and measurement of length is based on our intuitive appreciation of the notion of space and of our position in it in relation to other material things with which we share the space. Questions relating to spatial measurements are usually of two kinds: 'Where?' (in relation to some known reference point or a system of coordinates) and 'How long?' (referring to a distance between two points). Answers to both questions require a suitable measuring instrument and an agreed standard unit of length. The concept of **length** can be defined as a measure of distance between two points, one of which may be the origin of a coordinate system.

It should be noted that length, or linear dimension, is always measured along a single line. This line, however, does not have to be a straight line. Thus, while the diameter of a circle is a convenient description of the circle's size, the length of its circumference is an equally valid linear measure.

The base unit of length is the **metre**, originally conceived as one ten-millionth part of the distance from the equator to the pole along the meridian line through Paris, introduced in 1801 by law of the French National Assembly. Historically, the metre was the first unit in the metric system of units, both words having a common origin in the Greek word *metron* meaning 'measure'. As a result of the International Metric Convention of 1875, the metre was redefined as the distance between two defining marks on the International Prototype Metre, which became the physical embodiment of the unit. It is a standard bar of platinum–iridium alloy held at the specified temperature of 0°C on two roller supports in a horizontal position, and is kept at the International Bureau of Weights and Measures at Sèvres near Paris, France. Because of its inaccessibility, secondary standards, or copies of the primary standard, were made and distributed to the standardising agencies of various countries. The accuracy of measuring tapes and other length-measuring devices is usually fixed by manufacturing specifications derived from the secondary standards.

In 1960 the definition of the metre was further refined in terms of new techniques for measuring light waves, which made the standard more accurate and more readily reproducible. The metre is now defined as equal to exactly 1 650 763.73 wavelengths of the orange line in the spectrum of the krypton-86 atom in an electrical discharge. This incredibly precise definition was the result of a very careful comparison made between the light emitted from a krypton-86 lamp under specified laboratory conditions and the distance between the defining lines on the prototype metre. All precise measurements of length in scientific laboratories are now made with light waves.

In engineering practice, linear dimensions are the most common measurements made. The instruments used vary with the size and nature of the item and the degree of accuracy required. They range from a simple rule or tape to vernier calipers, precision micrometers, dial indicators and non-contacting electronic probes. The measurements can be made directly from a scale, or indirectly by transferring a calipered dimension to separate standards or gauge blocks. In transfer measurements with calipers used by a skilled mechanic, accuracies in the order of 0.01 mm can be achieved. However, direct micrometer and vernier caliper measurements are more reliable and are capable of accuracies from 0.01 to 0.001 mm.

For day-to-day work in engineering measurement which does not require a high degree of accuracy, a steel rule or tape, usually graduated in millimetres, is used. Dimensions of solid objects, to which an ordinary rule cannot be applied, are measured with the aid of engineer's calipers consisting of a pair of hinged steel jaws which are gradually closed until they touch both sides of the object in the desired position. The distance between the jaws is afterwards compared with an ordinary scale. For many purposes, slide or vernier calipers (Fig. 3.1), capable of a more accurate direct reading, are preferred. The vernier scale was named after Pierre Vernier, a 17th century French technician who patented this ingenious device. It enables an accurate reading of the last significant decimal place in the measurement without the necessity to estimate fractions of a division by eye. Besides calipers, many other measuring instruments are fitted with vernier scales.

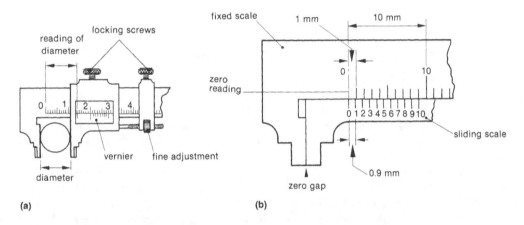

Fig. 3.1   (a) *Vernier calipers*   **(b)** *Principle of 0.1 mm vernier*

For measuring the diameter of a small shaft or a piece of wire, or similar small dimensions which nevertheless require a relatively high order of accuracy, a micrometer screw gauge as shown in Figure 3.2 may be used. The main feature of the instrument is a screwed spindle fitted with a graduated thimble. When the object is gripped gently between the spindle and the anvil, an accurate reading is obtained from the graduated sleeve and the thimble.

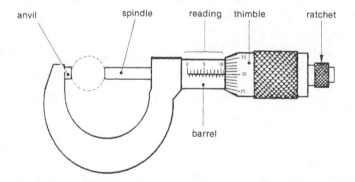

Fig. 3.2   *Micrometer screw gauge*

Many attachments are used to adapt these instruments to round, square, inside, outside, screw thread or other special measurements. Non-contacting electronic micrometers can be used for dynamic measurements, such as those on a rotating shaft. A digital display eliminates the need for skilful contact measurements.

Let us keep in mind that the most convenient unit of measurement for mechanical engineering components is the millimetre (= 0.001 m) and that it is common practice to show all dimensions on mechanical engineering drawings in millimetres. On the other hand, the requirement of dimensional homogeneity of physical equations often demands that base units be used for calculations, especially if quantities other than purely linear dimensions are involved. Therefore, care should be exercised in converting the measured dimensions from millimetres to the base unit, the metre, when necessary for calculations.

The concept and units of plane angle

In plane geometry, the **angle** is defined as the inclination of one straight line to another; i.e. the angle between two straight lines is determined by the amount of turning about the point of intersection required to bring one line into coincidence with the other.

The SI unit of angular measure is the **radian**, defined as the angle subtended at the centre of a circle by an arc of length equal to the radius. By virtue of this definition, the radian is a ratio of two equal lengths, and is therefore dimensionless. On the other hand, angular measure is a useful geometrical concept. The radian is therefore incorporated into the set of fundamental units with the status of a supplementary unit. There are 2π (= 6.283 approx.) radians in a circle.

In addition to the radian, the use of degrees, minutes and seconds is allowed for measurement of plane angle. The circle is divided into 360 degrees (°), the degree into 60 minutes (′) and the minute into 60 seconds (″). The degree is sometimes decimally divided. The degree–minute–second system was codified by Ptolemy of Alexandria (about AD 130) and has remained unchanged for almost 2000 years. Within Australia's metric system, degrees, minutes and seconds have been declared non-SI units which may be used without restriction. It should be remembered, however, that the radian is mathematically more fundamental than the degree, and must be used where dimensional homogeneity demands.

The relationship between degrees and radians follows from the fact that in a full circle there are 2π radians or 360 degrees:

$$1° = \frac{2\pi}{360} \text{ rad} \doteq 0.017\,45 \text{ rad}$$

$$1 \text{ rad} = \frac{360°}{2\pi} \doteq 57.3°$$

Example 3.1

In a centre lathe, the cutting-tool movement required for accurate control of the amount of metal removed is achieved by the use of a graduated indexing dial attached to the lead screw, so that for one complete revolution of the lead screw, the tool has a linear movement equal to the pitch of the lead screw thread.

Assuming that the lead screw shown in Figure 3.3 has a pitch of 2.5 mm and an indexing dial with 100 divisions, determine the angle in degrees through which the dial must turn to move the cutting tool a distance of 1 mm. What will this angle be if it is expressed in radians? What is the angle turned per division on the dial?

sliding member

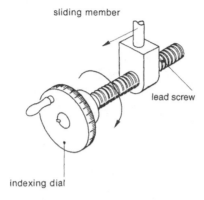

lead screw

indexing dial

Fig. 3.3

Solution

$$\begin{array}{l}\text{Distance moved by the tool}\\\text{for one revolution} = 2.5\ \text{mm}\end{array}$$

$$\begin{array}{l}\text{Distance moved by the tool}\\\text{for one division on the dial} = \dfrac{2.5\ \text{mm}}{100}\\\qquad\qquad\qquad\qquad\qquad\quad = 0.025\ \text{mm}\end{array}$$

$$\text{Number of divisions required} = \dfrac{1\ \text{mm}}{0.025\ \text{mm}}$$
$$= 40$$

$$\therefore \text{Angle of revolution required} = 360° \times \dfrac{40}{100}$$
$$= 144°$$

$$\text{Angle expressed in radians} = 2\pi \times \dfrac{40}{100}$$
$$= 2.51\ \text{rad}$$

$$\text{Angle per division on the dial} = \dfrac{360°}{100}$$
$$= 3.6°$$
$$= 3°36'$$

Plane angles are usually measured with a protractor. A plain protractor, as used by a draughtsperson or an engineer, is graduated in degrees, with an accuracy for each reading of plus or minus half a degree. When a protractor cannot be used directly to check the taper of a component, a bevel gauge is used to transfer the angle from the component to the protractor. For more accurate readings, a vernier scale can be incorporated into an instrument such as the universal bevel vernier protractor, shown in Figure 3.4, or a variety of optical instruments such as the surveyor's level.

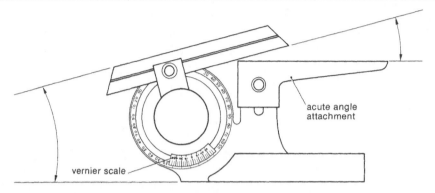

acute angle
attachment

vernier scale

Fig. 3.4   *Universal bevel vernier protractor*

The concept and units of mass

In engineering science we have to deal with material things, or matter. Steel, aluminium, concrete, water, oil, steam and air are typical examples of physical substances or matter. Some are used to manufacture engineering items, such as steel girders, concrete beams, and aluminium components for aircraft. Others, like steam and air, are used as working agents in turbines and compressors. Although all kinds of material substances occupy space, the volume they occupy is not a constant property, being subject to changes with variations in temperature and pressure.

However, we can think of **mass** as a non-varying quantity of matter in a body; it depends only on the constant number and kind of elementary particles that make up the substance of the body.[*]

The concept of mass was developed gradually over many centuries, with the greatest contributions coming from Leonardo da Vinci, Galileo and Newton. Our understanding of mass comes from the experience of observing how different objects respond to our attempts to set them in motion. The property of the object which determines its resistance to change of motion is called the **inertia** of the object. It is reasonable to assume that the inertia of a body depends on the quantity of matter contained in that body, i.e. the greater the quantity of matter, the greater the tendency to resist a change of motion. Unlike volume, the quantity of matter in a given body is constant, and not subject to variation with location, temperature or pressure, for as long as the body retains its physical identity.

This leads to the definition of **mass** as the measure of the quantity of matter in a body as evidenced by its inertia. In a scientific laboratory, a device called the **ballistic balance** is sometimes used to compare an unknown mass to a set of graded reference masses by means of observing their inertial effects during accelerated motion.

Another more convenient method by which one can determine the mass of an object is to compare its mass with that of known masses, using an equal-arm platform balance, shown in Figure 3.5. This method is based on the fact that equal masses experience equal attraction towards the Earth, a phenomenon discussed in more detail in Chapter 4. Unfortunately, in everyday conversation this procedure is called 'weighing', even though it is the mass, i.e. the quantity of matter, and not the weight which is being measured.[†]

---

[*] This is not quite true according to Einstein's theory of relativity, from which it follows that mass must be distinguished from the amount of substance. However, this fine distinction is not of any practical significance in ordinary engineering mechanics.

[†] In science, the word 'weight' means the force of gravity, which is quite different from the concept of mass, and we must carefully distinguish between them (see Ch. 4).

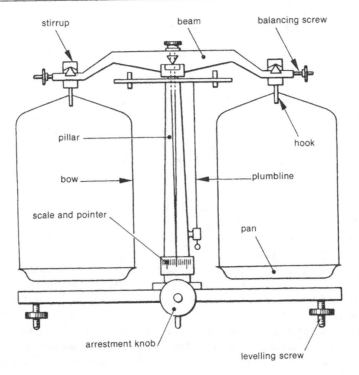

Fig. 3.5 *Equal-arm platform balance*

The base SI unit of mass is the **kilogram**. The kilogram is defined as the mass of the International Prototype Kilogram, kept at the International Bureau of Weights and Measures in Sèvres. The Prototype Kilogram is a cylinder 39 mm in diameter and 39 mm high, made from an alloy containing 90 per cent platinum and 10 per cent iridium.

Special attention should be drawn to the anomaly that, although it contains the prefix 'kilo' for historical reasons, the kilogram is nevertheless a base unit and not a multiple. In order to avoid the use of two consecutive prefixes, the SI prefixes are attached to the stem word 'gram' when forming units of mass larger and smaller than the kilogram. Furthermore, a unit equal to 1000 kg is often referred to as the **tonne**. The correct names, symbols and relationships between units of mass are summarised in Table 3.3.

Table 3.3 *SI units of mass*

Name	Symbol	Relationship to base unit
milligram	mg	$1\ mg = 10^{-6}\ kg$
gram	g	$1\ g = 10^{-3}\ kg$
kilogram	kg	SI base unit
tonne	t	$1\ t = 10^{3}\ kg$
kilotonne	kt	$1\ kt = 10^{6}\ kg$

(*Note:* 1 t = 1 Mg, and 1 kt = 1 Gg)

The concept and units of time

Of all the fundamental concepts of science, time is probably the most difficult to define satisfactorily. Time can be described as a measure of the sequence of events taking place in the physical world. It has to do with such notions as beginning and end, before and after, past and future, but thoroughly eludes any attempt at simple explanations. From ancient times in Egypt, Greece and Rome, time has been associated with periodic natural phenomena such as the duration of an average day and the duration of an average year, i.e. the movements of the Earth as observed by the astronomers. The problem with this approach is that such phenomena are not in fact constant in duration and have very inexact and awkward relationships between themselves. Besides, they do not provide the answer to the fundamental question 'What is time?'.

Our standard of time originated from the astronomical observation of the spinning of the Earth, as reflected in the apparent motion of the Sun. The duration of the mean solar day was divided into 24 hours. Each hour was itself divided into 60 minutes, each containing 60 seconds.*

In essence, this system of time measurement has not changed in the International System of Units and is still the legal standard. The SI base unit of time, however, is now the **second**, redefined in 1967 as the time interval occupied by 9 192 631 770 cycles of a specified energy change in the caesium atom, as measured by the caesium atomic clock. The atomic clock is similar to a radio transmitter giving out short waves, the frequency of which is controlled by energy changes of gaseous caesium atoms. The accuracy of such a clock is considerably better than the results of astronomical observations, even with the most modern instruments possible. The added advantage is that, unlike the movements of the Earth, the frequency of the atomic radiation is constant and can be accurately reproduced in any suitably equipped laboratory anywhere in the world.

Accordingly, the day, hour and minute have been redefined in terms of the standard second, as shown in Table 3.4. It should be remembered, however, that these units do not differ significantly from the mean solar day, hour, minute and second.

Table 3.4 *Units of time*

Name	Symbol	Definition in terms of SI unit
second	s	SI base unit
minute	min	1 min = 60 s
hour	h	1 h = 3.6×10^3 s
day	d	1 d = 86.4×10^3 s

Practical measurement of time, depending on the accuracy required, is made with pendulum clocks, chronometers, stopwatches, electric clocks and quartz clocks. All these instruments depend for their operation on some form of uniformly repeated motion: pendulums of suitable lengths swinging in the gravitational field of the Earth, vibrations of coiled springs, electric motors which operate synchronously on alternating current of standard frequency, or oscillating quartz crystals.

---

* The system of sexagesimal fractions ($\frac{1}{60}$, $\frac{1}{3600}$ etc.), which survived also in the angular measure, was derived from the Chaldeans, became standard in Greek arithmetic and continued in general use until the introduction of decimal notation about AD 1600.

The concept and units of temperature

Temperature can be described as a measure of the degree of hotness or coldness of a physical body or substance with respect to a fixed scale. There are two temperature scales used in science and engineering.

The first, and more fundamental of the two, is the **thermodynamic temperature scale**, also known as the **Kelvin scale**. The concept of thermodynamic temperature is closely associated with the idea that temperature is the property of a region in space that determines the rate at which heat energy will be transferred to or from it. The position of the origin, or zero point, on the scale is based on the notion of an unattainable lower limit of temperature, called **absolute zero**, at which all particles of matter are supposed to reach the lowest theoretically possible energy state. All this is the province of the branch of science called thermodynamics and is of little concern to us here in engineering mechanics, except for the fact that the SI unit of temperature is based on the thermodynamic scale.

The formal definition of the SI unit of temperature is not easy to understand. It states than one kelvin, symbol K,[*] is the fraction 1/273.16 of the thermodynamic temperature of the triple point of water. In simpler language, this means that ice melts at approximately 273 kelvins above absolute zero. You are probably already getting the feeling that the thermodynamic temperature scale is not very convenient for everyday use. Quite so! This is why an alternative scale has been allowed for ordinary practical applications.

This second temperature scale is known as the practical or **Celsius scale**. After the introduction of the SI metric units, the original Celsius scale was redefined in terms of kelvins and retained for general and engineering measurements because of its practical importance. The unit of temperature on the Celsius scale, equal in magnitude to one kelvin, is called 'degree Celsius', with the symbol °C. However, the zero point of the Celsius scale is different. It is approximately 273 kelvins above absolute zero. For our purposes, the Celsius scale is recognised as the scale on which the melting point of ice is 0°C and the boiling point of water at atmospheric pressure is 100°C.

The relationship between the two scales is represented graphically in Figure 3.6.

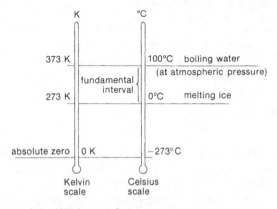

Fig. 3.6 *Comparison of the Kelvin and Celsius scales*

---

[*] Note that by international agreement since 1968, the unit of temperature is the kelvin (K), *not* the degree Kelvin.

The concept of temperature has very limited relevance in the context of engineering mechanics, applicable only to problems involving thermal expansion of solids, and stresses induced in the material of structural members and mechanical components subjected to variations in temperature. We will return to this in Chapter 26.

3.5   *DERIVED DIMENSIONS AND UNITS*

Dimensions and units of many physical quantities can be derived from those of the fundamental quantities length, angle, mass and time. In this section, we examine area, volume and some time-related quantities.

The concept and units of area

Area is defined as a measure of the extent of a surface. The SI unit of area is the **square metre**, defined as the area enclosed by a square each side of which is one metre in length. This can clearly be seen as a derived unit, having a dimension of a linear measure squared. The square metre does not have a special symbol and is an example of an SI unit with a compound name. The symbol used is m^2, which reflects its derivation.

The preferred multiples and submultiples of the square metre are formed from the preferred units of length, the millimetre and the kilometre, with compound symbols mm^2 and km^2. Due to the second-order relationship between the dimensions of length and area, the multiplier between one preferred unit of area and the next is 1000^2, or one million, as seen in Table 3.5.

In addition to the preferred units described above, there is the hectare, equal to 10 000 square metres (100 m × 100 m), which has been declared an acceptable unit. The use of the hectare, however, is generally limited to computation of land areas smaller than one square kilometre, and is not of particular interest to the mechanical or structural engineer.

Table 3.5   *SI units of area*

Name	Symbol	Relationship to the parent unit
square millimetre	mm^2	$1\ mm^2 = 10^{-6}\ m^2$
square metre	m^2	SI unit
hectare	ha	$1\ ha = 10^4\ m^2$
square kilometre	km^2	$1\ km^2 = 10^6\ m^2$

In general, areas are not measured directly, but are calculated from measured linear dimensions of surfaces, e.g. sides, diameters etc. The formulae for calculating the areas of most common geometrical shapes are given in Appendix B. One interesting exception is the planimeter, an instrument for measuring mechanically the area of a plane surface of irregular shape. A tracing point on an arm is moved around the closed curve, whose area is then given to scale by the revolutions of a small wheel supporting the arm.

Example 3.2

How many parquet flooring blocks, each measuring 125 mm × 25 mm, are required to cover an area 10 m × 8 m?

Solution

$$\text{Area to be covered} = 10 \text{ m} \times 8 \text{ m}$$
$$= 80 \text{ m}^2$$

$$\text{Area of each block} = 0.125 \text{ m} \times 0.025 \text{ m}$$
$$= 0.003\,125 \text{ m}^2$$

$$\text{Number of blocks required} = 80 \text{ m}^2 \div 0.003\,125 \text{ m}^2$$
$$= 25\,600$$

Note that where different units are involved, e.g. metres and millimetres, it is convenient to convert all given information to base units and then do the calculation. This is not an absolute rule, provided that consistent units are used.

The concept and units of volume

Volume is defined as a measure of the amount of space occupied by an object or matter. The SI unit of volume is the **cubic metre**, defined as the volume of a cube, each side of which is one metre in length. This is a derived unit having a dimension of linear measure cubed. The cubic metre does not have a special symbol and is another example of an SI unit with a compound name. The symbol used is m^3.

The preferred multiples and submultiples of the cubic metre are the cubic millimetre (mm^3) and the cubic kilometre (km^3). In this case, the third order relationship between the dimensions of length and volume make the multiplier equal to 1000^3, or one thousand million (10^9), which is a very large gap between one preferred unit and the next.

It is for this reason that an additional decimally related unit called the **litre**,* equal to one-thousandth part of a cubic metre, has been declared for general use. The SI prefixes can be attached to the word 'litre' to form its multiples and submultiples, resulting in some units having two possible names, as can be seen from Table 3.6.

Table 3.6 *SI units of volume*

Name	Symbol	Alternative name	Symbol	Relationship to the parent unit
cubic millimetre	mm^3	microlitre	μL	$=10^{-9}$ m^3
—	—	millilitre	mL	$=10^{-6}$ m^3
—	—	litre	L	$=10^{-3}$ m^3
cubic metre	m^3	kilolitre	kL	SI unit
—	—	megalitre	ML	$=10^3$ m^3
—	—	gigalitre	GL	$=10^6$ m^3
cubic kilometre	km^3	—	—	$=10^9$ m^3

The volumes of regular geometrical shapes such as prisms, spheres and cylinders are usually calculated from their linear dimensions. The appropriate formulae can be found in Appendix B. To determine the volumes of liquids and gases we use special calibrated

---

* Contrary to the general rule, the accepted symbol for the litre is the upper case 'L', in order to avoid confusion between lower case 'l' and the numeral '1', particularly in typewritten material.

containers. Volumes of solids of irregular shapes can be found indirectly by measuring the volume of the liquid displaced when the solid is immersed in a liquid, a method suggested by Archimedes.

Example 3.3

The pressure vessel shown in Figure 3.7 has hemispherical ends. Determine the volume of gas, in cubic metres and in litres, contained in the vessel.

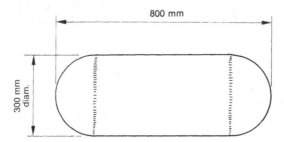

Fig. 3.7

Solution

The length of the cylindrical part:

$$l = 0.8 \text{ m} - 0.3 \text{ m}$$
$$= 0.5 \text{ m}$$

The volume of the cylindrical part:

$$V_1 = \frac{\pi D^2}{4} l$$

$$= \frac{\pi \times 0.3^2}{4} \times 0.5$$

$$= 0.035\,34 \text{ m}^3$$

The volume of the hemispherical ends:

$$V_2 = \frac{\pi D^3}{6}$$

$$= \frac{\pi \times 0.3^3}{6}$$

$$= 0.014\,14 \text{ m}^3$$

Total volume of the vessel:

$$V_1 + V_2 = 0.035\,34 \text{ m}^3 + 0.014\,14 \text{ m}^3$$
$$= 0.049\,48 \text{ m}^3$$

Converting to litres:

$$\text{Volume} = 0.049\,48\ \text{m}^3 \times \frac{1000\ \text{L}}{\text{m}^3}$$
$$= 49.48\ \text{L}$$
$$\doteqdot 49.5\ \text{L}$$

Proportional measure: rate and ratio

Many engineering concepts represent a comparative measure of one physical quantity or magnitude with another. The units used to express such proportional measures are formed by the division of the two different units, resulting in a new compound unit. There are too many derived concepts and units for all of them to be introduced here. However, a few examples are given below to illustrate the idea.

The term **rate** usually refers to a quantity or process in which time is a factor. For example, we speak of water flow rate in a pipe, which can be measured in litres per second, i.e. volume per unit time. Note that the quotient of two units is indicated by the word 'per' immediately in front of the unit forming the denominator. The symbol of the compound unit is formed using a horizontal line, an oblique stroke or the use of a negative index, for example:

$$\text{Litre per second is }\ \frac{\text{L}}{\text{s}},\ \text{L/s or L.s}^{-1}$$

Sometimes, the dimension of time does not appear in the definition of a concept, but a continuous process, such as travelling by car, is involved. Under these circumstances it is still appropriate to use the term 'rate' to describe relationships between physical quantities, e.g. petrol consumption rate in litres per kilometre, or volume of fuel used per unit distance travelled.

Quite often, a comparative measure does not involve or imply time or process, but describes some other physical property of a substance or object, in which case it is usually called a **ratio**. For example, volume occupied by a unit mass of substance can be described as the ratio of the volume occupied by a given quantity of matter to its mass. The appropriate derived unit could be litre per kilogram, or cubic metre per tonne, or some other quotient of volume and mass units.

If the quantities being compared are of the same kind and measured in the same units, the ratio is dimensionless and represents the relative sizes of the two quantities. For example, if the area of room x is 18 m^2 and the area of room y is 12 m^2, the ratio of the areas is:

$$\frac{A_x}{A_y} = \frac{18\ \cancel{\text{m}^2}}{12\ \cancel{\text{m}^2}}$$
$$= 1.5$$

meaning that room x is one-and-a-half times larger in area than room y. Notice how the units cancel out, making the ratio dimensionless.

Example 3.4

Determine the cost of heating oil used during the three winter months if the price of fuel is 35 cents/litre, the average fuel consumption rate is 2.3 litres/hour and the average duration of heater use is 5 hours/day.

Solution

The number of days in the three winter months:

$$\text{No. of days} = 30 \text{ (June)} + 31 \text{ (July)} + 31 \text{ (August)}$$
$$= 92 \text{ days}$$

The total number of hours the heater was in use:

$$\text{No. of hours} = 92 \cancel{d} \times \frac{5 \text{ h}}{\cancel{d}}$$
$$= 460 \text{ h}$$

$$\text{Total fuel consumption} = 460 \cancel{h} \times 2.3 \frac{\text{L}}{\cancel{h}}$$
$$= 1058 \text{ L}$$

$$\therefore \text{Total cost of fuel used} = 1058 \cancel{L} \times 0.35 \frac{\$}{\cancel{L}}$$
$$= \$370.30$$

Note that if consistent units are used and the relation between all variables is well understood, it is possible to combine the entire solution into one line of calculations, as follows:

$$\text{Cost of fuel used} = (30 + 31 + 31) \cancel{d} \times 5 \frac{\cancel{h}}{\cancel{d}} \times 2.3 \frac{\cancel{L}}{\cancel{h}} \times 0.35 \frac{\$}{\cancel{L}}$$
$$= \$370.30$$

Observe how the units cancel out, leaving the dollar as the unit of the final answer.

 Problems

3.1 Express the following quantities in SI base units:
(a) 0.4 km
(b) 53 g
(c) 75.3 mm
(d) 45 ms
(e) 357 000 mg
(f) 0.08 Mg
(g) 734 μm
(h) 0.54 Gs

3.2 Express the following quantities using the most appropriate multiples or submultiples of SI units:
(a) 12 300 m
(b) 7500 kg
(c) 0.079 s
(d) 0.0047 m
(e) 0.03 kg
(f) 85 × 10³ kg
(g) 3 × 10⁻⁶ m
(h) 4.7 × 10⁻⁶ kg

3.3 Convert:

   (a) 5 m² to mm²

   (b) 750 mm² to m²

   (c) 663 mm³ to L

   (d) 500 000 mm³ to m³

   (e) 1.35 m³ to L

   (f) 0.75 m³ to mm³

   (g) 632 L to m³

   (h) 47 L to mm³

3.4 Convert:

   (a) 35°15′ to degrees

   (b) 150° to radians

   (c) 1.57 rad to degrees

   (d) 2.75° to radians

   (e) 2.5 rad to degrees and minutes

   (f) 25°50′ to radians

3.5 Convert:

   (a) 3 days 5 hours to minutes

   (b) 0.75 hour to seconds

   (c) 5000 seconds to hours, minutes and seconds

   (d) 6.7 minutes to seconds

3.6 A heating panel is 750 mm long by 350 mm wide. What is its area in square metres?

3.7 It is proposed to replace a round duct of 400 mm diameter by a square duct of equivalent cross-sectional area. Determine the required dimensions of the new duct.

3.8 A cylindrical vessel is 340 mm in diameter and 900 mm long. Determine its total surface area in square metres and its volume in litres, neglecting the thickness of the walls.

3.9 A rotating shaft makes 2100 revolutions in one minute. How many times does it rotate per second?

3.10 During a test of a diesel engine, the average time to consume 50 mL of oil was 35.4 s. What was the fuel consumption in litres per hour?

3.11 A rectangular swimming pool is 50 metres long and 20 metres wide. The average depth of water is 3.6 metres. How long would it take to fill it with a pump which delivers 12 cubic metres of water per minute?

3.12 A car consumes 11 litres of petrol every 100 km. The price of petrol is 40 cents per litre. What is the cost of petrol for a 750 km trip?

3.13 A boiler generates 720 kg of steam per hour, while consuming 0.3 t of coal. What is the mass of steam generated per kilogram of coal?

3.14 A reciprocating pump has a 75 mm diameter piston with a 90 mm stroke, and produces 5 pumping strokes per second. Determine the number of litres of water pumped per hour.

Review questions

   1. What is a *physical quantity*? Give examples.

   2. Length, mass, time and temperature are regarded as fundamental quantities.

      (a) Explain the difference between plane angle and the four fundamental quantities.

      (b) Explain the difference between volume and the four fundamental quantities.

   3. What is a *unit of measurement*?

   4. What does SI stand for?

   5. List names and symbols of SI units of length, mass, time, temperature and plane angle.

6. Which of the following are SI base units?
metre, hour, tonne, square metre, kilogram, kilometre, litre, cubic metre, second, gram

7. Which of the following are non-preferred prefixes?
giga, mega, kilo, deci, centi, milli, micro

8. Is the litre a unit of mass or volume?

9. What is your mass in kilograms?

10. What is your height in metres?

STATICS

PART 2

STATIC FORCES

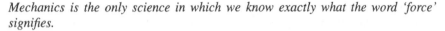

Mechanics is the only science in which we know exactly what the word 'force' signifies.

Friedrich Engels
Dialectics of Nature

CHAPTER 4

Force and gravity

The concept of force is one of the most important for the engineer. Design of structures and machine components requires a thorough understanding of the action and the effects of forces in the connected elements of bridges, engines, machine tools and other structures and mechanical devices. In this chapter the concept of force is introduced, together with some important principles associated with the mathematical and graphical analysis of forces.

Expected learning outcomes

After carefully studying the material presented in this chapter, working through all numerical examples, and successfully completing all practice problems, students should be able to:
1. state the name and symbol of the SI unit of force, and describe the basic characteristics of force;
2. resolve a given force into two components;
3. find the resultant of several forces mathematically and graphically;
4. calculate the weight of a body of known mass.

4.1   *THE CONCEPT OF FORCE*

In everyday experience we usually associate force with muscular effort. When weightlifters try to lift heavy barbells, they use the strength of their arms to first move the barbell and then to hold it above the head by applying a force. Other athletes use force to throw a heavy metal sphere in a shot-put event, while in a circus a 'strongman' amuses by bending a metal rod with the force of his hands. Note that in all of these instances a force is applied, but the results are quite different: lifting, holding, throwing and bending. In all cases, however, the **force** can be described as **push** or **pull**, i.e. effort. Furthermore, force is seen to be an interaction between two bodies.

In mechanics, forces are usually associated with any action that tends to *maintain* the position of a body (e.g. support, hold), to *alter* the position of a body (e.g. lift, throw), or to *distort* it (e.g. stretch, bend). It is very useful to continue thinking about forces in terms of push or pull, or effort.

There are many kinds of forces—muscular effort, tractive effort at the wheels of a vehicle, cutting force at the tip of a cutting tool, forces due to gas pressure in an engine cylinder—which are a type of push or pull, produced as a result of interaction between two or more bodies.

Another very important type of force is **gravity**, or gravitational attraction, i.e. the pull exerted by the Earth on every physical object located on or near its surface. The law of gravity is discussed in some detail in Sections 4.8–4.10.

There is also a group of forces which are the result of friction (see Ch. 10), often described as resistance to motion, e.g. friction between two surfaces sliding relative to each other, air resistance etc. These forces usually oppose some applied force such as tractive effort, which tends to cause or maintain motion.

Forces are often discussed and classified with respect to their effects. The science of **statics** deals with the equilibrium of bodies at rest, under the combined action of several balanced forces. When all forces acting on a body are balanced, the body is said to be in **static equilibrium**. A bridge is a structure subjected to a large number of forces. The forces acting on the bridge at any one time must be in equilibrium for the bridge to remain in its position. The equilibrium of forces is discussed further in Chapters 5 and 6.

If a force, or forces, acting on a body are *not* balanced, the condition of rest or motion will be altered, and the body will accelerate or slow down. For example, an object dropped from a height falls with an increasing speed under the influence of the unbalanced force of gravity. The part of mechanics dealing with the analysis of bodies in motion under the influence of various forces acting on the bodies is called **kinetics**. The relations used to predict the motion caused by given forces or to determine the forces required to produce a given motion will be explained in Chapter 13.

Another possible effect of a force is to deform the material of the body to which the force is applied. The branch of engineering science which studies the relations between forces and the amount of deformation produced in the material is called **strength of materials**. An introduction to the concept of elasticity, which is fundamental to strength of materials, is given in Chapter 25. In many practical problems, the ability of a material to withstand the applied forces without failure or excessive deformation is of paramount importance. Selection of suitably sized components by the designer is usually based on the analysis of these forces and of the elastic behaviour of the material.

However, as the first step towards understanding the concept of force, let us start with the proposition that **force** is a push or a pull exerted on a body.

4.2   *MEASUREMENT AND UNIT OF FORCE*

The SI unit of force is the **newton**, symbol N. The exact definition of the newton is based on the ability of a unit force, when applied to a unit mass, to impart to that mass a unit of acceleration. In order to fully appreciate the significance of this definition, one must have a good understanding of the laws of linear motion. We shall therefore return to this definition in Chapter 13, where the exact relationship between the forces acting on a body, the mass of the body and the motion of the body will be discussed. It is more important for the sutdent at this time to get some idea of the actual magnitude of the unit force, rather than to grapple with its definition.

Force is a push or pull. The unit of force, the newton, is therefore a push or pull of a particular strength, which is taken to be the standard for comparing forces. It is not possible to describe the newton as we can describe some material object such as an apple. However, a small apple is a convenient aid with which to provide a recognition point for learning purposes. If you put an apple on the palm of your hand, you can feel a small downward force acting on your hand (Fig. 4.1(a)). The force is approximately one newton. The newton, as can be seen from this simple demonstration, is not very large. In engineering practice, a force one thousand times stronger than the newton is a more useful unit. With the aid of SI decimal prefixes, such a unit is called the **kilonewton**, with the symbol kN. The force required to support 26 common house bricks is approximately equal to one kilonewton (Fig. 4.1(b)).

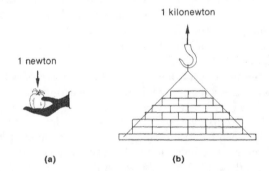

Fig. 4.1

Forces are usually measured by means of a spring dynamometer.[*] A simple dynamometer (Fig. 4.2) consists of a spring with one end restrained and the other movable, where the force is applied. The extension of the spring is proportional to the applied force within the normal operating range of a given instrument. As the force stretches the spring, a marker attached to the movable end of the spring moves along a graduated scale, indicating the magnitude of the applied force. As a force-measuring device, a spring is calibrated in the units of force, i.e. newtons or kilonewtons.[†] Simple spring dynamometers are not very accurate. More sophisticated ones are capable of accuracies in the order of ±0.2 per cent.

---

[*] The term 'dynamometer' is also used for a device for testing power output of engines or electric motors.

[†] It should be noted that many mass-measuring devices, e.g. butcher's scales, are in fact spring balances calibrated to read in kilograms and/or grams. However, their use is limited to measuring a quantity, e.g. meat, and should not be extended to measures of force. Always remember: kilogram is *not* a unit of force.

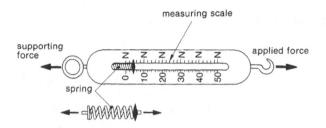

Fig. 4.2   *A simple dynamometer*

Other force-measuring instruments, such as hydraulic and pneumatic load cells and strain gauges, are used extensively in many varied engineering applications, particularly where the force cannot be transmitted through a dynamometer, and must therefore be measured indirectly. Strain gauge load cells are probably the most common devices for measuring forces of large magnitudes. Accuracies of ±0.1 per cent are common. Strain gauges are devices built around a strip of elastic material that is attached to a member of structure or machine which is being compressed or stretched by a force. Connected to this strip is an electric device that measures its compression or elongation but provides an output calibrated in units of force.

4.3   *CHARACTERISTICS OF A FORCE*

A force is characterised by its magnitude, its direction and its point of application.

The **magnitude** of a force is a measure of the strength of the pushing or pulling effort, expressed in standard units of force, usually newtons or kilonewtons, e.g. 560 N, 3.75 kN.

The **direction** of a force is defined by the line of action and the sense of the force. The **line of action** is a straight line along which the force acts. It can be described by the angle it forms with some reference axis, e.g. 30° to the horizontal. The **sense** can be indicated descriptively as to the right or to the left, up or down, etc.

Finally, when applied to an actual object or component, a force must have a **point of application**, e.g. the point at which a cable is attached to a mast.

Graphically, a force can always be represented by a straight line, drawn to scale, in the direction corresponding to that of the force. For example, a force of 50 N acting down and to the right along a line inclined to the horizontal at 30° and applied at point A can be represented by a line 50 mm long, i.e. a scale of 1 mm = 1 N, as shown on Figure 4.3. Notice how the graphical representation is clear and precise in comparison with the somewhat unwieldy verbal description.

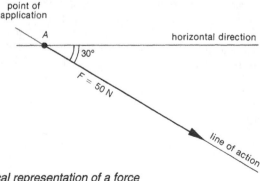

Fig. 4.3   *Graphical representation of a force*

Those familiar with the mathematical concept of vectors will easily recognise that, as a quantity which has both magnitude and direction, force is a vector. As a vector, a force can not only be represented graphically as shown, but can also be manipulated according to the rules of vector algebra.

4.4   *BASIC PRINCIPLES*

There are three main principles concerning forces which are necessary for the solution of problems involving forces in mechanics. These are action and reaction, transmissibility of a force and the parallelogram of forces.

Action and reaction forces

The **principle of action and reaction** was formulated by Sir Isaac Newton, who pointed out that whenever a force acts on a body, there must be an equal and opposite force or reaction acting on some other body. Expressed in a simple form, this principle states that action and reaction are equal and opposite.

To take an example, a car is pulling a trailer with a force of 0.2 kN (Fig. 4.4). The trailer experiences a pull of 0.2 kN exerted by the car. At the same time, the trailer exerts an equal, but opposite, reaction equal to 0.2 kN on the car. Note that if a spring dynamometer were fitted between the car and the trailer, it would indicate a force of 0.2 kN.

action = 0.2 kN = reaction

Fig. 4.4   *Action and reaction forces*

Action and reaction forces are always collinear (i.e. they act along the same line), are always equal in magnitude, but are opposite in sense. However, they are applied to two different bodies and should always be considered with respect to the body to which they are applied. In the example above, the action is applied to the trailer and the reaction is applied to the car.

Transmissibility of a force

The **principle of transmissibility** states that the effect of a force on a body to which it is applied is not altered when the point of application of the force is moved to some other position on the line along which the force acts. In short, a force can be moved along its line of action without changing its effect. For example, a locomotive can *pull* a train, or it can *push* a train with equal force and equal effect (Fig. 4.5).

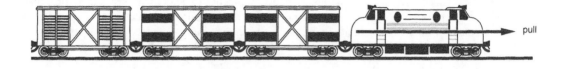

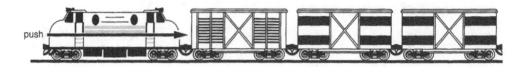

Fig. 4.5 *Equal push and pull forces*

Parallelogram of forces

The **principle of the parallelogram of forces**, formulated by Stevin in 1586, states that if two forces intersecting at a point are represented in magnitude and direction by the adjacent sides of a parallelogram, their combined action is equivalent to the action of a single force, represented both in magnitude and direction by the diagonal of the parallelogram. The single force, which has exactly the same effect as the two given forces, is called the **resultant** force.

Example 4.1

Two cables are attached to a bracket as shown in Figure 4.6(a).

Determine the resultant force acting on the bracket if the force in the horizontal cable is 4.2 kN and that in the inclined cable is 2.4 kN.

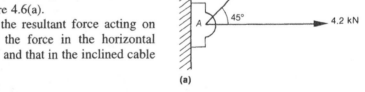

(a)

Solution

The resultant can be found by constructing the parallelogram of forces, as shown in Figure 4.6(b), using a suitable scale such as 10 mm = 1 kN.

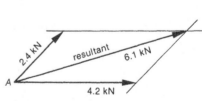

(b)

The answer is a force of 6.1 kN acting at 16° to the horizontal, as shown in Figure 4.6(c).

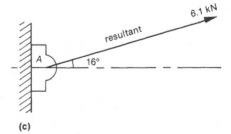

(c)

Note that in a simple example like this, the three diagrams can be superimposed to form a combined diagram as in Figure 4.6(d). However, as the complexity of force systems increases in other problems, it is more convenient to draw separate force diagrams as in Figure 4.6(b).

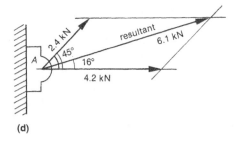

(d)

Fig. 4.6

4.5   *RECTANGULAR COMPONENTS OF A FORCE*

In the previous section it was shown how two forces could be combined into a single resultant force. The reverse problem, called **resolution** of a force into components, is the separation of a single force into two component forces acting in different directions on the same point. The previous example could be reversed as follows.

Example 4.2

It is necessary to exert a force of 6.1 kN on the bracket at 16° to the horizontal by means of two cables, one horizontal, the other at an angle of 45° to the first cable. What should the force in each cable be?

Solution

The information given in the problem is sufficient for us to construct a parallelogram of forces, which in fact is identical to that in Figure 4.6(b). The answers now are 4.2 kN in the horizontal cable and 2.4 kN in the other cable. Note that the forces in the two cables are the components of the required force of 6.1 kN.

In many engineering problems, the two components into which a single force must be resolved are at right angles to each other. Such components are called **rectangular components** of a force, as the parallelogram of forces becomes a rectangle. Problems involving the resolution of a force into rectangular components can either be solved graphically by constructing a rectangle of forces, or mathematically by using trigonometric relationships.

Example 4.3

A force of 5 kN is acting up and to the right at 30° to the horizontal. Determine its horizontal and vertical components.

Solution

(a) *Graphical method*

Draw the parallelogram, i.e. rectangle of forces, to scale, as in Figure 4.7. The horizontal component is 4.3 kN to the right and the vertical component is 2.5 kN upwards.

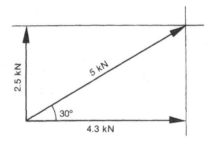

Fig. 4.7

(b) *Mathematical method*

Sketch a diagram similar to that in Figure 4.7, but not necessarily to scale. From the geometry of the triangles involved, the horizontal component is:

$$F_H = 5 \times \cos 30°$$
$$= 5 \times 0.866$$
$$= 4.33 \text{ kN} \rightarrow$$

and the vertical component is:

$$F_V = 5 \times \sin 30°$$
$$= 5 \times 0.5$$
$$= 2.5 \text{ kN} \uparrow$$

The relationships between a force F and its rectangular components in the mutually perpendicular x and y directions, F_x and F_y, are:

$$\boxed{F_x = F \cos \theta} \quad \text{and} \quad \boxed{F_y = F \sin \theta}$$

where θ is the angle between the force and the x-direction.[*]

Conversely, if the two components F_x and F_y are known, the force itself and the angle it makes with the x-direction can be calculated from:

$$\boxed{F = \sqrt{F_x^2 + F_y^2}} \quad \text{and} \quad \boxed{\tan \theta = \frac{F_y}{F_x}}$$

The x and y directions are usually horizontal and vertical, but they may also be chosen in any two mutually perpendicular directions, as in Figure 4.8.

---

[*] θ (theta) is a letter of the Greek alphabet often used to represent angular measurement.

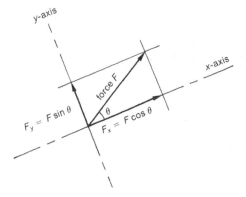

Fig. 4.8

Example 4.4

If the components of the force acting on a gear tooth are 54.6 N in the radial direction and 150 N in the tangential direction, determine the total force and the angle between the force and tangential direction (Fig. 4.9).

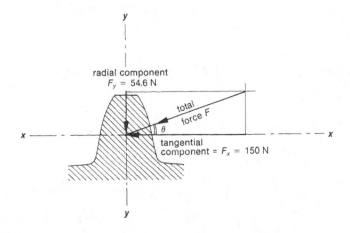

Fig. 4.9

Solution

Total force:

$$F = \sqrt{F_x^2 + F_y^2}$$
$$= \sqrt{150^2 + 54.6^2}$$
$$= 159.6 \text{ N}$$

Angle:

$$\tan \theta = \frac{F_y}{F_x}$$
$$= 54.6 \div 150$$
$$= 0.364$$
$$\therefore \ \theta = 20°$$

Problems

4.1 What will the reading be on each of the spring dynamometers shown in Figure 4.10, if a force, or forces, are applied as indicated?

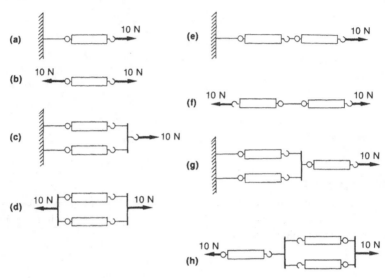

Fig. 4.10

4.2 Determine the resultant of the two forces acting on point *A* in each of the examples in Figure 4.11.

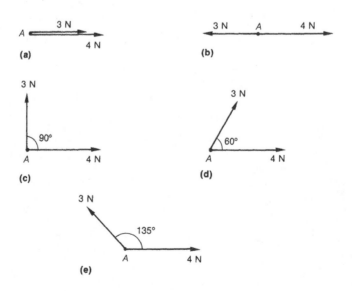

Fig. 4.11

4.3 Two cranes are attempting to right an overturned truck by each applying a pull of 50 kN. The two cables form an angle of 30° between each other. What is the resultant force on the truck?

4.4 A man pulls a loaded wagon with a force of 300 N acting along the handle of the wagon which makes an angle of 30° with the ground. What are the magnitudes of the horizontal and vertical components?

4.5 Four forces are acting at a point as shown in Figure 4.12. For each of the forces, determine the horizontal and vertical components:
(a) graphically
(b) mathematically

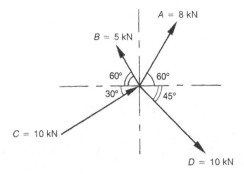

Fig. 4.12

4.6 A wagon is being pushed up an incline by a horizontal force of 500 N as shown in Figure 4.13.

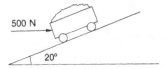

Fig. 4.13

Determine the components of the force acting along the plane and perpendicular to the plane.

4.7 A structural member is subjected to a load of 10 kN as shown in Figure 4.14. Determine the components of the load:
(a) horizontal and vertical
(b) along and perpendicular to the axis of the member

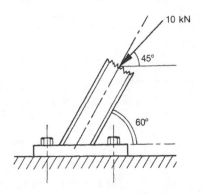

Fig. 4.14

4.8 Resolve the 800 N force, acting on a cutting tool, into components in directions perpendicular and parallel to the axis of the tool (Fig. 4.15).

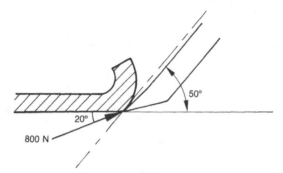

Fig. 4.15

4.6 *GRAPHICAL ADDITION OF FORCES*

In many engineering problems there are systems of forces consisting of more than two forces. If a resultant of such a system is required, it is possible to apply the parallelogram of forces principle to find the resultant of any two forces in the system, then combine that resultant with another force, and then repeat the procedure until all forces have been included. As one can imagine, the construction required is rather complicated, involving too many construction lines.

The solution can be simplified by introducing the triangle of forces rule, and then extending this rule to the polygon of forces method.

The triangle of forces rule

The **triangle of forces rule** (or force triangle rule) is derived simply from the parallelogram principle. When two forces are added by the parallelogram method, the opposite sides of the parallelogram are always equal, and any one of the two opposite sides can represent a force in magnitude, resulting in two possible triangles of forces, as in Figure 4.16.

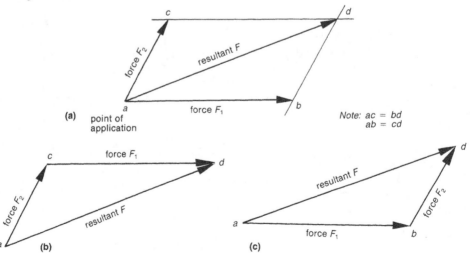

Fig. 4.16   **(a)** *Parallelogram of forces*   **(b)** *Triangle of forces (first alternative)*
(c) *Triangle of forces (second alternative)*

The force triangle rule states that by arranging given forces F_1 and F_2 in a tip-to-tail fashion, taking into account the scaled magnitude and direction of the forces, the resultant of the two forces is found by connecting the tail of one with the tip of the other, as shown in Figure 4.16.

Although closely related, the parallelogram of forces and the triangle of forces differ in one very important respect: the parallelogram of forces shows all given and resultant forces as passing through a common point, the actual point of application of the forces. In this respect, the parallelogram is a geometrical representation of a fundamental principle showing true relationships between all forces. In a triangle of forces, one of the forces does not pass through the point of application. The triangle construction can only be regarded as a graphical rule or method for determining the magnitude and direction of the resultant force. As such, the triangle of forces should always be drawn as a separate 'force diagram', and not superimposed on the diagram showing the actual layout of forces in relation to the point of application.

Example 4.5

A damaged vehicle is being pulled by means of two ropes as shown in Figure 4.17(a). Determine the resultant force in magnitude and direction, using the triangle of forces rule.

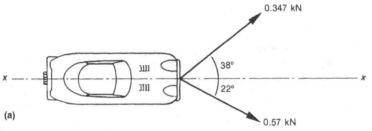

Solution

To solve the problem, construct the triangle of forces according to the rule, and scale-off the magnitude and direction (angle).

The answer is 0.8 kN in the *x*-direction, acting to the right.

Fig. 4.17

The polygon of forces method

The rule or method for finding resultants of force systems involving more than two forces consists of repeated applications of the force triangle rule to successive pairs of forces, until all the given forces are reduced to a single resultant force. This is called the **polygon of forces method**.

Example 4.6

Determine the resultant force on the eye bolt used to anchor four guy wires as shown in Figure 4.18(a).

Solution

Construct the polygon of forces by successive addition of forces, as shown in Figure 4.18(b) on page 62.

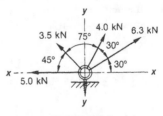

Fig. 4.18(a)

The answer is a vertical resultant force of 9.1 kN.

Note that the results of intermediate addition steps shown by dotted lines are usually omitted; only the outline of the force polygon is needed.

Fig. 4.18(b)

Before leaving the subject, be warned against losing sight of the real physical significance of the answers obtained. This should not be obscured by the method or construction used to obtain an answer. Thus in Example 4.5, the combined effect of the applied forces is that the vehicle is being pulled forward with a force equal to 0.8 kN. Likewise, in Example 4.6, the combined pull of the guy wires produces an upward force of 9.1 kN acting on the bolt. Always try to visualise the answers in tangible physical terms, as illustrated in Figure 4.19, and do not think of the method or construction used to obtain the answers as the end in itself.

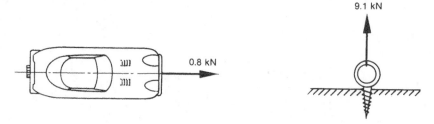

Fig. 4.19

4.7 *MATHEMATICAL ADDITION OF FORCES*

Addition of forces, i.e. solving for the resultant of a system of forces, can also be achieved mathematically by summing their x and y components. The method of solution consists of several steps, as follows:

Step 1
Resolve given forces into x and y components (usually horizontal and vertical), using $F_x = F \cos \theta$ and $F_y = F \sin \theta$, where θ is the acute angle between each force and the x-axis (horizontal axis).

Step 2
Assign positive and negative signs to each component, according to the usual mathematical sign convention, i.e. to the right—positive; to the left—negative; upwards—positive; downwards—negative.

Step 3
Add all x-components, ΣF_x,* taking into account the positive and negative signs, then add all y-components, ΣF_y, taking signs into account.

---

* Σ is a mathematical sign meaning 'the sum of', pronounced 'sigma'. It is a letter of the Greek alphabet.

Step 4

The two sums can now be used to determine the resultant force, using

$F = \sqrt{(\Sigma F_x)^2 + (\Sigma F_y)^2}$.

Step 5

Determine the angle the resultant makes with the x-direction, using $\tan \theta = \dfrac{\Sigma F_y}{\Sigma F_x}$.

In summary, this method involves resolving forces into rectangular components (steps 1 and 2), reducing all components to a single force in each of the two directions x and y (step 3), and adding these two remaining forces into the resultant (steps 4 and 5), according to the rules explained in Section 4.5.

The procedure is greatly assisted by the use of a table for recording all intermediate results, as illustrated in the following example.

Example 4.7

Determine the resultant force acting on the barge due to the combined effort of four tugboats as shown in Figure 4.20(a).

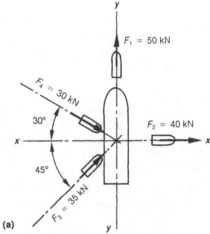

(a)

Solution

The solution is carried out according to the procedure outlined above, with the aid of a diagram as illustrated in Figure 4.20(b), if necessary.

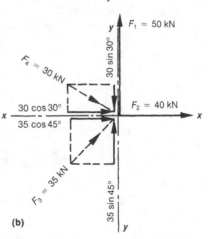

(b)

Fig. 4.20   *(continued)*

The results are tabulated in Table 4.1.

Table 4.1

Force	Magnitude	x-component	y-component
F_1	50	0	50.0
F_2	40	40.0	0
F_3	35	24.7	24.7
F_4	30	26.0	−15.0
		$\Sigma F_x = \overline{90.7}$	$\Sigma F_y = \overline{59.7}$

Resultant:

$$F = \sqrt{90.7^2 + 59.7^2}$$
$$= 108.6 \text{ kN}$$

Angle:

$$\tan \theta = \frac{59.7}{90.7}$$
$$= 0.658$$
$$\therefore \ \theta = 33.4°$$

The resultant force is as shown in Figure 4.20(c).

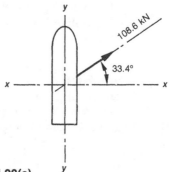

Fig. 4.20(c)

Finally, it is always a good idea to cross-check the solution using an independent alternative method. A mathematical solution can be verified graphically, and vice versa. Can you solve this example graphically? Example 4.6 mathematically?

Problems

4.9 For each of the systems of forces acting at a point in Figure 4.21, determine the magnitude and direction of the resultant by constructing polygons of forces.

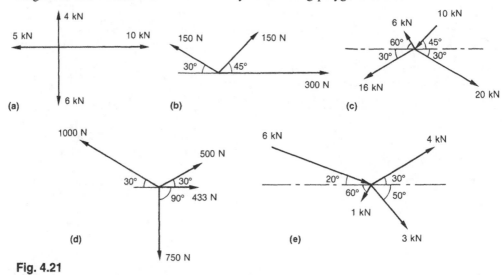

Fig. 4.21

4.10 Solve problem 4.9 mathematically.

4.8   *UNIVERSAL GRAVITATION*

Gravity is one of the most common physical phenomena seen in nature; it manifests itself as a **force** of mutual attraction between masses.

Historically, our understanding of the law of universal gravitation was a result of many centuries of astronomical observations culminating in the work of Johannes Kepler (1571–1630) who discovered important regularities in the motion of the planets, lending support to the Copernican heliocentric theory of the solar system.

However, it was the mathematical insight of Isaac Newton that enabled the formulation of the law relating the force of mutual attraction between two bodies of known masses to the square of the distance between them.

Newton's **law of universal gravitation** states that the force of gravitational attraction (F_g) between two bodies having masses m_1 and m_2 separated by a distance d is given by:

$$F_g = G \frac{m_1 m_2}{d^2}$$

It is important to understand that the constant G, known as the **universal gravitational constant,** has a numerical value which is the same for any pair of bodies, regardless of their masses or the distance separating them. The actual value of G was measured experimentally by Henry Cavendish in 1798. Significant improvements in the accuracy of the measurement were achieved in the 19th and 20th centuries. The present accepted value, determined at the United States National Bureau of Standards, is:

$$G = 66.7 \times 10^{-12} \ \frac{N.m^2}{kg^2}$$

Example 4.8

Determine the force of mutual attraction between the following pairs of bodies:
(a) the Earth and the Moon, given the mass of Earth as 5.97×10^{24} kg, the mass of the Moon as 73.7×10^{21} kg and the distance between them as 0.38×10^6 km;
(b) two ships, 30 000 t each, at a centre-to-centre distance of 50 m;
(c) two 1 kg masses, at a distance of 1 m.

Solution

(a) $F_g = 66.7 \times 10^{-12} \ \dfrac{N.m^2}{kg^2} \times \dfrac{5.97 \times 10^{24} \ kg \times 73.7 \times 10^{21} \ kg}{(0.38 \times 10^9 \ m)^2}$

$= 0.203 \times 10^{21}$ N

This is the huge pull between the Earth and the Moon which keeps the Moon in its orbit.

(b) $F_g = 66.7 \times 10^{-12} \ \dfrac{N.m^2}{kg^2} \times \dfrac{30 \times 10^6 \ kg \times 30 \times 10^6 \ kg}{(50 \ m)^2}$

$= 24$ N

In comparison with other forces acting on the ships, a force of 24 N is totally insignificant.

(c) $$F_g = 66.7 \times 10^{-12} \; \frac{\text{N.m}^2}{\text{kg}^2} \times \frac{1 \text{ kg} \times 1 \text{ kg}}{(1 \text{ m})^2}$$

$$= 66.7 \times 10^{-12} \text{ N}$$

This is a tiny force which is of no particular interest to the engineer.

It may appear at this point that the law of universal gravitation is of no particular practical use to the engineer, since forces of attraction between objects, even as large as ships, are very small indeed. However, as we shall see in the next section, the understanding of the law is necessary for developing the concept of weight.

4.9 *WEIGHT OF A BODY*

The engineer is concerned with structures and machines which are located on, or very near, the surface of the Earth, i.e. at nearly constant distance from the centre of the Earth, equal to its mean radius of 6370 km. If this distance is taken as constant for all such objects, the law of universal gravitation can be reduced to a special case applicable at or near the Earth's surface, as follows:

$$F_g = G \frac{m_e \times m_o}{r_e^{\,2}}$$

where m_e is the mass of the Earth $= 5.97 \times 10^{24}$ kg
  r is the mean radius of the Earth $= 6.37 \times 10^6$ m
  m_o is the mass of a given object

Substitution and combining of constants yields:

$$F_g \approx 66.7 \times 10^{-12} \; \frac{\text{N.m}^2}{\text{kg}^2} \times \frac{5.97 \times 10^{24} \text{ kg} \times m_o \text{ kg}}{(6.37 \times 10^6 \text{ m})^2}$$

$$\approx 9.81 \; \frac{\text{N}}{\text{kg}} \times m_o \text{ kg}$$

The new constant we have obtained is called the **local gravitational constant** with the symbol *g*.

In science and engineering, the force of gravity exerted by the Earth on an object is often referred to as the **weight of the object**, F_w, given by:

$$\boxed{F_w = mg}$$

where *m* is the mass of the object, and:

$$\boxed{g = 9.81 \; \frac{\text{N}}{\text{kg}}}$$

Being a force, weight is measured in units of force, i.e. newtons. In this regard, one must not be misled by the common usage of the word 'weight' as equivalent to mass, or quantity. To the engineer, mass is a measure of quantity expressed in kilograms, and *weight is a measure of gravitational force expressed in newtons.*

The force of gravity on an object, or weight, is always acting towards the centre of the Earth, i.e. vertically downwards, and is applied to the object at the centre of its mass distribution known as its **centre of gravity**. When the mass of an object is distributed uniformly throughout its volume, the centre of gravity coincides with the geometrical centre of the shape, e.g. the centre of gravity of a uniform solid sphere is at its geometrical centre.

Example 4.9

A man has a mass of 79 kg. What is his weight?

Solution

$$F_w = mg$$
$$= 79 \text{ kg} \times 9.81 \; \frac{\text{N}}{\text{kg}}$$
$$= 775 \text{ N}$$

Example 4.10

What is the force in a cable supporting a load of 1.5 tonnes?

Solution

$$F = F_w$$
$$= mg$$
$$= 1500 \text{ kg} \times 9.81 \; \frac{\text{N}}{\text{kg}}$$
$$= 14 \; 715 \text{ N}$$
$$= 14.7 \text{ kN}$$

4.10 *LOCAL VARIATIONS IN GRAVITY*

In the previous section the value of the local gravitational constant g was assumed to be the same for all places on the Earth's surface. However, this is true only if the distance to the centre of the Earth is the same for all locations, which is not strictly correct.

Because the Earth is slightly ellipsoidal in shape and not a perfect sphere, there is a small gradual variation in g with latitude, as can be seen from Table 4.2. These values are average values only, because there are other local influences owing to the nature of the underlying rocks etc. at different locations along the same latitude.

Table 4.2   *Variation of g at sea-level*

Latitude	equator	10°	20°	30°	40°	50°	60°	70°	80°	90°
g (N/kg)	9.780	9.782	9.786	9.793	9.802	9.811	9.819	9.826	9.831	9.832

Gravity also varies with altitude above sea-level, as shown in Table 4.3 for a latitude of 45°.

Table 4.3   *Variation of g with altitude*

Altitude (km)	sea-level	1	2	5	10	30	100
g (N/kg)	9.806	9.803	9.800	9.791	9.776	9.714	9.598

The overall average, known as the International Standard value, has been defined as $g = 9.806\ 65$ N/kg at sea-level. The local value for Sydney (Australia) is 9.796 83 N/kg.

It is common practice to use the value of 9.81 N/kg for most engineering calculations.* It can easily be seen that the error involved in using this value instead of the more accurate local value is quite insignificant, except for very high altitudes.

 # Problems

4.11 Determine the magnitude of the gravitational attraction between the Sun and the Earth, given that the mass of the Sun is 1.99×10^{30} kg, that of the Earth is 5.97×10^{24} kg, and the distance between them is 1.5×10^8 km.

4.12 Determine the weight of the following:
(a) a one gram mass
(b) a one kilogram mass
(c) a one tonne mass
(d) a brick of mass 4 kg
(e) a 50 g egg

4.13 Determine the weight of a truck which has a mass of 2.3 tonnes.

4.14 Determine the reading in newtons on each of the spring balances shown in Figure 4.22.

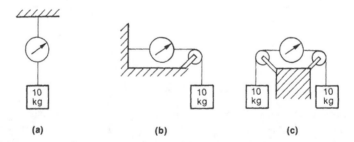

Fig. 4.22

4.15 A vehicle having a mass of 1.35 tonnes is standing on an incline at 15° to the horizontal. Determine the components of its weight parallel and perpendicular to the road surface.

4.16 If the tractive effort applied at the wheels of the vehicle in problem 4.15 is equivalent to a force of 4 kN acting up along the incline, determine the resultant of the weight and the tractive effort.

---

* In engineering practice, an occasional approximation to 9.8 N/kg may be quite acceptable when dealing with other parameters which are not capable of being measured with a high degree of precision. However, the use of the 'nice round figure' of 10 N/kg, as found in some secondary school physics texts, is strongly discouraged because it tends to give the erroneous impression that the gravitational constant is an exact decimal factor, which it is not. Throughout this book, 9.81 N/kg is used consistently.

4.17 In a laboratory experiment, forces are applied to a bar by means of attaching different masses as shown in Figure 4.23. The mass of the bar itself is 600 g.

Draw a diagram of the bar showing all forces acting on the bar, including the weight of the bar. Indicate the correct magnitude and direction of each force.

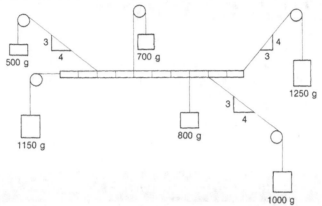

Fig. 4.23

Review questions

1. What is meant by *force*?
2. What is the SI unit of force?
3. Which one of the following cannot be measured in newtons?
   muscular effort, gravitational attraction, quantity of matter, frictional resistance.
   Why?
4. Describe a simple spring dynamometer.
5. Name the three main characteristics of a force.
6. What is the principle of action and reaction?
7. What is meant by *transmissibility* of a force?
8. Explain the principle of the parallelogram of forces.
9. What is meant by the *resultant* of two or more forces?
10. State the formulae for resolving a force into rectangular components.
11. State the formulae for combining two forces at 90° to each other into a resultant force.
12. Explain how the force triangle and the force polygon can be used for graphical addition of forces.
13. Outline the steps for the mathematical addition of forces.
14. If the mathematical solution to a problem is compared with the graphical solution, would you expect the answers to be the same?
15. State the *law of universal gravitation*.
16. What is the value of the universal gravitational constant?
17. Which is greater: the attraction of the Earth for the Moon, or the attraction of the Moon for the Earth?
18. What is meant by *weight*?
19. What is the SI unit of weight?
20. If a body is taken from the equator to the South Pole, what will be the effect on its mass and on its weight?
21. What is the value of the local gravitational constant generally used in practical engineering calculations? Are there any limitations on its accuracy?
22. Is it possible to use a spring balance to measure mass? Explain.

CHAPTER 5

Concurrent forces

When a structure or a machine component is at rest, it is said to be in **static equilibrium**. In general, the state of **equilibrium** can be defined as a state of rest or balance under the action of forces which counteract each other.

This chapter is concerned with conditions of equilibrium of **concurrent forces**. A system of forces is called concurrent when the lines of action of all forces intersect at a common point, the point of concurrency.

Expected learning outcomes

After carefully studying the material presented in this chapter, working through all numerical examples, and successfully completing all practice problems, students should be able to:
1. state the conditions of equilibrium for a system of concurrent forces;
2. define and determine, mathematically and graphically, the equilibrant of a system of concurrent forces;
3. determine unknown forces in a system of concurrent forces when the lines of action of all forces are known;
4. locate the point of concurrency and determine non-parallel reaction forces using the three-force principle.

5.1 CONDITIONS OF EQUILIBRIUM

For a system of concurrent forces to be in equilibrium, the resultant force, i.e. the result of summation of all forces acting through that point, must be equal to zero. In other words, all forces balance each other in such a way that there is no resultant push or pull acting on the body at the point of application of the forces.

To prove that a given system of forces is in fact in equilibrium, we must demonstrate, graphically or mathematically, that the forces add up to zero. Naturally, since forces are vector quantities, the addition must be vectorial addition as explained in Chapter 4. Consider the following example.

Example 5.1

A joint of a pin-jointed truss is as shown in Figure 5.1(a) and has internal and external forces as shown. Prove that the joint is in equilibrium.

Solution

In order to prove the equilibrium of this system of concurrent forces *graphically*, it is necessary to construct the polygon of forces, using all of the applied forces, external (F_1) as well as internal $(F_2, F_3$ and $F_4)$ forces, and to show that the starting point of the construction coincides with the end point, i.e. the polygon must close, leaving no room for a gap representing a resultant force. (*Remember*: The resultant must be zero.) See Figure 5.1(b).

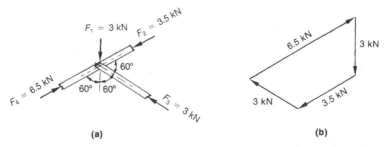

(a)   **(b)**

Fig. 5.1

Mathematically, the summation of forces is signified by ΣF, meaning that the resultant force is the sum of all the applied forces. For a system in equilibrium, the resultant is equal to zero:

$$\Sigma F = 0$$

This is the very useful equation of forces in equilibrium. However, this equation is often interpreted in terms of the perpendicular components of the resultant force. If the resultant is equal to zero, its components must also be equal to zero, i.e. no force—no components. It follows that, for a given system of forces, the sum of all components in any direction must be zero. This is usually expressed in terms of mutually perpendicular directions x and y, often horizontal and vertical.

$$\Sigma F_x = 0 \qquad \text{and} \qquad \Sigma F_y = 0$$

Now let us consider a mathematical solution to the previous example.

Alternative solution

To prove the equilibrium of forces *mathematically*, it is necessary to demonstrate that all vertical components and all horizontal components both add up to zero. We can use a table similar to Table 4.1 to record all the force components, including their positive and negative signs. The solution will appear as in Table 5.1.

Table 5.1

Force	Magnitude	x-component	y-component
F_1	3.0	0	–3.0
F_2	3.5	–3.03	–1.75
F_3	3.0	–2.60	1.50
F_4	6.5	5.63	3.25
		$\Sigma F_x = 0$	$\Sigma F_y = 0$

The sum of the horizontal, or *x*-direction, components and the sum of the vertical, or *y*-direction, components have been shown to be zero, i.e. the two equations $\Sigma F_x = 0$ and $\Sigma F_y = 0$ have each been satisfied. The conclusion is that the system of forces is in equilibrium, as we would expect in a structure such as a roof truss.

5.2 THE EQUILIBRANT FORCE

If a system of concurrent forces is not balanced, i.e. not in equilibrium, it is possible to determine the additional force required to produce equilibrium. Such a force is called the **equilibrant**. In other words, the equilibrant of a system of concurrent forces is that force, which when added to the system, produces equilibrium. In the event of a balanced system of forces, the equilibrant force would be equal to zero.

Example 5.2

Determine the equilibrant force (F_e) for the system of forces shown in Figure 5.2(a).

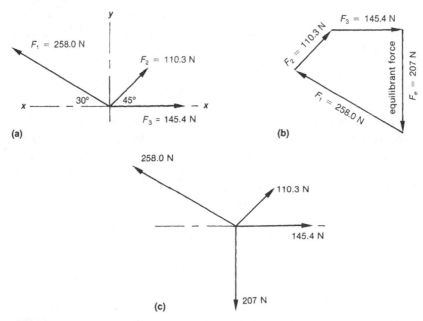

Fig. 5.2

Solution

To solve this problem *graphically*, we must recall that for a system of forces in equilibrium, the force polygon must close. If we attempt to construct the force polygon, using only the given forces F_1, F_2 and F_3 and arranging them in head-to-tail order, we find that the first and last points do not coincide. In order to close the gap an additional line is required, as shown in Figure 5.2(b). This line will close the polygon and represent the required equilibrant force, both in magnitude and direction. The system of forces in equilibrium is shown in Figure 5.2(c).

It is important to remember that the closed polygon of forces has all its forces, including the equilibrant force, follow the head-to-tail order. This enables us to determine the correct direction of the equilibrant force; in this example the equilibrant is a vertical downward force.

It should also be understood that the equilibrant and the resultant forces are always equal in magnitude and opposite in direction.

Alternative solution

The equilibrant force can also be found *mathematically* by means of addition of rectangular components, including those for the unknown equilibrant force.

The solution is best set out in tabular form as in Table 5.2:

Table 5.2

Force	Magnitude	x-component	y-component
F_1	258.0	−223.4	129.0
F_2	110.3	78.0	78.0
F_3	145.4	145.4	0
equilibrant	F_e	F_x	F_y

For equilibrium:

$$\Sigma F_x = -223.4 + 78.0 + 145.4 + F_x = 0$$
$$\therefore F_x = 0$$

Also:

$$\Sigma F_y = 129.0 + 78.0 + 0 + F_y = 0$$
$$\therefore F_y = -207 \text{ N}$$

The magnitude of the equilibrant force is therefore:

$$F_e = \sqrt{F_x^2 + F_y^2}$$
$$= \sqrt{0 + 207^2}$$
$$= 207 \text{ N}$$

Consideration of the directions of components F_x and F_y indicates that, in this case, the equilibrant is acting vertically downwards.

 Problem

5.1 Determine, graphically and mathematically, the equilibrant for each of the systems of forces in Figure 5.3.

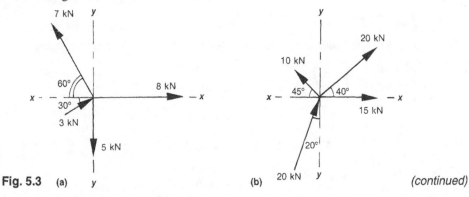

Fig. 5.3   **(a)**   **(b)**   *(continued)*

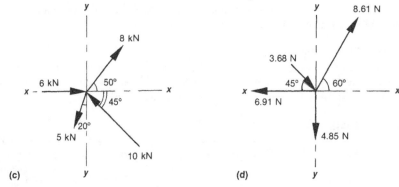

(c)   (d)

Fig. 5.3

5.3 *FREE-BODY DIAGRAMS*

When problems involving the equilibrium of bodies under the action of force systems are being solved, a method of setting out the essential details in the form of a diagram is necessary.

We often start with a semipictorial sketch or space diagram showing the physical conditions of the problem, i.e. a layout of the mechanical or structural components such as pulleys, supports, cables, rollers etc.

The next step is to isolate the essential facts about the forces involved and to draw a **free-body diagram**. Such a diagram usually shows the point of concurrency acted upon by all the forces, indicating the magnitudes and directions of the forces.

A separate force polygon is then constructed, based on the information contained in the free-body diagram.

Example 5.3

Two cranes are supporting a 5 tonne mass as shown in Figure 5.4(a). Consider the equilibrium of forces and hence determine the force in each cable.

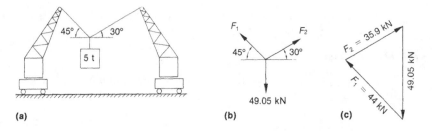

(a)   (b)   (c)

Fig. 5.4   (a) *Space diagram*   **(b)** *Free-body diagram*   **(c)** *Force triangle*

Solution

The weight of the load is:

$$F_w = mg$$
$$= 5000 \text{ kg} \times 9.81 \ \frac{\text{N}}{\text{kg}}$$
$$= 49\,050 \text{ N}$$
$$= 49.05 \text{ kN}$$

The free-body diagram, Figure 5.4(b), is drawn to represent all forces acting on the point of concurrency. Note that each force is represented as a vector, showing its magnitude and direction in relation to the point.

Knowing one force (i.e. the weight) and the direction of the other two forces enables us to construct the force triangle. (Note that a triangle is just a special case of force polygon where the number of forces involved is three.) If the triangle of forces is drawn to scale, the answers can easily be scaled off.

Alternatively, the sine rule can be used to calculate the unknown forces as side lengths of the force triangle. In order to achieve this, all three internal angles of the force triangle must be calculated first. Be careful here — internal angles of the force triangle are related to the angles shown on the free-body diagram, but are not *directly* equal to them. Examine Figure 5.5 carefully and make sure that you understand how the internal angles were obtained.

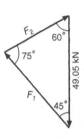

Fig. 5.5

The sine rule yields:

$$\frac{49.05}{\sin 75°} = \frac{F_1}{\sin 60°} = \frac{F_2}{\sin 45°}$$

Hence:

$$F_1 = 49.05 \times \frac{\sin 60°}{\sin 75°}$$
$$= 43.98 \text{ kN}$$

$$F_2 = 49.05 \times \frac{\sin 45°}{\sin 75°}$$
$$= 35.91 \text{ kN}$$

Example 5.4

An elastic member ABC is stretched as shown in Figure 5.6(a) by three forces F_1, F_2 and F_3.

Determine the forces in AB and BC.

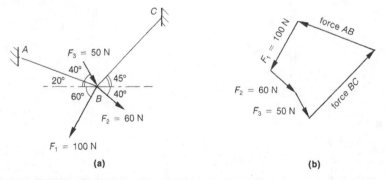

Fig. 5.6 (a) *Free-body diagram* (b) *Force polygon*

Solution

In this example more than three forces are involved. Therefore, a polygon of forces, rather than a triangle, must be constructed, as in Figure 5.6(b).

The procedure is to construct as much of the polygon as possible using all known forces and then to complete it by drawing two lines, parallel to the unknown forces, through the first and last points.

The answers are scaled off the force polygon and are:

$$F_{AB} = 148 \text{ N}$$
$$F_{BC} = 167 \text{ N}^*$$

 Problems

5.2 A light fitting of mass 1.5 kg is hanging from the ceiling on wire *AB* and is tied to the wall by string *BC* (Fig. 5.7). Determine the forces in *AB* and *BC*.

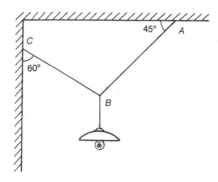

Fig. 5.7

5.3 A streetlight of mass 15 kg is supported at midpoint between two poles by a cable *ABC* (Fig. 5.8). If the length of the cable *ABC* is 20 m and distance *BD* at midpoint is 2 m, determine the force in the cable.

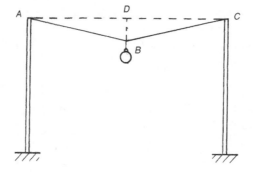

Fig. 5.8

---

* An alternative mathematical solution is also possible, but is rather complex and involves two simultaneous equations in terms of components of unknown forces. It is not recommended.

5.4 The jib-crane in Figure 5.9 carries a load of 1.4 tonnes. Determine the forces in the jib and the tie.

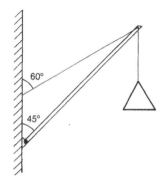

Fig. 5.9

5.5 Determine the forces in *AB*, *BC*, *CD* and *BE* (Fig. 5.10) if the mass *m* = 500 kg.

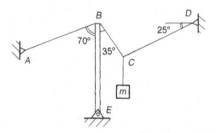

Fig. 5.10

5.6 Determine the force acting through the axis of the pulley shown (Fig. 5.11). Does the magnitude or direction of the force depend on pulley diameter?

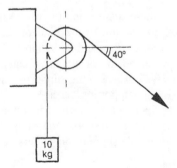

Fig. 5.11

5.7 Determine the mass lifted and the forces in members *AB* and *BC* (Fig. 5.12) when the tension in the cable is 20 kN.

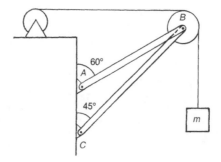

Fig. 5.12

5.8 Figure 5.13 represents joints in a simple pin-jointed roof truss. Using the information available for each joint, determine the unknown forces by constructing a separate polygon of forces for each joint.

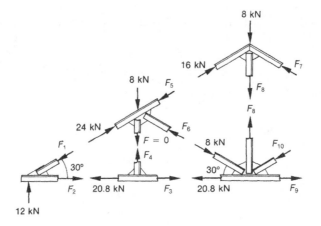

Fig. 5.13

5.4   *SUPPORT REACTIONS*

Before proceeding to the next section, it is necessary to discuss different types of contact surfaces or supports in relation to the kind of reaction force that may exist between a body, such as a beam or a truss, and its supporting surface.

In this chapter we need to consider two categories of supports: those at which the reaction force is always *normal* to the supporting surface, and those which can support a force at *any angle* to the supporting surface.

The first category includes a *smooth*, frictionless surface contact between a body and a supporting surface, such that the body can slide along the surface at the point of contact without any resistance.[*] A support on rollers, e.g. as seen under some long bridge

---

[*] Friction is discussed in Chapter 10.

sections, has a similar force action, i.e. it provides normal reaction only.* The rule to remember is that at a smooth surface or at a roller support, the reacting force is always perpendicular to the supporting surface.

The second category includes a *rough* surface contact between a body and a supporting surface in which friction prevents relative sliding motion, producing a reaction force in any direction. A fixed hinge or bearing has a similar force action capable of supporting a force at any angle to the supporting surface.

In addition to the above-mentioned reaction forces, it is helpful to recognise that forces in links pivoted at each end and in cables connecting two points of a structure always act along the axes of such members (see Table 5.3).

Table 5.3   *Classification of supports, connections and contact surfaces*

1. *Normal reaction only*

   (a) smooth contact surface   (b) ball or roller support

2. *Force along axis of member*

   (a) cable   (b) link

3. *Reaction at any angle*

   (a) rough contact surface   (b) pin or hinge

5.5   *THE THREE-FORCE PRINCIPLE*

A particular case of equilibrium which is of considerable interest is that of a body in equilibrium under the action of three non-parallel forces. Such a body is usually called a **three-force body**.

---

* The normal to a line or surface is a direction perpendicular to it, e.g. normal force is perpendicular to the supporting surface.

It can be shown that if a three-force body is in equilibrium, the lines of action of the three forces must intersect at a common point.[*] This principle, known as the **three-force principle**, is very useful in the solution of many engineering problems as it helps to determine easily the direction of an unknown force without recourse to more complex mathematical methods.

Example 5.5

A 2.5 m long ladder of mass 10 kg rests on the floor and against a smooth wall as shown in Figure 5.14(a) below. Determine the reactions.

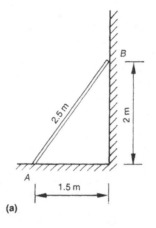

(a)

Solution

Draw a free-body diagram (Fig. 5.14(b)) showing weight:

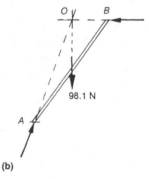

$$F_w = mg$$
$$= 10 \times 9.81$$
$$= 98.1 \text{ N}$$

(b)

acting through the midpoint, and a horizontal reaction on the smooth wall at *B*.

Extend their lines of action to intersect at point *O* and join point *O* with point *A*. This determines the direction of the reaction at *A*.

A triangle of forces can now be constructed (Fig 5.14(c)) and the reaction forces found:

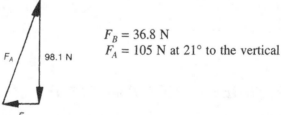

$F_B = 36.8$ N
$F_A = 105$ N at 21° to the vertical

Fig. 5.14   (c) F_B

Example 5.6

The truss shown in Figure 5.15 is subjected to wind loads equivalent to a force of 10 kN applied at 90° to the member at the joint as shown. Determine the reactions at *A* and *B*.

---

[*] The only exception to this principle is when the forces are parallel.

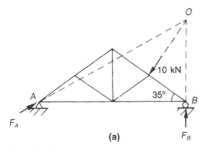

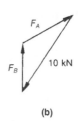

Fig. 5.15

Solution

The normal reaction will be at the roller support at *B*. Therefore, draw a vertical line through *B* to intersect with the line of action of the load at point *O*.

Join *O* and *A* and construct the force triangle. From the force triangle:

$$F_A = 6.5 \text{ kN}$$
$$F_B = 5.14 \text{ kN}$$

When solving problems using the three-force principle, always follow this sequence of steps:

Step 1
Draw a free-body diagram of the structure or component in essential outline only. All dimensions and angles must be drawn to scale.

Step 2
Draw the line of action of the known force and identify its magnitude and direction. (Force representing weight should always be drawn vertically downwards. Any other force must be drawn as given in the problem.)

Step 3
Consider the nature of supports and draw the line of action of the second force *either* normal to the smooth surface or roller support surface *or* along a link or a cable, as the case may be.

Step 4
Extend the two lines of action drawn in steps 2 and 3 as far as necessary until they intersect at a point, i.e. locate the point of concurrency.

Step 5
Join the point of concurrency with the point of application of the third force, which is usually at the other support. This defines the line of action of the third force, and hence its angle to the horizontal.

Step 6
Draw the force triangle as a separate diagram. (Sides representing forces must be to scale and parallel to the lines of action of the forces.)

Step 7
Scale off the force magnitudes from the force triangle. Directional sense is established by the head-to-tail sequence of arrows, starting with the known force.

Now you are ready to tackle the following problems, but do not forget to refer back to the seven steps outlined above.

Problems

5.9 A 120 N force is required to operate the foot pedal shown in Figure 5.16. Determine the force on the connecting link and the reaction at the bearing.

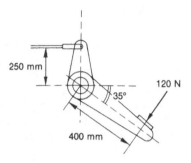

Fig. 5.16

5.10 From Figure 5.17, determine the force in the supporting cable *AB* and the reaction at *C* for the following conditions of loading on the jib-crane shown.
(**a**) $m = 1$ t, $x = 1$ m   (**c**) $m = 2$ t, $x = 2$ m
(**b**) $m = 1$ t, $x = 4$ m   (**d**) $m = 2$ t, $x = 3$ m

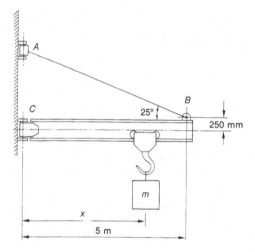

Fig. 5.17

5.11 A beam is subjected to a load of 10 kN as shown in Figure 5.18. Determine the support reactions at each end for:
(**a**) $\theta = 20°$
(**b**) $\theta = 40°$
(**c**) $\theta = 60°$

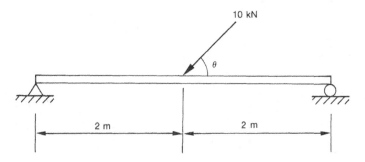

Fig. 5.18

5.12 A streetlight has a mass of 20 kg and is supported as shown in Figure 5.19. Determine the reactions at *A* and *B*.

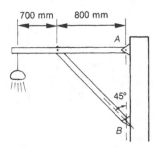

Fig. 5.19

5.13 An advertising sign is supported as shown in Figure 5.20. Determine the force in *AB* and the reaction at *C* if the mass of the sign is 30 kg.

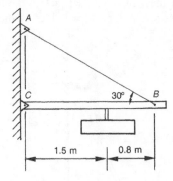

Fig. 5.20

5.14 Determine the force *F* required to start to tip the cabinet shown (Fig. 5.21) about *A*, assuming it is not going to slide. Determine also the reaction at *A*. The mass of the cabinet is 50 kg.

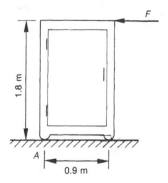

Fig. 5.21

5.15 A pin-jointed frame supporting a scoreboard is subjected to a horizontal wind load as shown in Figure 5.22. Determine reactions at *A* and *B*.

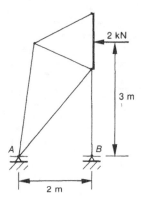

Fig. 5.22

5.16 A ladder rests on a rough floor surface and against a smooth wall as shown in Figure 5.23. Neglecting the mass of the ladder, determine the reactions at *A* and *B* when a man of mass 85 kg climbs to a point on the ladder as shown.

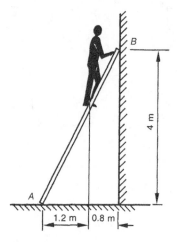

Fig. 5.23

5.17 Determine the force F required to pull a 500 mm diameter roller, having a mass of 100 kg, over a step 60 mm high, and the reaction at the point of contact (Fig. 5.24).

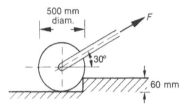

Fig. 5.24

5.18 Find the force in cable AB and the reaction at C when a load of 500 kg is supported by a flexible rope passing over a sheave as shown in Figure 5.25.

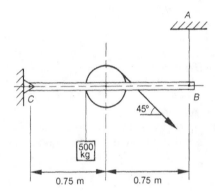

Fig. 5.25

5.19 The door (viewed from above) in Figure 5.26 is held in a closed position by tension in the spring equal to 70 N. Determine the reaction at the hinge.

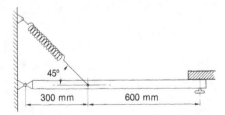

Fig. 5.26

5.20 For each metre of its length, the retaining wall shown in Figure 5.27 has a mass of 1.8 t, and the pressure of the earth behind it is equivalent to a force of 7.57 kN. The point at which the reaction at the base can be regarded as applying is point A. Determine the reaction, in magnitude and direction, and the position x of point A.

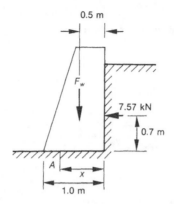

Fig. 5.27

Review questions

1. What is meant by *equilibrium of forces*?
2. Define *concurrent forces*.
3. What is the condition for equilibrium of a concurrent force system?
4. State two equations commonly used to express the conditions of equilibrium.
5. What is a *free-body diagram*?
6. Define *equilibrant force*.
7. What is a *three-force body*?
8. State the *three-force principle*.
9. What is the only exception to the three-force principle?
10. Explain the difference between smooth and rough surface contact.

CHAPTER 6

Non-concurrent forces

This chapter is about non-concurrent force systems. A system of forces is called **non-concurrent** if it does not have a single point of concurrency, i.e. the lines of action of the forces do not all meet at a common point.

When forces are non-concurrent, there is always a possibility that they may have a turning effect with respect to the structure or mechanical component to which they are applied. The concept of **moment of a force** is introduced; it is a measure of a force's turning effect about some reference point. Equilibrium of non-concurrent forces requires that, in addition to the force balance, the moments of all forces must also be in balance.

Expected learning outcomes

After carefully studying the material presented in this chapter, working through all numerical examples, and successfully completing all practice problems, students should be able to:

1. define *moment of a force* and calculate moments of non-concurrent forces about a chosen reference point;
2. determine the resultant of several non-concurrent forces and locate the resultant relative to a specified reference point;
3. define a *force couple* and express the effect of a given force acting on a body as an equivalent force-couple system acting elsewhere on the body.

6.1 *MOMENT OF A FORCE*

At the very dawn of physical science, Archimedes is reputed to have declared: 'Give me a place to stand, and I can move the Earth'. This was not just an idle boast, but a poetical assertion of the law of the lever, which Archimedes understood and developed mathematically. The lever is a simple device, one of the earliest ever used. Levers enable us to multiply the force available in our own hands to move very large loads. Today, the lever, in various forms and combinations, is the most common machine element.

The lever is only one example of the application of the principle of moments which is the subject of this chapter. Every time we drive a car, tighten up a nut, ride a bicycle or open a door, we make use of the turning effect of a force or forces, applied at some distance from the axis or fulcrum about which turning takes place.

The **moment of a force** about a point is defined as the product of the force and the perpendicular distance of its line of action from the point:

$$M = Fd$$

The perpendicular distance d is referred to as the **moment arm**.

The moment is a measure of the turning effect of the force acting on a body, relative to a specified point. As a product of force and distance, the moment of a force is measured in units derived from those of force and length. The SI unit of moment of force is the product of the newton and the metre, called **newton metre**, with a compound symbol N.m. The multiples and submultiples of the newton metre are formed by using decimal prefixes in front of the new unit, e.g. 1 kilonewton metre (kN.m) is equal to 1000 newton metres (N.m).

Example 6.1

If a force of 65 N is applied to the lever shown in Figure 6.1, and the length of the moment arm is 0.75 m, determine the moment of the force about the pivot point.

Solution

Moment is the product of the force and the perpendicular distance called the moment arm:

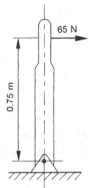

Fig. 6.1

$$M = Fd$$
$$= 65 \text{ N} \times 0.75 \text{ m}$$
$$= 48.75 \text{ N.m clockwise}$$

Note that the answer also indicates the directional sense of the moment, clockwise in this case. In this example, the actual length of the lever is in fact the moment arm at right angles to the force. In many problems, care must be exercised in using correct perpendicular distance with each force, as is illustrated by the following example.

Example 6.2

Determine the magnitude and sense of the moments of forces F_1, F_2 and F_3 about point A in Figure 6.2.

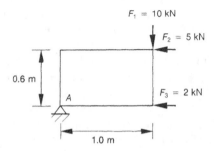

Fig. 6.2

Solution

Moment of force F_1:

$$M_1 = F_1 d_1$$
$$= 10 \text{ kN} \times 1.0 \text{ m}$$
$$= 10 \text{ kN.m clockwise}$$

Moment of force F_2:

$$M_2 = F_2 d_2$$
$$= 5 \text{ kN} \times 0.6 \text{ m}$$
$$= 3 \text{ kN.m anticlockwise}$$

Moment of force F_3:

$$M_3 = F_3 d_3$$
$$= 2 \text{ kN} \times 0$$
$$= 0$$

Note that force F_3 does not have a moment about point A due to the fact that its line of action passes through the point, i.e. its perpendicular distance from A is equal to zero. Students should study this example very carefully in order to understand which distance is in fact the moment arm for each of the forces in question. The diagrams in Figure 6.3 may be helpful in this regard.

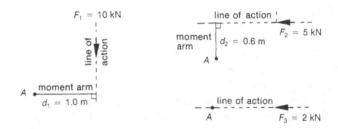

Fig. 6.3

If a force is inclined to the convenient principal directions, such as the horizontal and vertical directions, the correct perpendicular distance must be determined graphically or trigonometrically, as in the following example.

Example 6.3

Determine the moment of force $F = 50 \text{ N}$ about point A if the distance AB is 800 mm (Fig. 6.4).

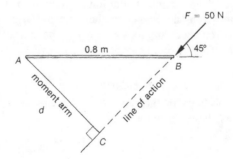

Fig. 6.4

Solution

In this case, the given distance AB is not the moment arm because AB is not perpendicular to the line of action of the force. To solve the problem, we need distance AC. This can be found graphically by drawing the diagram to scale and measuring the required distance *AC*. Alternatively, from the right-angled triangle *ABC*:

$$\text{Moment arm } d = AB \sin 45°$$
$$= 0.8 \times 0.707$$
$$= 0.566 \text{ m}$$

The moment of the force is therefore:

$$M = Fd$$
$$= 50 \text{ N} \times 0.566 \text{ m}$$
$$= 28.3 \text{ N.m clockwise}$$

6.2   ADDITION OF MOMENTS

If more than one force is acting on a body, as in Example 6.2, there is a corresponding number of moments of force, each tending to produce a turning effect about the point. The total turning effect, or resultant moment, is the algebraic sum of the moments of all the forces acting on the body. 'Algebraic sum' means that the different sense of the various moments must be taken into account when the moments are being added. A sign convention usually used is that clockwise moments are taken to be positive, and anticlockwise moments to be negative.

Returning to Example 6.2, the total moment about point *A* of all the forces is:

$$M = M_1 + M_2 + M_3$$
$$= +10 \text{ kN.m} - 3 \text{ kN.m} + 0$$
$$= 7 \text{ kN.m clockwise}$$

The Varignon theorem

An extension of the principle of addition of moments is known as the **Varignon theorem**, originally proposed by the French mathematician Varignon (1654–1722). It states that the moment of a force about any axis is equal to the sum of the moments of its components about that axis. This theorem can be used to solve Example 6.3 mathematically.

Mathematical solution of Example 6.3

To solve this problem mathematically, resolve force *F* into its horizontal and vertical components, as shown in Figure 6.5.

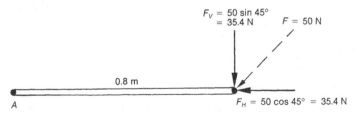

Fig. 6.5   *Resolving a force into its horizontal and vertical components*

As the moment of the horizontal component is zero, due to the line of action passing through point A, the total moment about A is:

$$M = F_V \times d + F_H \times 0$$
$$= 35.4 \text{ N} \times 0.8 \text{ m} + 0$$
$$= 28.3 \text{ N.m}$$

It is important to realise that for a given system of forces, there is no single answer for the moment of forces unless the reference point is specified. A moment must always be calculated with respect to a particular reference point. It must also be understood that in many problems there is no rotation actually taking place. The moment represents only the tendency for rotation under the influence of a force or forces, and not actual rotation. The following example should help to illustrate these points.

Example 6.4

A horizontal beam rests on two supports A and B, and supports three forces as shown in Figure 6.6. Calculate the total moment due to the applied forces:
(a) about the left-hand support A
(b) about the right-hand support B

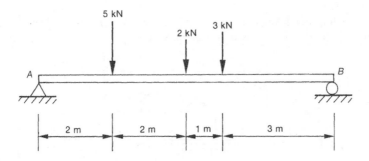

Fig. 6.6

Solution

(a) The total moment about A is the sum of the moments of all forces about A.
   (*Note*: All distances are measured from point A.)

$$M_A = \Sigma(F \times d)$$
$$= 5 \times 2 + 2 \times 4 + 3 \times 5$$
$$= 33 \text{ kN.m clockwise}$$

(b) The total moment about point B (distances measured from B) is:

$$M_B = \Sigma(F \times d)$$
$$= 5 \times 6 + 2 \times 4 + 3 \times 3$$
$$= 47 \text{ kN.m anticlockwise}$$

It is obvious that the moments about A and B are different, not only in magnitude, but also in sense. There is no actual rotation taking place. However, if the right-hand support B is suddenly removed, then rotation is produced by the turning moment M_A about A in the clockwise direction. If instead the left-hand support A is removed, then the beam turns anticlockwise about B under the influence of the moment M_B.

6.3 *EQUILIBRIUM OF TURNING MOMENTS*

When a structure or a machine component is in equilibrium, the moments, as well as forces, must be in a state of balance, otherwise the unbalanced resultant moment would cause rotation of the body. Thus for a body to be in rotational equilibrium, the resultant moment about any point must be equal to zero, i.e. there must be no resultant turning effect. Mathematically this can be stated as:

$$\boxed{\Sigma M = 0}$$

Alternatively this principle can be expressed in terms of equivalence of the sum of the clockwise moments about any point and the sum of anticlockwise moments about the same point. That.is, for equilibrium:

Sum of clockwise moments = sum of anticlockwise moments

The simplest case of this relation is the equivalence of moments of two forces in static equilibrium about some pivot point, as in the following example.

Example 6.5

The wheelbarrow shown in Figure 6.7 carries a load of 58 kg. Calculate the effort required to hold the handles in a stationary position.

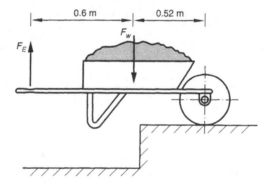

Fig. 6.7

Solution

First, calculate the weight of the load:

$$F_w = mg$$
$$= 58 \text{ kg} \times 9.81 \text{ N/kg}$$
$$= 569 \text{ N}$$

Now, equate the moments of the two forces about the pivot point, which is the axis of the wheel, and solve for the unknown force:

$$569 \times 0.52 = F_E \times 1.12$$
$$\therefore F_E = 264 \text{ N}$$

Hence the total effort required is 264 N, or 132 N per handle.

To prove that a given system of forces is in rotational equilibrium, we must demonstrate that the algebraic sum of all moments about any point is zero.

Example 6.6

A bridge structure is subjected to forces as shown in Figure 6.8. Prove that the structure is in rotational equilibrium.

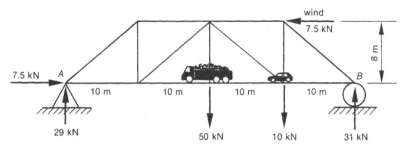

Fig. 6.8

Solution

Take the algebraic sum of all moments about the left-hand support A:

$$\Sigma M_A = 50 \times 20 + 10 \times 30 - 7.5 \times 8 - 31 \times 40 = 0$$
$$\therefore \ \Sigma M_A = 0$$

Therefore the moments are balanced about point A.

Take the algebraic sum of all moments about the right-hand support B:

$$\Sigma M_B = 29 \times 40 - 50 \times 20 - 10 \times 10 - 7.5 \times 8 = 0$$
$$\therefore \ \Sigma M_B = 0$$

Therefore the moments are balanced about point B.

6.4 GENERAL CONDITIONS OF EQUILIBRIUM

It has already been established that for a system of forces to be in equilibrium, the resultant force, i.e. the result of vectorial summation of all forces, must be equal to zero. That is, there must be no resultant push or pull.

This statement is often interpreted mathematically in terms of the perpendicular, usually horizontal and vertical, components of the resultant force, and can be stated simply as:

$$\Sigma F_x = 0 \text{ and } \Sigma F_y = 0$$

Only when both of these conditions are satisfied is the force equilibrium established.

These two conditions were found sufficient for the study of equilibrium of concurrent force systems, as discussed in Chapter 5. However, under certain conditions rotation may occur.

An additional condition for equilibrium of a system of non-concurrent forces must, therefore, ensure the absence of a turning moment, stated as:

$$\Sigma M = 0$$

Collectively, the three conditions of equilibrium are known as the **three equations of statics:**

1. The sum of x-components of all forces must equal zero:

$$\boxed{\Sigma F_x = 0}$$

2. The sum of y-components of all forces must equal zero:

$$\boxed{\Sigma F_y = 0}$$

3. The sum of moments of all forces about any point must equal zero:

$$\boxed{\Sigma M = 0}$$

These equations may be used to prove that a particular structure is in static equilibrium under the combined action of a system of non-concurrent forces.

Example 6.7

The structure in Figure 6.9 is subjected to forces as shown. Prove that the structure is in equilibrium.

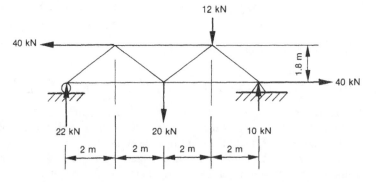

Fig. 6.9

Solution

Sum of horizontal forces:

$$\Sigma F_x = 40 \text{ kN} - 40 \text{ kN} = 0$$

Sum of vertical forces:

$$\Sigma F_y = 22 \text{ kN} + 10 \text{ kN} - 20 \text{ kN} - 12 \text{ kN} = 0$$

Sum of moments about left-hand support:[*]

$$\Sigma M = 20 \times 4 - 10 \times 8 + 12 \times 6 - 40 \times 1.8 = 0$$

All three equations are satisfied. The conclusion is that the structure is in static equilibrium.

Proving equilibrium of a static structure, such as a truss or a bridge span, when all forces (i.e. all loads and all reactions) acting on it are known and taken into account, does not in itself seem to be much more than a mathematical exercise of confirming by calculation the obvious fact that any static structure must always be in equilibrium.

It is quite a different matter when some of the forces, usually support reactions, are not given. In this case the three equations of statics become a powerful tool, which enables us to calculate up to three unknown reaction forces for any structure which is presumed to be in equilibrium. This will be made use of for the analysis of beams, frames and trusses in Chapters 7 to 9.

 ## Problems

6.1 A force at the end of a spanner applied 300 mm from the centre of a nut is 45 N. What is the maximum turning moment produced by this force? What should be the direction of the force to achieve the maximum turning effect?

6.2 A winding hoist has a drum of diameter 300 mm. Determine the moment about its centreline produced by tension in the cable equal to 0.5 kN.

6.3 A horizontal beam 2 m long is supported at its ends. Determine the moments about each of the supports due to the following loads:
(a) a downward force of 3 kN at midpoint
(b) a downward force of 3 kN at a point 0.5 m from the left support
(c) three downward forces, 1 kN each, located at 0.5 m, 1 m and 1.5 m from the left support
(d) a downward force of 1 kN and a downward force of 2 kN located at 0.5 m and 1.5 m respectively from the left support.
(e) a force of 3 kN applied at midpoint and inclined at 60° to the horizontal

6.4 The beam in Figure 6.10 is built into the wall and carries a load of 10 kN as shown.

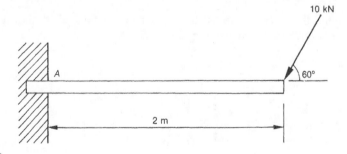

Fig. 6.10

---

[*] It is suggested that you should try calculating moments about some alternative point, e.g. the right-hand support, to be satisfied that the sum of the moments is equal to zero irrespective of the position of the reference point.

Determine the moment of the force about the support *A* by using:

(a) perpendicular distance to the force

(b) rectangular components of the force

6.5 The truss shown in Figure 6.11 is subjected to three loads as indicated. Determine the total moment, due to the applied forces, about:

(a) left-hand support *A*

(b) right-hand support *B*

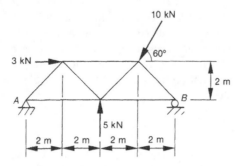

Fig. 6.11

6.6 The arm shown in Figure 6.12 is keyed to a 200 mm diameter shaft. Assuming that the turning effort is transmitted by the key only, determine the force on the key if the load at the end of the arm is 250 N.

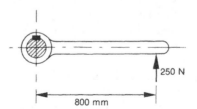

Fig. 6.12

6.7 Determine the force that must be applied to the foot pedal shown in Figure 6.13 to produce a force in the connecting link of 200 N.

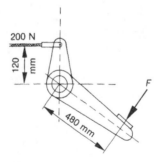

Fig. 6.13

6.8 Determine the force F required to start to tip the cabinet shown in Figure 6.14, if the mass of the cabinet is 50 kg.

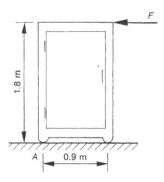

Fig. 6.14

6.9 The triangular roof truss in Figure 6.15 is subjected to a number of loads which include various weights and wind loads as shown. The horizontal and vertical reaction forces at the supports are also given. Use the three equations of statics to demonstrate that the system of non-concurrent forces acting on the truss is in equilibrium.

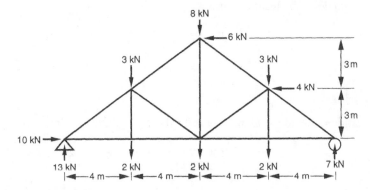

Fig. 6.15

6.5   *RESULTANT OF NON-CONCURRENT FORCES*

We have previously explained the resultant force as that single force which has exactly the same effect as a given system of forces, and learned how to determine its magnitude and direction by the force polygon method or by mathematical addition of forces.

So far, this is applicable to the case of non-concurrent forces acting on a rigid body, except that each force in such a system will also tend to rotate the body upon which it acts about some axis. It is therefore necessary to consider the rotational effect of the force system in addition to the linear push–pull action of the forces.

When solving for the resultant of a non-concurrent force system, the problem usually consists of two steps:

1. finding the magnitude and direction of the resultant force by mathematical or graphical addition of all forces;
2. finding the location of the resultant relative to an arbitrary reference point, usually by applying the principle of moments.

Example 6.8

Determine the magnitude, direction and location of the resultant of the system of forces in Figure 6.16.

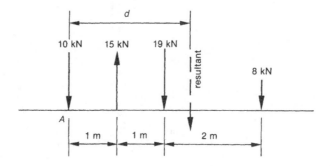

Fig. 6.16

Solution

The magnitude of the resultant force:

$$F = \Sigma F$$
$$= -10 + 15 - 19 - 8$$
$$= -22 \text{ kN}$$

i.e. 22 kN down.

The location is found by taking moments about an arbitrary point, such as point *A*:

$$\Sigma M = 10 \times 0 - 15 \times 1 + 19 \times 2 + 8 \times 4$$
$$= 55 \text{ kN.m}$$

This sum must be equal to the moment of the resultant about the same point:

$$F \times d = 55 \text{ kN.m}$$
$$22 \text{ kN} \times d = 55 \text{ kN.m}$$
$$\therefore d = \frac{55 \text{ kN.m}}{22 \text{ kN}}$$
$$= 2.5 \text{ m}$$

It is therefore possible to replace the given system of forces by a single downward force of 22 kN, located at 2.5 m from point *A*, as shown.

Example 6.9

A concrete foundation for a brick wall must support two vertical loads and a horizontal load due to soil pressure on one side, as shown in Figure 6.17(a). Its own weight is also shown. Determine the resultant and check if it passes through the base of the foundation, as required for stability.

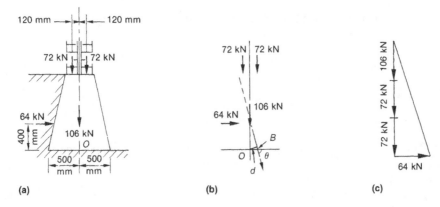

Fig. 6.17

Solution

Sum of the vertical forces:

$$\Sigma F_V = 106 + 72 + 72$$
$$= 250 \text{ kN down}$$

Sum of the horizontal forces:

$$\Sigma F_H = 64 \text{ kN to the right}$$

The resultant is therefore:

$$F = \sqrt{250^2 + 64^2}$$
$$= 258.1 \text{ kN}$$

The angle to the horizontal:

$$\theta = \tan^{-1} \left(\frac{250}{64} \right)$$
$$= 75.6°$$

To locate the resultant, take moments about a convenient point, such as midpoint of the base:

$$\Sigma M = 64 \times 0.4 - 72 \times 0.12 + 72 \times 0.12$$
$$= 25.6 \text{ kN.m}$$

Distance of the resultant from the midpoint:

$$d = \frac{\Sigma M}{F}$$
$$= \frac{25.6 \text{ kN.m}}{258.1 \text{ kN}}$$
$$= 0.0992 \text{ m}$$
$$= 99.2 \text{ mm}$$

The point at which the resultant intersects with the base can be calculated from:

$$OB = \frac{d}{\sin 75.6°}$$
$$= \frac{99.2}{\sin 75.6°}$$
$$= 102.4 \text{ mm}$$

Parts of this solution can also be done graphically. For example, the magnitude and direction of the resultant can very conveniently be found by constructing the polygon of forces as in Figure 6.17(c).

However, the principle of moments, for determining the distance *d*, is best dealt with mathematically.[*]

The last part of the question can also be answered by using a diagram such as Figure 6.17(b) if drawn to scale.

 # Problems

6.10 For each of the systems of forces in Figure 6.18, determine and locate the resultant force, stating the distance along the horizontal line from the first force on the left.

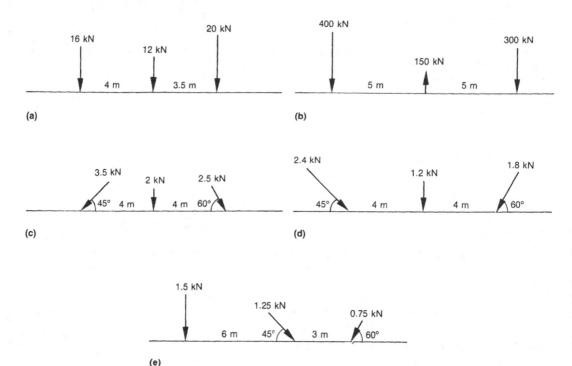

Fig. 6.18

---

[*] The alternative graphical method, known as the **funicular** (from *funis* meaning 'rope') **polygon,** involves a special construction procedure which locates a point on the line of action of the resultant force, thus locating the force itself. It is not discussed in this book.

6.11 Determine and locate the resultant of the three forces acting on the ladder as shown in Figure 6.19.

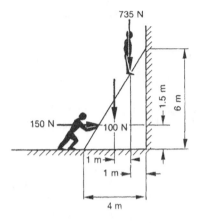

Fig. 6.19

6.12 Three forces acting on a 300 mm diameter pulley are as shown in Figure 6.20. These include belt tensions and the weight of the pulley. Determine the resultant in magnitude, direction, and distance from the centreline of the pulley.

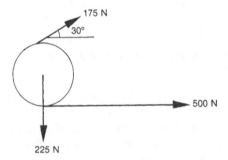

Fig. 6.20

6.13 Determine the resultant of the loads acting on the structure shown in Figure 6.21, and locate it relative to the left-hand support.

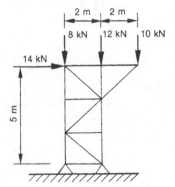

Fig. 6.21

6.6 *MOMENT OF A COUPLE*

So far we have discussed turning effects of single forces. We now turn our attention to a very special combination of forces called a **couple.** A couple consists of two forces which have:

1. the same magnitude
2. parallel lines of action
3. opposite sense

When your hands are on two opposite points of the steering wheel of a car, one hand pushing up and the other pulling down with equal but opposite forces, the result is a couple (Fig. 6.22).

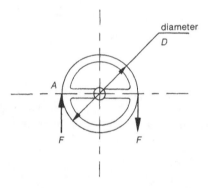

Fig. 6.22 *Forces on a steering wheel illustrating a couple*

Clearly the algebraic sum of the two forces is equal to zero, i.e. there is no net push or pull in any direction. However, there is a turning effect, which can be calculated relative to any point. The obvious point of reference for calculating the moments is the centre point of the wheel, which is the axis of its rotation.

The total moment is:

$$M = F \times \frac{D}{2} + F \times \frac{D}{2}$$
$$= F\left(\frac{D}{2} + \frac{D}{2}\right)$$
$$= F \times D$$

where F is the magnitude of each of the forces in the couple, and D is the distance between them, in this case the wheel diameter. The sense of the total moment is clockwise for the forces shown.

The product of one of the forces and the distance between them is called the **moment of the couple.** One important characteristic of a couple is that its moment, $M = F \times D$, does not depend on the choice of the reference point. For example, the moment about point A is:

$$M_A = F \times D + 0$$
$$= F \times D, \text{ clockwise as before}$$

This can be checked by repeating the same calculation about any other point on, or even outside, the wheel.

Example 6.10

Determine the moment of the couple acting on the steering wheel in Figure 6.22 if the wheel diameter is 350 mm and the forces applied as shown are each 5 N. Determine also the forces required to produce the same moment if the hands are held on the spokes, halfway between the axis and the rim of the wheel.

Solution

Moment of the couple:

$$M = F \times D$$
$$= 5 \text{ N} \times 0.35 \text{ m}$$
$$= 1.75 \text{ N.m clockwise}$$

To apply the same moment when $D = \dfrac{0.35 \text{ m}}{2} = 0.175$ m, the forces required can be calculated from $M = F \times D$:

$$1.75 \text{ N.m} = F \times 0.175 \text{ m}$$
$$\therefore F = \frac{1.75 \text{ N.m}}{0.175 \text{ m}}$$
$$= 10 \text{ N each}$$

This example illustrates the idea of equivalent couples, i.e. couples which involve different forces but produce equal turning moments. Next time you drive a car, turn off a tap or turn a key in a lock, think of the two forces applied by your hands or fingers, producing a turning moment.

Let us summarise some important points, which are often misunderstood.

1. A **couple** is a pair of actual forces applied to a structure or component at specified points. A couple always consists of equal and opposite forces acting along parallel lines.
2. **Moment of a couple** is a pure turning effect produced by the couple, expressed in newton metres. A couple has no resultant force in any direction; the pair of forces cancel each other out.
3. The magnitude of the moment of a couple is independent of the reference point about which it is calculated. Therefore, the simplest way to determine it is by multiplying one of the forces by the perpendicular distance between them.
4. The magnitude of the moment is also independent of the actual location of the couple on the structure or component to which it is applied. Therefore, a couple can be relocated to any other position without affecting the overall balance of moments.
5. An **equivalent couple** is any other couple with the same magnitude of turning moment and the same directional sense. A given couple can be replaced by an equivalent couple applied elsewhere on the body without changing its turning effect.

6.7 *EQUIVALENT FORCE-AND-COUPLE SYSTEM*

When a force F is applied at a point on a rigid body, such as point A on the spanner shown in Figure 6.23(a), its effect can be viewed from some other significant point, such as the centreline of the bolt, point B. In this case, the bolt (point B) experiences a downward push equal to the applied force F, as well as a clockwise turning moment M caused by force F about point B.

If we were to draw a separate diagram of the bolt itself, showing the effect of the applied force as viewed from B, the force F would be represented by the usual arrow symbol, and the turning moment indicated by the symbol $\curvearrowright$ placed around point B, as shown in Figure 6.23(b).

It is usually sufficient to leave the turning moment symbol as it is, without any further interpretation. However, it should be understood that in reality, this 'pure turning moment' is a result of a force couple acting on the head of the bolt, as shown in Figure 6.23(c). In fact, if the distance between the flats of the hexagonal bolt head is known, it is possible to work out the magnitudes of the forces that constitute this couple.

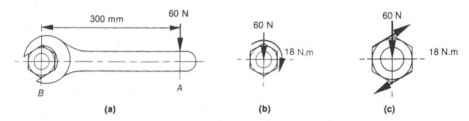

Fig. 6.23   **(a)** *Force applied to a spanner to turn a bolt*   **(b)** *The head of the bolt showing the turning moment*   **(c)** *The force-and-couple acting on the head of the bolt*

Example 6.11

A bolt is being tightened by a force of 60 N applied to a spanner at a distance of 300 mm, as shown in Figure 6.23(a). What is the force and what is the turning moment acting on the bolt at point B?

Solution

The force–moment system at B comprises a force of 60 N and a turning moment equal to:

$$M = 60 \text{ N} \times 0.3 \text{ m}$$
$$= 18 \text{ N.m clockwise}$$

Diagrams 6.23(b) and 6.23(c) can each be taken to represent the 'force-and-couple system' acting at point B, which is equivalent to the action of force F applied at point A. Note that on both diagrams only the magnitude and clockwise directional sense of the turning moment are stated, without spelling out the individual force magnitudes in the pair.

There are situations, such as that illustrated by the following example, where it is necessary to know the individual forces in the couple, and the perpendicular distance between them is well defined.

Example 6.12

For the hook shown in Figure 6.24(a), replace the force of 200 N acting in plane A by an equivalent force-and-couple system at B.

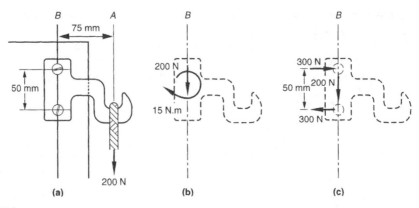

Fig. 6.24

Solution

The equivalent force–moment system at B consists of a force of 200 N and a clockwise turning moment equal to:

$$M = 200 \text{ N} \times 0.075 \text{ m}$$
$$= 15 \text{ N.m}$$

as shown in Figure 6.24(b).

In this case, where the moment is taken by the two screws, it is meaningful to interpret the moment as an equivalent couple with a distance of 50 mm between forces:

$$M = Fd$$
$$15 \text{ N.m} = F \times 0.05 \text{ m}$$
$$\therefore F = \frac{15}{0.05}$$
$$= 300 \text{ N, as in Figure 6.24(c)}$$

 Problems

6.14 A screw jack is operated by means of 40 N forces applied at each end of a double arm, at right angles to it. Determine the magnitude of the turning moment if the total length of the double arm is 700 mm.

6.15 A 400 mm × 300 mm plate is subjected to forces as shown in Figure 6.25. Show that the turning effects of the forces add up to zero at any point.

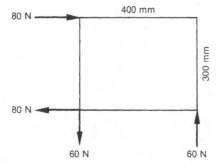

Fig. 6.25

6.16 A horizontal bar is subjected to a number of forces as shown in Figure 6.26. Prove that the bar is in rotational equilibrium:
(a) by adding the moments of all forces about any point on the bar and then repeating for one other point;
(b) by recognising pairs of forces as couples and adding their moments.

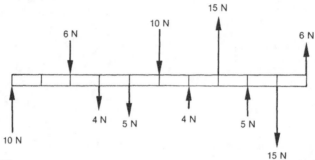

Fig. 6.26

6.17 Determine the force F at each end of the bar AB (Fig. 6.27) required to maintain equilibrium, if the force in the string which is passed around the pulleys is 160 N.

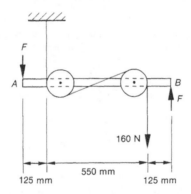

Fig. 6.27

6.18 A column 0.5 m wide carries a 2 kN force as shown in Figure 6.28. Reduce the force to an equivalent axial load and a moment.

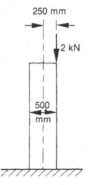

Fig. 6.28

6.19 Determine the eccentricity *d* (i.e. the perpendicular distance from the axis) of a load which can be represented as an axial load of 1 kN and a moment of 15 N.m acting on a column.

6.20 Determine the equivalent force–moment system at the built-in end of the cantilever beam shown in Figure 6.29.

Fig. 6.29

6.21 If the mass supported by the bracket in Figure 6.30 is 5 kg, interpret the load as a force and a couple acting through the bolts.

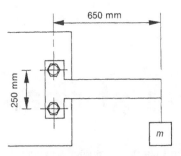

Fig. 6.30

 Review questions

1. Define a *non-concurrent force system.*
2. Define *moment of a force.*
3. What is the SI unit of moment of a force?
4. Does moment of a force have directional sense?
5. What is the principle known as the *Varignon theorem*?
6. What is the condition for rotational equilibrium of a system of forces?
7. What are the three conditions for equilibrium of non-concurrent forces?
8. Explain the method of determining the magnitude and direction of the resultant of a non-concurrent force system.
9. How can the line of action be located?
10. Define *moment of a couple.*
11. What is an *equivalent couple*?
12. What is meant by an *equivalent force–moment system*?
13. What is an *equivalent force-and-couple system*?

PART 3

STRUCTURAL ANALYSIS ≡

The Sydney Harbour Bridge is a steel arch of the two-hinged type. A steel deck hangs from the arch and five steel truss approach spans lead to each side of the arch. The arch is hinged at the base on each side of the harbour. These hinges, or bearings, support the full weight of the bridge and spread their load through large concrete skewbacks into a foundation of solid sandstone.

The Story of the Sydney Harbour Bridge

CHAPTER 7

Reactions at beam supports

Beams are structural members designed to support loads applied at various points along their lengths. In this chapter we learn the procedure for calculating reactions at beam supports. This is one of the most common practical applications of the three equilibrium equations, which were introduced in the previous chapter.

Expected learning outcomes

After carefully studying the material presented in this chapter, working through all numerical examples, and successfully completing all practice problems, students should be able to:
1. describe a simply supported beam, an overhanging beam and a cantilever beam;
2. distinguish between concentrated loads and uniformly distributed loads;
3. describe different types of beam supports and state the kind of reactions that may exist at each type of support;
4. calculate reactions at beam supports using the equations of statics.

7.1 BEAMS, LOADS AND SUPPORT REACTIONS

Beams are usually long, straight, solid bars, having a uniform cross-sectional shape and a constant cross-sectional area. They are made from rolled structural steel, reinforced concrete, timber, or similar materials.

In most cases, beams are placed in a horizontal position and carry vertical loads, i.e. forces which are perpendicular to the axis of the beam. The majority of loads supported by horizontal beams are usually weights, i.e. forces directed vertically downwards due to gravity acting on the various masses resting on the beam. Occasionally some loads, which are not weights, are directed vertically upwards, or are inclined to the axis of a horizontal beam.

Beams are generally classified in accordance with the way in which they are supported. A **simply supported beam** is pinned at one end and roller supported at the other (Fig. 7.1(a)). An **overhanging beam** also rests on two similar supports, but extends beyond one or both of its supports (Fig. 7.1(b)). A **cantilever beam** is rigidly fixed at one end and free at the other (Fig. 7.1(c)).

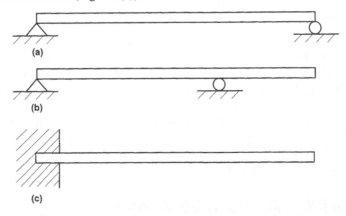

Fig. 7.1   **(a)** *Simply supported beam*   **(b)** *Overhanging beam*   **(c)** *Cantilever beam*

A beam may be subjected to **concentrated loads**, to **distributed loads**, or to a **combination of both**. It is common practice in engineering design to ignore the weight of the beam itself as relatively insignificant in comparison with all other forces acting on the beam. However, if it is deemed inappropriate to neglect the weight of the beam, it can be accounted for as an additional concentrated load acting through the midpoint of the beam.

When a beam carries loads and rests on its supports, the supports react to provide the necessary static balance which keeps the beam a rest, i.e. in a state of static equilibrium. The kinds of reactions that may exist at any given support depend on the nature of the support.

A roller support, or any support that offers negligible resistance to small displacements along the axis of the beam, can only provide a reaction force which is perpendicular to the supporting surface, i.e. a vertical reaction to a beam resting on a horizontal surface, as shown in Figure 7.2(a). Thus there can be only one unknown reaction force at a roller support.

The possible direction of the reaction at a pinned support is not immediately obvious. This unknown reaction is usually analysed in terms of its vertical and horizontal components, thereby simplifying mathematical solutions. Therefore, there can be two unknown reaction forces at a pinned support, one of them vertical and the other horizontal, as shown in Figure 7.2(b).

The cantilever beam is a special case. Its only supported end is embedded rigidly into an unyielding solid abutment, which has to provide the total reaction against any movement or rotation. The total reaction at the fixed end may contain a horizontal reaction force, a vertical reaction force, and a reaction moment that prevents rotation of the cantilever beam about its support, as shown in Figure 7.2(c). Thus there can be as many as three unknowns at the fixed end of a cantilever beam.

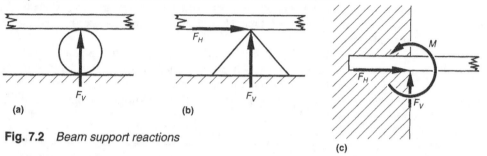

Fig. 7.2 *Beam support reactions*

It should be noted that since we only have the three equations of statics to work with, namely $\Sigma F_H = 0$, $\Sigma F_V = 0$ and $\Sigma M = 0$, the methods of statics are not sufficient by themselves to determine beam reactions if more than three unknown quantities are involved. For this reason, all discussion and all problems covered in this chapter are strictly limited to examples of statically determinate beams, i.e. beams supported in such a way that the number of unknown reactions does not exceed three.[*]

7.2 SIMPLY SUPPORTED BEAMS

Let us begin by learning how to calculate reaction forces for a simply supported beam which carries concentrated vertical loads only.

In the absence of any horizontal forces, the first equation, $\Sigma F_H = 0$, is only a formality and not very helpful. However, since there are only two unknown forces, one at each support, the other two equations are sufficient, i.e. $\Sigma F_V = 0$ and $\Sigma M = 0$.

The method illustrated by the following example consists of two steps:

Step 1
Take moments about one of the supports and equate the algebraic sum of all moments to zero. This gives an equation in which the reaction force at the opposite support is the only unknown. Solve for this unknown reaction force.

Step 2
Equate the algebraic sum of all vertical forces, including reactions, to zero, and solve for the second unknown reaction force.

Example 7.1
Calculate the reactions for the simply supported beam shown in Figure 7.3.

---

[*] The solution of problems containing more than three unknown reactions, such as the case of a beam rigidly fixed at both ends, requires additional consideration to be given to the resistance of the beam material to bending. Statically indeterminate structures are outside the scope of this book.

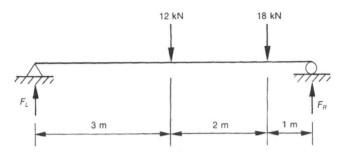

Fig. 7.3

Solution

Let F_L and F_R stand for the reaction forces at the left-hand and right-hand supports respectively.

Taking moments about the left-hand support:

$$\Sigma M_L = 12 \text{ kN} \times 3 \text{ m} + 18 \text{ kN} \times 5 \text{ m} - F_R \times 6 \text{ m} = 0$$
$$36 + 90 - F_R \times 6 = 0$$
$$\therefore F_R = 21 \text{ kN}$$

Note that the moment of F_L about the left-hand support is zero because the force passes through the point there. Remember also that by convention, clockwise moments are taken to be positive and anticlockwise moments are negative.

Summation of vertical forces:

$$\Sigma F = F_L - 12 \text{ kN} - 18 \text{ kN} + F_R = 0$$

But $F_R = 21$ kN. Hence:

$$F_L - 12 \text{ kN} - 18 \text{ kN} + 21 \text{ kN} = 0$$
$$\therefore F_L = 9 \text{ kN}$$

Here again the sign convention is observed—upward forces are positive and downward forces are negative.

Alternatively, F_L could also be found by taking moments about the right-hand support, as follows:

$$\Sigma M_R = -12 \text{ kN} \times 3 \text{ m} - 18 \text{ kN} \times 1 \text{ m} + F_L \times 6 \text{ m} = 0$$
$$\therefore F_L = 9 \text{ kN}$$

In this case the result may be checked by the summation of vertical forces:

$$\Sigma F = F_L - 12 \text{ kN} - 18 \text{ kN} + F_R$$
$$= 9 \text{ kN} - 12 \text{ kN} - 18 \text{ kN} + 21 \text{ kN}$$
$$= 0$$

7.3 OVERHANGING BEAMS

If a beam has an overhanging portion extending beyond the supports on one or both sides, the procedure to be followed is exactly the same as described above. Extra care should be exercised in distinguishing between positive (clockwise) moments and negative (anticlockwise) moments relative to the selected reference point, which is usually the left-hand support.

Example 7.2

Calculate the reactions for the beam shown in Figure 7.4.

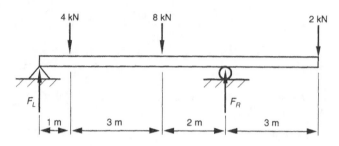

Fig. 7.4

Solution

Take moments about the left-hand support:

$$\Sigma M_L = 4 \text{ kN} \times 1 \text{ m} + 8 \text{ kN} \times 4 \text{ m} - F_R \times 6 \text{ m} + 2 \text{ kN} \times 9 \text{ m} = 0$$
$$\therefore F_R = 9 \text{ kN}$$

Summation of vertical forces:

$$\Sigma F_V = F_L - 4 \text{ kN} - 8 \text{ kN} + 9 \text{ kN} - 2 \text{ kN} = 0$$
$$\therefore F_L = 5 \text{ kN}$$

 Problem

7.1 For each of the beams in Figure 7.5, calculate reactions at the supports.

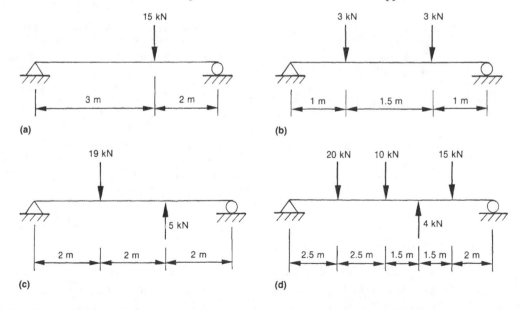

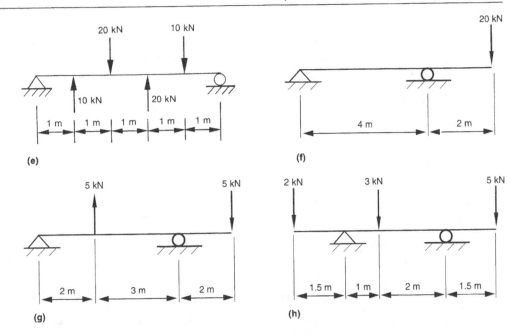

Fig. 7.5

7.4 CANTILEVER BEAMS

Let us now consider cantilever beams with concentrated vertical loads only. In this case there is only one support at the fixed end of the beam, where the total reaction is a combination of a vertical reaction force (F) and a reaction moment (M) exerted by the support.

Example 7.3

Calculate the reactions for the cantilever beam shown in Figure 7.6.

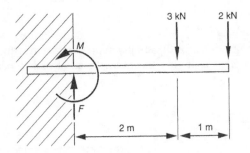

Fig. 7.6

Solution

Summation of moments taken about the support:

$$\Sigma M = 3 \text{ kN} \times 2 \text{ m} + 2 \text{ kN} \times 3 \text{ m} - M = 0$$

When setting up this equation, the unknown reaction moment M must be included in the total sum of all moments acting on the beam at the support. Note carefully how forces are multiplied by their respective distances from the support, while the reaction moment is entered as a single entity represented by the symbol M.[*] Note also that the negative sign denotes the anticipated anticlockwise sense of the reaction moment, as shown on the diagram.

Solving for M establishes the magnitude and units of the reaction moment:

$$M = 12 \text{ kN.m}$$

Summation of vertical forces:

$$\Sigma F = F - 3 \text{ kN} - 2 \text{ kN} = 0$$
$$\therefore F = 5 \text{ kN}$$

Hence the reaction force at the support is 5 kN.

7.5 *UNIFORMLY DISTRIBUTED LOADS*

Quite frequently, forces acting on a beam are applied over an extended portion of a beam, such as the weight of goods distributed over the floor of a warehouse, which in turn is supported by the beams.

When the intensity of load distribution has a constant value per unit length of a beam between two points along the beam, the load is said to be **uniformly distributed** over that part of the beam. Such loads are expressed in units of force per unit length of beam, i.e. in newtons per metre (N/m), or kilonewtons per metre (kN/m). Only uniformly distributed loads are considered in this book.

When it is necessary to determine reactions which are due, in whole or in part, to a uniformly distributed load, it is possible to replace the distributed load by an equivalent concentrated load located at the midpoint of the load distribution. The solution is considerably simplified by such substitution, the remaining steps being the same as in all previous examples.

Example 7.4

Determine the reactions for the simply supported beam with a uniformly distributed load of 5 kN/m as well as a single concentrated load of 22 kN located as shown in Figure 7.7.

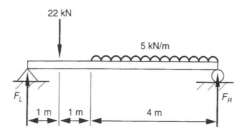

Fig. 7.7

---

[*] The reaction moment M stands for the resultant product of forces and distances acting on the built-in end of the beam.

Solution

Here the intensity of the load distribution is given as 5 kN/m over 4 m. This means that the total magnitude of the distributed load is:

$$5 \text{ kN/m} \times 4 \text{ m} = 20 \text{ kN}$$

The midpoint of this distributed load is located 2 m from the right-hand support. Therefore, the problem can be restated in terms of the actual concentrated load plus the equivalent concentrated load, as shown in Figure 7.8.

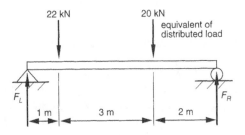

Fig. 7.8

The usual method of solution can now be followed:

$$\Sigma M_L = 22 \text{ kN} \times 1 \text{ m} + 20 \text{ kN} \times 4 \text{ m} - F_R \times 6 \text{ m} = 0$$
$$\therefore F_R = 17 \text{ kN}$$
$$\Sigma F_V = F_L - 22 \text{ kN} - 20 \text{ kN} + 17 \text{ kN} = 0$$
$$\therefore F_L = 25 \text{ kN}$$

Example 7.5

Determine the reactions for a cantilever beam with a uniformly distributed load of 3 kN/m as well as a single concentrated load of 4 kN located as shown in Figure 7.9.

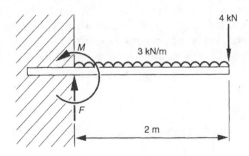

Fig. 7.9

Solution

The solution follows similar lines. The total magnitude of the distributed load is:

$$3 \text{ kN/m} \times 2 \text{ m} = 6 \text{ kN}$$

The midpoint of this distributed load is located 1 m from the support. Therefore, the problem becomes as shown in Figure 7.10.

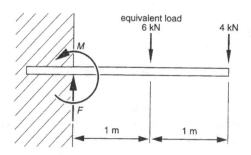

Fig. 7.10

The usual solution for a cantilever beam follows:

$$\Sigma M = 6 \text{ kN} \times 1 \text{ m} + 4 \text{ kN} \times 2 \text{ m} - M = 0$$
$$\therefore M = 14 \text{ kN.m}$$
$$\Sigma F = F - 6 \text{ kN} - 4 \text{ kN} = 0$$
$$\therefore F = 10 \text{ kN}$$

7.6 AXIAL LOADS

Occasionally a load is inclined to the axis of a beam at an angle other than 90°. Such a load can be resolved into two components, one at right angles to the beam and the other along the axis of the beam. In the case of a *horizontal* beam, the **transverse component** of an inclined load is vertical and the **axial component** is horizontal.

The transverse (vertical) component of the load can be included with any other vertical loads when calculating moments about supports and when working out vertical reaction forces, as discussed previously. The axial (horizontal) component is transmitted along the axis of the beam to that support which is capable of resisting horizontal forces, i.e. to the pinned support of a simply supported beam or to the built-in end of a cantilever beam, and gives rise to a reaction force acting in the axial (horizontal) direction.

Example 7.6
Determine all support reactions for the beam shown in Figure 7.11.

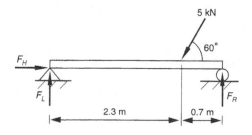

Fig. 7.11

Solution

Resolve the inclined force into two components:

$$\text{Vertical component} = 5 \times \sin 60°$$
$$= 4.33 \text{ kN}$$
$$\text{Horizontal component} = 5 \times \cos 60°$$
$$= 2.50 \text{ kN}$$

Now the problem is reduced to vertical and horizontal forces only, as shown in Figure 7.12.

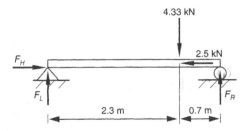

Fig. 7.12

Summation of horizontal forces:

$$\Sigma F_H = F_H - 2.5 \text{ kN} = 0$$
$$\therefore F_H = 2.5 \text{ kN}$$

Hence the horizontal reaction at the pinned support is 2.5 kN.

Summation of moments about the left-hand support:

$$\Sigma M_L = 4.33 \text{ kN} \times 2.3 \text{ m} - F_R \times 3 \text{ m} = 0$$
$$\therefore F_R = 3.32 \text{ kN}$$

Hence the vertical reaction force at the right-hand support is 3.32 kN.

Summation of vertical forces:

$$\Sigma F = F_L - 4.33 \text{ kN} + 3.32 \text{ kN} = 0$$
$$\therefore F_L = 1.01 \text{ kN}$$

Therefore the vertical reaction force at the left-hand support is 1.01 kN.

 Problems

7.2 For each cantilever beam in Figure 7.13, calculate the reaction force and the reaction moment at the support.

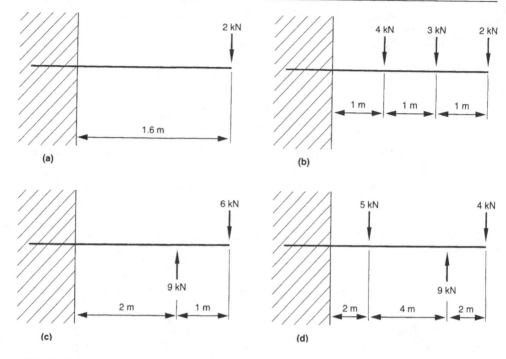

Fig. 7.13

7.3 For each beam in Figure 7.14, replace uniformly distributed loads by equivalent concentrated loads, and hence determine reactions at the supports.

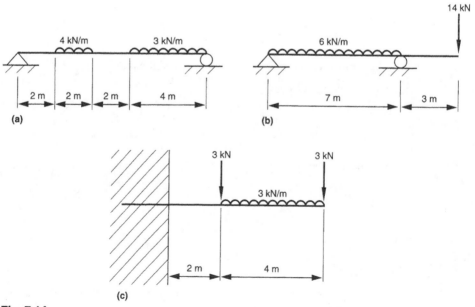

Fig. 7.14

7.4 For the beam in Figure 7.15, determine all reaction forces.

(*Hint*: The resultant of the rope tensions passes through the centre of the pulley bisecting the angle between the ropes.)

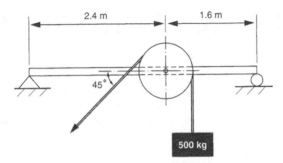

Fig. 7.15

Review questions

1. Describe a *simply supported beam*.
2. Describe an *overhanging beam*.
3. Describe a *cantilever beam*.
4. Distinguish between *concentrated loads* and *uniformly distributed loads*.
5. Describe three different types of beam supports.
6. What kinds of reactions apply to:
   (a) a simply supported beam?
   (b) an overhanging beam?
   (c) a cantilever beam?

CHAPTER 8

Pin reactions in frames

A frame is a common type of structure composed of several straight members pin-joined together. Frames are generally stationary and are used to support loads. Provided a frame is properly supported and contains no more joints and supports than is necessary to prevent collapse, then the reaction forces at each joint can be determined mathematically, using the three equations of statics:

$$\Sigma F_H = 0, \quad \Sigma F_V = 0 \quad \text{and} \quad \Sigma M = 0$$

In this chapter, we study the analytical method for calculating pin reactions in frame structures, based on these three equations.

Expected learning outcomes

After carefully studying the material presented in this chapter, working through all numerical examples, and successfully completing all practice problems, students should be able to:
1. draw separate free-body diagrams for each member of a given frame;
2. identify all force components acting on each member;
3. write a related set of equilibrium equations and solve for all pin-reaction forces in a frame.

8.1 *OUTLINE OF THE METHOD OF MEMBERS*

In a stationary pin-jointed frame, designed and used to support a load, forces produced by the load are transmitted through members to the pins at the joints, and then by the pins to other connected members. The forces at the joints are referred to as **pin reactions**.

If we are to use the three equations of statics, it is necessary to resolve all known forces into their horizontal and vertical components. Likewise, unknown forces should be represented by horizontal and vertical components at the joints.

When calculating pin reactions between members of a given frame, we assume the pins to be perfectly smooth, i.e. any possible friction at the joints is ignored. We also neglect the weights of the members, regarding them as being relatively insignificant in comparison with the magnitude of the load supported by the frame.

The mathematical method for determining pin reactions in a frame is sometimes called the **method of members**, since it consists of examining the conditions of static equilibrium of each member individually. It contains several steps which must generally be followed in a logical sequence. These are now summarised.

Step 1
Considering the frame as a whole, determine all support reactions in terms of their horizontal and vertical components.

Step 2
Identify any two-force members in the frame. It is important to recognise these as having two equal and opposite forces applied through the pins and acting along the axis of such members. This helps to recognise the lines of action of forces acting on other connected members.

Step 3
Isolate each separate member of the frame and draw it as a free body, showing all known and unknown forces acting on the member. As usual, the sense of an unknown force may be assumed, and then revised in the light of subsequent calculations.

Step 4
Select a member, with at least one known force, which contains no more than three unknowns. use the three equations of statics to calculate the unknown pin-reaction forces acting on the member.

Step 5
With the knowledge gained from the previous step, select another member and solve for more unknowns. It is essential to remember that a reaction force acting on any given member at a joint is the pulling or pushing action from another member, transmitted through the joint. Between any two members connected through a pin, action and reaction forces are always equal and opposite.

Step 6
Repeat step 5 until all components of pin-reaction forces have been determined.

Step 7
As a final step, if desired, combine horizontal and vertical components of pin-reaction forces at each joint into their resultant in order to determine the total magnitude of the reaction force at the joint.

8.2   *ILLUSTRATIVE EXAMPLE*

The following example illustrates the application of these steps in solving a typical problem.

Example 8.1

Determine the horizontal and vertical components of pin reactions at all joints of the frame shown in Figure 8.1, as well as the total magnitude of the reaction force at each joint.

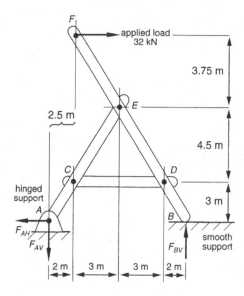

Fig. 8.1

Solution

Step 1

Considering the frame as a whole, determine support reactions F_{AH}, F_{AV} and F_{BV}, as follows.

Take moments about the hinged support A:

$$\Sigma M = 32 \text{ kN} \times 11.25 \text{ m} - F_{BV} \times 10 \text{ m} = 0$$
$$\therefore F_{BV} = 36 \text{ kN}$$

Hence reaction at the smooth support *B* is 36 kN upwards.

Summation of forces:

$$\Sigma F_H = 32 \text{ kN} - F_{AH} = 0$$
$$\therefore F_{AH} = 32 \text{ kN to the left}$$
$$\Sigma F_V = F_{BV} - F_{AV}$$
$$= 36 \text{ kN} - F_{AV} = 0$$
$$\therefore F_{AV} = 36 \text{ kN downwards}$$

Step 2

Identify member *CD* as a two-force member in the frame. As such it must have equal and opposite forces applied through the pins and acting along the axis *CD*. Therefore, the lines of action of forces acting through joints *C* and *D*, and their interaction with members *ACE* and *BDEF*, must be horizontal (Fig. 8.2). Directional sense of these forces may be assumed, but there is no absolute certainty at this stage until further calculations have been performed.

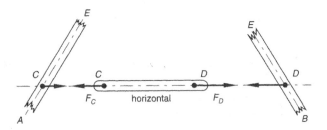

Fig. 8.2

Step 3
Isolate each separate member of the frame and draw it as a free body, showing all known and unknown forces acting on the member (Fig. 8.3).

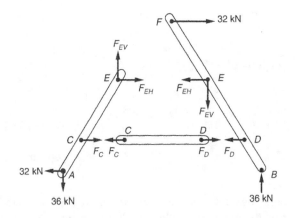

Fig. 8.3

Here again, the directional sense of unknown forces may be assumed and shown by an arrowhead, subject to revision in the light of subsequent calculations. However, it should be carefully noted that for each pair of action-and-reaction forces interacting at the same joint, the arrowheads must be consistently shown as opposing each, other in the two images of each 'separated' joint. Examine Figure 8.3 very carefully to observe this feature for joint *C*, joint *D*, and especially for joint *E*.

Step 4
Look for a member, with at least one known force, which contains no more than three unknowns. Member *ACE* is one such member (alternatively *BDEF* could be used). Now consider the equilibrium of member *ACE* (Fig. 8.4), and use the three equations of statics to calculate the unknown pin-reaction forces acting at joints *C* and *E*.

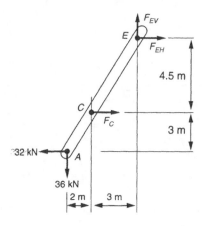

Fig. 8.4

Taking moments about point E:

$$\Sigma M = 32 \text{ kN} \times 7.5 \text{ m} - 36 \text{ kN} \times 5 \text{ m} - F_{CH} \times 4.5 \text{ m} = 0$$
$$\therefore F_{CH} = 13.3 \text{ kN}$$

Summation of forces:

$$\Sigma F_H = F_{EH} + F_{CH} - 32 \text{ kN}$$
$$= F_{EH} + 13.33 \text{ kN} - 32 \text{ kN} = 0$$
$$\therefore F_{EH} = 18.67 \text{ kN to the right}$$
$$\Sigma F_V = F_{EV} - 36 \text{ kN} = 0$$
$$\therefore F_{EV} = 36 \text{ kN upwards}$$

Step 5
From this point onwards, the rest follows like dominoes, each subsequent step being made easier by the information obtained from previous steps. Consider member *CD* as a free body (Fig. 8.2). Now that we know $F_{CH} = 13.3$ kN, it is easy to see that force F_D must be equal and opposite to F_C.

$$\therefore F_{DH} = 13.3 \text{ kN}$$

Step 6
A very similar consideration applies to member *BDEF* with respect to force interactions at joint E. From Figure 8.3 it can be seen that $F_{EH} = 18.67$ kN to the left, and $F_{EV} = 36$ kN downwards.

Step 7
As a final step, we can combine horizontal and vertical components of the pin-reaction force at joint E in order to determine its total magnitude.

$$F_E = \sqrt{F_{EH}^2 + F_{EV}^2}$$
$$= \sqrt{18.67^2 + 36^2}$$
$$= 40.6 \text{ kN}$$

Similarly, the total magnitude of reaction force at A is:

$$F_A = \sqrt{F_{AH}^2 + F_{AV}^2}$$
$$= \sqrt{32^2 + 36^2}$$
$$= 48.2 \text{ kN}$$

Summarising the answers for the total pin-reaction force at each joint:

Joint A: Pin reaction 48.2 kN ($F_{AH} = 32$ kN, $F_{AV} = 36$ kN)
Joint C: Pin reaction 13.3 kN ($F_{CH} = 13.3$ kN, $F_{CV} = 0$)
Joint D: Pin reaction 13.3 kN ($F_{DH} = 13.3$ kN, $F_{DV} = 0$)
Joint E: Pin reaction 40.6 kN ($F_{EH} = 18.67$ kN, $F_{EV} = 36$ kN)

 Problems

8.1 Determine the horizontal and vertical components of pin reactions at all joints of the frame shown in Figure 8.5.

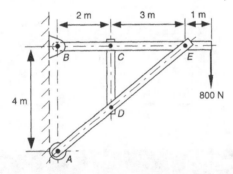

Fig. 8.5

8.2 Determine the horizontal and vertical components of pin reactions at all joints of the frame shown in Figure 8.6.

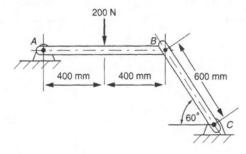

Fig. 8.6

8.3 Determine the horizontal and vertical components of pin reactions at all joints of the frame shown in Figure 8.7.

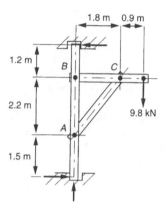

Fig. 8.7

8.4 Determine the horizontal and vertical components of pin reactions at all joints of the frame shown in Figure 8.8.

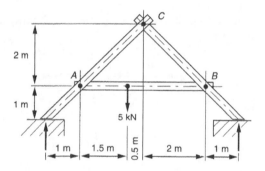

Fig. 8.8

 Review questions

1. What is a *pin-jointed frame*?
2. Explain what is meant by *pin-reaction forces*.
3. Briefly outline the steps for calculating pin reactions.
4. What is the method for calculating pin reactions called?
5. Which quantities are ignored when pin reactions are calculated?

CHAPTER 9

Internal forces in trusses

In this chapter we use the concepts and methods of static equilibrium to solve for internal forces in truss members.

A truss is one of the major types of structures used in engineering, mainly in building and bridge construction. Truss analysis and design is based on a certain number of definitions and assumptions, which distinguish trusses from beams and frames.

By definition, a **truss** is a structure composed of slender members joined together at their ends. No member is taken to be continuous through a joint. Because members of a truss are slender, they cannot support significant transverse loads between joints. All loadings are considered to be applied at the joints. Although in reality the joints are usually formed by bolted or welded connections, individual members of a truss are assumed to be joined by smooth pins. This is valid provided the centrelines of the joining members are concurrent, i.e. the centrelines intersect at the same point.

On the basis of these assumptions, each member of a truss can be treated as a two-force body. It follows then that forces at the ends of any member must always be directed along the axis of that member. Therefore, an individual member can only be stretched or compressed. Hence, for the purposes of force analysis, a truss is always considered to be an assembly of two-force members, in axial tension or compression, joined by frictionless pins.

On the next few pages we will apply several different methods to the solution of the same problem in order to illustrate the relative merits of these methods.

Expected learning outcomes

After carefully studying the material presented in this chapter, working through all numerical examples, and successfully completing all practice problems, students should be able to:
1. use Bow's notation to identify members of a truss;
2. analyse simple trusses by the method of joints;
3. analyse simple trusses by the combined force diagram (Maxwell's diagram);
4. solve for selected members of a simple truss by the method of sections.

9.1 BOW'S NOTATION

Let us begin by selecting a typical truss problem to be used as an illustrative example throughout this chapter:

Determine the force in each member of the truss shown in Figure 9.1.

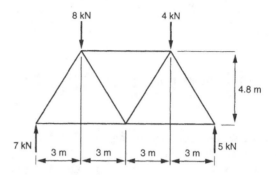

Fig. 9.1   *External forces on a truss*

In this example, all external forces (loads and reactions) are known. Often, only the loads are given. Finding the reactions becomes the first, preliminary step when solving a truss problem. Bow's notation is then used to label all individual members and joints.

Bow's notation is a well-established conventional method for clear and unambiguous identification of various elements of a truss. The idea is very simple indeed! Just place different capital letters in each of the spaces between external forces, and in each of the triangular panels between various members of the truss, as shown in Figure 9.2.

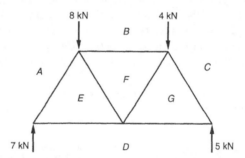

Fig. 9.2   *Bow's notation*

Make sure that the number of letters used is correct. There must be as many letters around the truss as there are external forces (4), and as many letters inside the truss as there are triangular spaces (3).

Each external force and each member will then lie between two letters and is identified by those letters, e.g. the 8 kN load is force *AB*, and the horizontal member at the top is member *BF* (or *FB*).

A joint is identified by the cluster of letters surrounding it, taken in a clockwise sequence, e.g. the joint at the left-hand support is joint *DAE*.

We are now well organised and ready to tackle the problem.

9.2   *THE METHOD OF JOINTS (GRAPHICAL)*

Truss analysis by the **method of joints** consists of considering successively the equilibrium of each joint,[*] always selecting a joint that contains no more than two unknown forces. Solution can be graphical or mathematical.

---

[*] It may be useful now to revise problem 5.8 from Chapter 5.

Example 9.1

Determine *graphically* the force in each member of the truss shown in Figure 9.1, using the method of joints.

Solution

Step 1   Consider joint DAE

Its free-body diagram and force polygon are shown side by side in Figure 9.3(a). The force polygon for this joint happens to be a triangle because there are only three forces meeting at this joint. When the force polygon is drawn accurately and to a sufficiently large scale, the magnitudes of forces in members *AE* and *ED* can be determined graphically. Hence the force in member *AE* = 8.25 kN and the force in member *ED* = 4.38 kN.

Note very carefully how the force polygon is labelled with the same letters, but in lower case, as those identifying the forces acting on the joint. Furthermore, the 'from–to' order of arrows on the force polygon corresponds to the clockwise sequence of letters around the joint. There is a special significance to this.

Take the 7 kN force, for example. Relative to the joint, it is identified on the free-body diagram as force *DA* (clockwise *D* to *A*). It is an upward force, labelled on the force polygon as 'from tail *d* to arrowhead *a*', i.e. force '*da*'.

If we now relate the directions of the other two forces back from the force polygon '*dae*' to the free-body diagram of the joint, we see that force *AE* is directed from *a* to *e*, i.e. towards the joint, and force *ED* is directed from *e* to *d*, i.e. away from the joint.

Forces directed *away* from the joint are **tensile**, and those directed *towards* the joint are **compressive**. Therefore, member *AE* is in compression and member *ED* is in tension. Let us make a note of this:

Force in member *AE* = 8.25 kN (compression)
Force in member *ED* = 4.37 kN (tension)

Step 2   Consider joint EABF

We already know that Member *EA* carries a compressive force of 8.25 kN. Notice that we have just switched from calling it member *AE* to member *EA*. This is because we are now considering it from joint *EABF*, in which the clockwise sequence of letters is *EA*. Compressive forces are the ones directed towards the joint. Therefore, when drawing the free-body diagram for this joint, show the arrowhead of force *EA* pointing towards the joint.

Now start with the known forces *EA* and *AB* and construct the force polygon '*eabf*', as shown in Figure 9.3(b). Notice again how the force polygon is labelled with the lower case version of the letters which identify forces acting on the joint. The 'from–to' order of arrows on the force polygon again corresponds to the clockwise sequence of letters around the joint we are considering at present. Notice also that the directional arrow of force '*ea*' is the reverse of force '*ae*' from step 1.

Now we can relate the directions of the previously unknown forces back from the force polygon '*eabf*' to the free-body diagram of the joint. Force *BF* is directed from *b* to *f*, i.e. towards the joint, and force *FE* is directed from *f* to *e*, also towards the joint. Therefore both forces are compressive. Scale off the magnitudes from the force polygon and record the results:

Force in member *BF* = 3.75 kN (compression)
Force in member *FE* = 1.18 kN (compression)

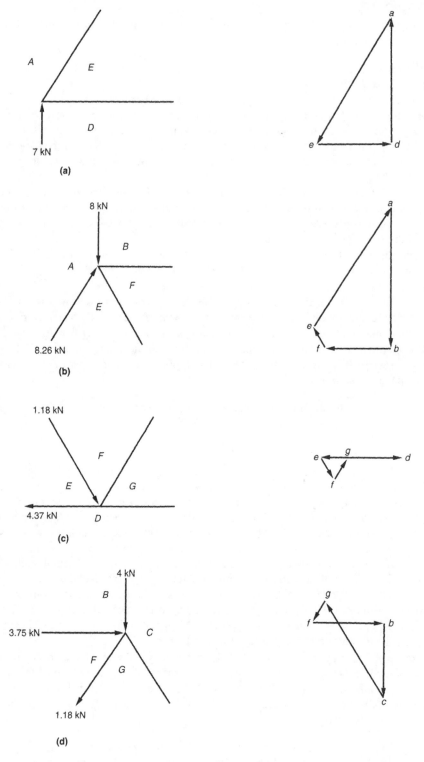

Fig. 9.3 (a) *Joint DAE* (b) *Joint EABF* (c) *Joint DEFG* (d) *Joint GFBC*

Step 3   Consider joint DEFG

From steps 1 and 2 it follows that force *DE* = 4.37 kN (tension), and force *EF* = 1.18 kN (compression). The free-body diagram for the joint can now show these two forces with arrowheads pointing appropriately in relation to this joint: force *DE* away from the joint and force *EF* towards the joint, as in Figure 9.3(c).

Once again, we start with the known forces *DE* and *EF* and construct the force polygon '*defg*'. When the magnitudes and directions of the previously unknown forces are related back from the force polygon to the free-body diagram of the joint, we find that:

Force in member *FG* = 1.18 kN (tension)
Force in member *GD* = 3.12 kN (tension)

Step 4   Consider joint GFBC

There is only one unknown force remaining, which is the force in member *CG*. Force *GF*, equal to 1.18 kN, is shown on the free-body diagram with the arrowhead pointing away from the joint, indicating tension, as in Figure 9.3(d).

After the force polygon '*gfbc*' has been constructed to scale and measured, we find that:

Force in member *CG* = 5.90 kN (compression)

It is customary to summarise the results obtained in tabular form:

Table 9.1   *Summary of results*

Member	Force (kN)	Direction
AE	8.25	compression
ED	4.37	tension
BF	3.75	compression
FE	1.18	compression
FG	1.18	tension
GD	3.12	tension
CG	5.90	compression

9.3   *THE METHOD OF JOINTS (MATHEMATICAL)*

The alternative to drawing force polygons is to consider each free-body diagram and solve for the unknown forces mathematically, in terms of their horizontal and vertical components.

Those who are mathematically minded may prefer this alternative, which lends itself more readily to computer-assisted methods of analysis and design. Others will probably choose the graphical method, which gives a better visual representation of the actual relationship between forces.

Example 9.2

Determine *mathematically* the force in each member of the truss shown in Figure 9.1, using the method of joints.

Solution

We need to know the angles which the truss members make with the horizontal so that we can use the formulae for resolving forces into their horizontal and vertical components:

$$F_x = F \cos \theta \qquad \text{and} \qquad F_y = F \sin \theta$$

The angle to the horizontal in our truss is:

$$\theta = \tan^{-1} \frac{4.8}{3.0}$$
$$= 58°$$

Step 1   Consider joint DAE

Refer to the free-body diagram in Figure 9.3(a).
From the summation of vertical components $\Sigma F_y = 0$:

$$7 + F_{AE} \sin 58° = 0$$
$$\therefore \; F_{AE} = 8.25 \text{ kN towards the joint, i.e. compression}$$

From the summation of horizontal components $\Sigma F_x = 0$:

$$F_{ED} - 8.25 \cos 58° = 0$$
$$\therefore \; F_{ED} = 4.37 \text{ kN away from the joint, i.e. tension}$$

Step 2   Consider joint EABF

Refer to the free-body diagram in Figure 9.3(b).
From the summation of vertical components $\Sigma F_y = 0$:

$$8.25 \sin 58° - 8 + F_{FE} \sin 58° = 0$$
$$\therefore \; F_{FE} = 1.18 \text{ kN towards the joint, i.e. compression}$$

From the summation of horizontal components $\Sigma F_x = 0$:

$$8.25 \cos 58° - 1.18 \cos 58° - F_{BF} = 0$$
$$\therefore \; F_{BF} = 3.75 \text{ kN towards the joint, i.e. compression}$$

Step 3   Consider joint DEFG

Refer to the free-body diagram in Figure 9.3(c).
From the summation of vertical components $\Sigma F_y = 0$:

$$F_{FG} \sin 58° - 1.18 \sin 58° = 0$$
$$\therefore \; F_{FG} = 1.18 \text{ kN away from the joint, i.e. tension}$$

From the summation of horizontal components $\Sigma F_x = 0$:

$$-4.37 + 1.18 \cos 58° + 1.18 \cos 58° + F_{GD} = 0$$
$$\therefore \; F_{GD} = 3.12 \text{ kN away from the joint, i.e. tension}$$

Step 4   Consider joint GFBC
Refer to the free-body diagram in Figure 9.3(d).
From the summation of horizontal components $\Sigma F_x = 0$:

$$3.75 - 1.18 \cos 58° + F_{CG} \cos 58° = 0$$
$$\therefore F_{CG} = 5.90 \text{ kN towards the joint, i.e. compression}$$

These answers are the same as those listed in Table 9.1 for the graphical method of solution.

 Problems

9.1 Determine the force in each member of the truss shown in Figure 9.4:
(a) graphically
(b) mathematically

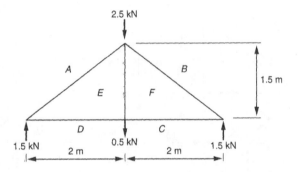

Fig. 9.4

9.2 Determine the force in each member of the truss shown in Figure 9.5:
(a) graphically
(b) mathematically

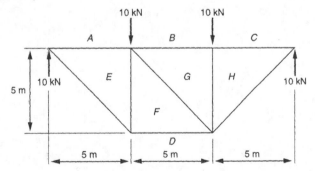

Fig. 9.5

9.3 Determine the force in each member of the truss shown in Figure 9.6:
(a) graphically
(b) mathematically

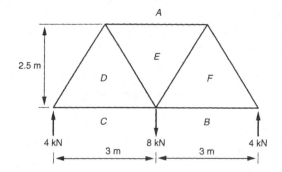

Fig. 9.6

9.4 Determine the force in each member of the truss shown in Figure 9.7:
(a) graphically
(b) mathematically

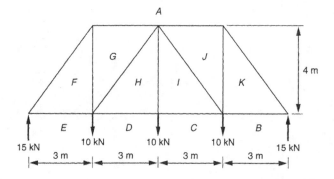

Fig. 9.7

9.5 Determine the force in each member of the truss shown in Figure 9.8:
(a) graphically
(b) mathematically

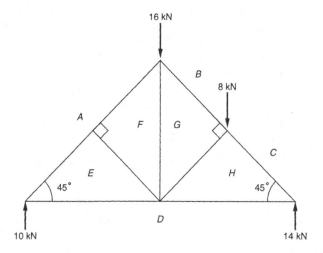

Fig. 9.8

9.4   *MAXWELL'S DIAGRAM METHOD*

The method of graphical analysis of trusses known as **Maxwell's diagram** is an extension of the graphical method of analysis by joints. Maxwell's diagram is a composite diagram which combines into one all the separate force polygons as previously drawn for the individual joints.

If we go back to the original force polygons shown in Figure 9.3 and superimpose them on each other, making sure that all corresponding intersection points coincide, we will get the combined diagram as shown in Figure 9.9.

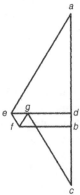

Fig. 9.9   *Maxwell's diagram*

You may have noticed that the arrowheads are missing on the combined diagram. The reason is that they appear to be in conflict between each pair of separate force diagrams containing the same force. This is not really a problem, as Bow's notation comes to the rescue and helps us to interpret the meaning of Maxwell's diagram.

Example 9.3

Construct Maxwell's diagram for the truss shown in Figure 9.1, and hence determine the force in each member of the truss.

Solution

Step 1

On a suitably large sheet of paper, draw the truss to scale. Show all loads and reaction forces and use Bow's notation to designate each region between forces outside the truss and each triangular panel inside the truss (Fig. 9.10(a)).

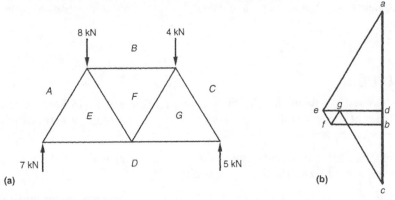

(a)

(b)

Fig. 9.10

Step 2

Take the sequence of letters around the truss in clockwise order and commence drawing the force diagram by taking each external force in turn and connecting them in a head-to-tail fashion. A vertical line results, which is a directional plot of force vectors joined together as follows:

$$a \text{ to } b = 8 \text{ kN down}$$
$$b \text{ to } c = 4 \text{ kN down}$$
$$c \text{ to } d = 5 \text{ kN up}$$
$$d \text{ to } a = 7 \text{ kN up}$$

This line serves as a skeleton on which to build all other force polygons.

Step 3

Consider joint *DAE* as before (refer to Fig. 9.3). Line *da* has already been drawn. Therefore, to locate point *e*, draw a line through point *a* parallel to member *AE*, and another line through point *d* parallel to member *ED*. Mark the intersection between these two new lines with the letter *e*. This completes force triangle *dae*. Do not show any arrows on the diagram you are constructing.

Repeat the process by considering joint *EABF*, and drawing lines *bf* and *fe* parallel to members *BF* and *FE* respectively.

Repeat with joints *DEFG* and *FBCG*. The diagram is now complete (Fig. 9.10(b)).

Notice that all lines previously drawn are being reused. Therefore it requires only two additional lines to effectively complete each new polygon. This is the main advantage of the Maxwell diagram method.

Step 4

The magnitudes of the forces in all members of the truss can now be measured from the combined diagram.

In order to determine whether a member is in tension or in compression, we have to decide whether the member pulls or pushes on the joint to which it is attached. For example, consider member *CG* in relation to joint *FBCG*. Reading the letters in the clockwise order around the joint, we read *CG*. The direction of the force exerted on the joint is found by reading the letters on the Maxwell force diagram in the same order, i.e. the direction of the force is 'from *c* to *g*'. This indicates that the force pushes on the joint and that the member is in compression.

Needless to say, the answers should be the same as previously obtained by the method of joints.

 Problems

9.6 Draw Maxwell's diagram for the truss shown in Figure 9.4 and hence determine the magnitudes of the forces in all members of the truss. Distinguish between tension and compression, and tabulate results.

9.7 Determine the magnitude and nature of the forces in members of the truss shown in Figure 9.5, using Maxwell's diagram method.

9.8 Determine the magnitude and nature of the forces in members of the truss shown in Figure 9.6, using Maxwell's diagram method.

9.9 Determine the magnitude and nature of the forces in members of the truss shown in Figure 9.7, using Maxwell's diagram method.

9.10 Determine the magnitude and nature of the forces in members of the truss shown in Figure 9.8, using Maxwell's diagram method.

9.5   THE METHOD OF SECTIONS

The **method of sections** is a mathematical method based on the principle that if the entire truss is in equilibrium, any part of the truss must also be in equilibrium. This method is particularly useful when forces in only a small number of nominated members need to be determined, because it is not necessary to solve the entire truss.

Example 9.4

Determine forces in members *BF*, *FG* and *GD* of the truss shown in Figure 9.1, using the method of sections.

Solution

Draw a section line *x–x* through the entire truss, and then consider the left-hand part of the truss as a free body in equilibrium, as shown in Figure 9.11.

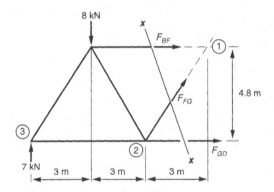

Fig. 9.11

We must consider all forces acting on this part of the truss only, and they include internal forces in the members cut by the sectioning line: F_{BF}, F_{FG} and F_{GD}.[*] Since we do not know at this stage whether these internal forces are tension or compression, tension is always assumed initially and arrowheads indicated accordingly.

Step 1
Select a point where two of the unknown forces meet, such as point 1. This eliminates these two forces from the equation of moments. Take moments of all other forces about this point and equate to zero.

$$\Sigma M = 7 \times 9 - 8 \times 6 - F_{GD} \times 4.8 = 0$$
$$\therefore F_{GD} = 3.125 \text{ kN}$$

---

[*] It is not essential to use Bow's notation with the method of sections. It could even be simpler to just number the three unknown forces F_1, F_2 and F_3. However, we have retained the Bow's notation letter subscripts to make comparisons with previous examples easier.

This answer is positive. Therefore, our initial assumption of the member being in tension was correct.

Step 2

Select another point where a pair of the unknown forces meet, such as point 2. This eliminates these two forces from the equation of moments. Take moments of all other forces about this point and equate to zero:

$$\Sigma M = 7 \times 6 - 8 \times 3 + F_{BF} \times 4.8 = 0$$
$$\therefore F_{BF} = -3.75 \text{ kN}$$

This answer is negative. Therefore, our initial assumption of the member being in tension was incorrect. The member is in compression.

Step 3

Now any convenient point may be chosen, provided it does not lie on the line of action of the remaining unknown force. The left-hand support, point 3, is as good as any other. Take moments about this point and equate to zero:

$$\Sigma M = 8 \times 3 + F_{BF} \times 4.8 - F_{FG} \times 6 \sin 58° = 0$$

Substitute $F_{BF} = -3.75$ kN:

$$\Sigma M = 8 \times 3 + (-3.75) \times 4.8 - F_{FG} \times 6 \sin 58° = 0$$
$$\therefore F_{FG} = 1.179 \text{ kN}$$

This answer is positive. Therefore, our initial assumption of the member being in tension was correct.

 Problems

9.11 Determine the forces in the members cut by line *x–x* as shown in Figure 9.12.

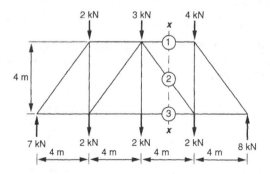

Fig. 9.12

9.12 Determine the forces in the members cut by line *x–x* as shown in Figure 9.13.

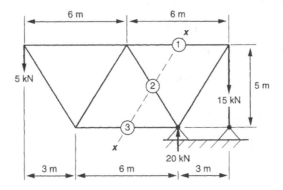

Fig. 9.13

9.13 As an additional exercise, select any members of any truss in problems 9.1 to 9.5 and determine the magnitudes and nature of the internal forces in these members using the method of sections.

Review questions

1. Briefly outline the method of truss analysis known as the *method of joints*.
2. What are the similarities and the differences between the mathematical and graphical solutions by the method of joints?
3. Why is Maxwell's diagram also called the *combined diagram*?
4. Which method would you use if you were interested in the force in only one member of a truss?
5. In general, which of the available methods of truss analysis do you prefer? Why?

PART 4

SLIDING FRICTION

Friction is what prevents perpetual motion from becoming a reality.

Anonymous

CHAPTER 10

Dry sliding friction

It is common experience that when one object is pushed or pulled along the surface of another object, there is resistance to such motion which must be overcome by applying an external force. The property of the surface that causes this resistance is called **friction.** More precisely, friction is a force that resists the sliding of one solid object over another.

The phenomenon of friction is one of the most fundamental and most common encountered by the engineer. It has been studied as a branch of mechanics where the methods of statics are best suited for estimating the magnitude and effects of friction.

A new science called **tribology** is now emerging which studies friction and its related subjects of wear and lubrication.

Expected learning outcomes

After carefully studying the material presented in this chapter, working through all numerical examples, and successfully completing all practice problems, students should be able to:

1. define *coefficient of friction*;
2. solve problems involving friction on a horizontal surface by applying the mathematical expression of the law of dry sliding friction;
3. define *angle of friction*;
4. solve problems involving friction on a horizontal surface by applying the concept of angle of friction.

10.1 *FRICTIONAL RESISTANCE*

Friction is not a simple phenomenon. Its laws have been studied for hundreds of years, and reasonably satisfactory methods for calculating friction in simple cases have been known for at least two centuries. However, the exact mechanism by which frictional forces are formed is not yet completely understood.

Friction is believed to be caused by the adhesion of or interference between microscopic irregularities of the surfaces in the areas of contact. It depends on the nature of the surfaces in contact, the force pressing one of the surfaces onto the other, and to some extent on whether the surfaces are already in relative motion or at rest.

Some frictional effects are beneficial, such as the traction needed to walk without slipping. The operation of brakes, clutches and power transmission belts depends on the presence of friction. Friction also makes the wheels of a locomotive grip the rails. On the other hand, much of the power used to drive machines of all kinds is consumed in overcoming friction between moving parts. Friction also tends to produce heat, which causes the surfaces to wear, necessitating costly repairs and replacements. Lubrication tends to reduce frictional resistance and wear between moving machinery parts and is therefore widely used.

Another form of resistance similar to friction, known as **rolling resistance,** is produced when a deformable tyre rolls over a surface. The main reason for rolling friction is deformation which occurs when the tyre is slightly flattened and the road surface is somewhat indented at the point of contact. The resistance to motion due to rolling friction is generally considerably less than that of sliding friction. A common engineering application which combines the effects of rolling friction and lubrication is the ball or roller bearings used to reduce frictional resistance to rotational motion of shafts and wheels.

Resistance of air to the motion of vehicles and projectiles through it also represents a friction-like force. However, mathematical treatment of air resistance is quite complex and different from that of sliding friction. Except for the laws of dry sliding friction discussed in some detail in this chapter, we will simply regard all forms of friction as forces resisting motion.

10.2 *THE LAWS OF DRY SLIDING FRICTION*

The first research on the laws governing sliding friction was carried out by Leonardo da Vinci in the 15th century. However, the credit for discovering the laws of friction is given to Guillaume Amontons, French physicist and inventor of scientific instruments, who published his investigations on friction in 1699. Another early worker was Charles Coulomb, also a French scientist and a military engineer best known for his work in electrostatics. He conducted the first complete investigation of dry friction in 1781. The theory of dry friction often bears the name of Coulomb friction.

Sliding friction is characterised by two simple experimental facts. First, it is nearly independent of the area in contact. Second, friction is proportional to the normal force that presses the surfaces together.

Consider a solid block, such as a brick, resting on a horizontal table surface (Fig. 10.1). The weight (F_w) acting down and the normal reaction (F_n) from the table, acting up, are equal and opposite. If a small push (F_p) is applied to the brick and the brick does not move, there must be a friction force (F_f), equal and opposite to the push, at the surface of contact between the brick and the table. If the push is increased gradually, it will eventually cause the brick to slide on the table. There is, therefore, a limiting value of the force of friction beyond which it cannot increase.

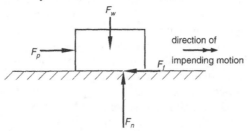

Fig. 10.1   *Friction force F_f on a horizontal plane*

Experiments show that the limiting value of the friction force is the same whether the brick is lying down or standing on its end, i.e. it is independent of the area in contact. Furthermore, if a pile of three bricks is pushed along the table, the friction is three times greater than when one brick is pushed, i.e. friction is proportional to the normal force.

At the moment of impending motion, the value of F_f is always a fixed ratio of the normal force F_n, which depends on the materials and the roughness of the contacting surfaces. The constant value of this ratio is called the **coefficient of static friction** and is usually symbolised by the Greek letter μ (mu).

$$\mu = \frac{F_f}{F_n}$$

Because both friction and normal forces are measured in units of force, i.e. newtons, the coefficient of static friction is dimensionless.

Table 10.1   *Approximate values of the coefficient of static friction*

Surfaces	Typical value	Usual range
metal on metal (greasy)	0.1	0.08–0.2
hardwood on metal (greasy)	0.2	0.15–0.3
metal on metal (dry)	0.2	0.15–0.35
wire rope on metal pulley	0.2	0.15–0.4
hemp rope on metal pulley	0.3	0.2–0.5
wood on wood	0.35	0.25–0.5
hardwood on metal (dry)	0.35	0.2–0.6
rubber or leather on metal	0.4	0.3–0.6
brake lining on metal	0.4	0.3–0.7
metal on stone	0.4	0.3–0.7
wood on stone	0.4	0.3–0.7
masonry on brickwork	0.6	0.55–0.7
rubber tyre on concrete	0.8	0.6–1.0

Owing to the variation in friction with the condition of the rubbing surfaces, it is impossible to specify exact values for each pair of materials in contact. Typical clean, unlubricated surfaces have friction coefficients in the range 0.2 to 0.4. Table 10.1 gives generally accepted approximate values.

The presence of a liquid lubricant entirely alters the character of friction and often has a profound effect on the friction coefficient. Good lubricants are able to reduce the coefficient of friction to as low as 0.05. A typical range for lubricated polished surfaces is 0.08 to 0.15.

The friction described up until now is that which arises between surfaces at rest with respect to each other. At the moment of impending motion, the smallest force (F_p) required to overcome static friction and to start motion is:

$$F_p = F_f = \mu F_n$$

Once the motion begins, the force required to continue the motion, i.e. to maintain motion against continuous frictional resistance, called **kinetic friction**, is usually less than static friction by about 25 per cent.

From these conclusions, the **laws of friction** can be summarised as follows:

1. Friction always acts in a direction opposite to impending or actual motion.
2. Static friction has a limiting value beyond which it cannot increase.
3. The limiting value of static friction is given by $F_f = \mu F_n$.
4. The value of the coefficient of static friction (μ) depends on the nature and condition of the surfaces in contact, but is independent of the areas in contact.
5. In general, kinetic friction is less than the limiting static friction.

Example 10.1

A body of mass 5 kg rests on a horizontal surface and the coefficient of friction between the two surfaces is 0.33. What horizontal force will be required to start the body moving?

Solution

Refer to Figure 10.1.
Weight of body:

$$F_w = mg$$
$$= 5 \text{ kg} \times 9.81 \ \frac{\text{N}}{\text{kg}}$$
$$= 49.05 \text{ N}$$

Normal force:

$$F_n = F_w$$
$$= 49.05 \text{ N}$$

Limiting friction:

$$F_f = \mu F_n$$
$$= 0.33 \times 49.05 \text{ N}$$
$$= 16.2 \text{ N}$$

Therefore, the force required to just start the body moving is:

$$F_p = F_f$$
$$= 16.2 \text{ N}$$

Example 10.2

In an experiment to determine the coefficient of static friction, a horizontal force of 50 N was required to start a 10 kg block moving on a horizontal surface. What was the value of the coefficient?

Solution

Refer to Figure 10.1.

$$F_n = F_w$$
$$= mg$$
$$= 10 \text{ kg} \times 9.81 \frac{\text{N}}{\text{kg}}$$
$$= 98.1 \text{ N}$$

$$F_f = F_p$$
$$= 50 \text{ N}$$

$$\mu = \frac{F_f}{F_n}$$

$$= \frac{50 \text{ N}}{98.1 \text{ N}}$$
$$= 0.51$$

Example 10.3

A 100 kg block rests on a plate as shown in Figure 10.2. The coefficient of friction between all surfaces is 0.2. Determine the force required to pull the plate from under the block.

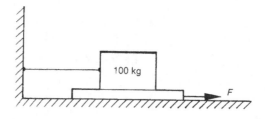

Fig. 10.2

Solution

Normal force:

$$F_n = F_w$$
$$= mg$$
$$= 100 \text{ kg} \times 9.81 \frac{\text{N}}{\text{kg}}$$
$$= 981 \text{ N}$$

Friction force:

$$F_f = \mu F_n$$
$$= 0.2 \times 981 \text{ N}$$
$$= 196.2 \text{ N}$$

The applied force F_p must overcome friction between two pairs of surfaces in contact.

Therefore:
$$F_p = 2F_f$$
$$= 2 \times 196.2 \text{ N}$$
$$= 392.4 \text{ N}$$

 ## Problems

10.1 A block of mass 2 kg rests on a horizontal table, and the coefficient of static friction between the surfaces is 0.28. What horizontal force will be required to start the block moving?

10.2 A block of wood having a mass of 5 kg rests on a horizontal table. A horizontal force of 12 N is just sufficient to cause it to slide. What is the coefficient of static friction between the surfaces?

10.3 A 2 tonne girder is pulled along a horizontal floor by a winch. The coefficient of friction between the girder and the floor is 0.35. What is the tension in the horizontal rope between the girder and the winch?

10.4 A 20 kg block rests on a horizontal surface and is attached to a mass of 3.5 kg by a cable, as shown in Figure 10.3. The pulley is frictionless and the coefficient of static friction is 0.2.

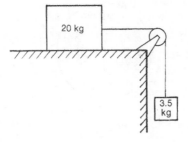

Fig. 10.3

Prove that the mass is not sufficient to start the block moving. What is the additional mass that would be required to start motion?

10.5 Each of the two blocks in Figure 10.4 has a mass of 10 kg, and the coefficient of friction between all surfaces is 0.3. Determine the force F_p required to pull one block from under the other.

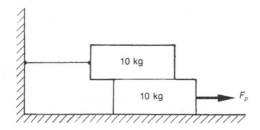

Fig. 10.4

10.6 In an automatic materials-handling operation, metal blocks 1.5 kg each are pushed one at a time from the bottom of a stack six blocks high as shown in Figure 10.5. If the coefficient of friction is 0.25, determine the horizontal force required.

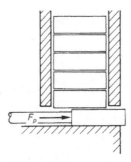

Fig. 10.5

10.7 What is the braking torque* applied to a brake drum of diameter 300 mm (Fig. 10.6) if the brake shoes are pressed to the drum with a force of 800 N each? The coefficient of friction is 0.6.

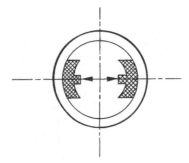

Fig. 10.6

10.8 In the brake shown in Figure 10.7, the coefficient of friction between the brake shoe and the drum is 0.45. Find the smallest value of force F required to prevent rotation of the drum against an applied torque of 75 N.m.

---

* Torque is a turning effect of one or several forces about the axis of a rotating component that tends to maintain or to resist rotation. (See Section 14.3.)

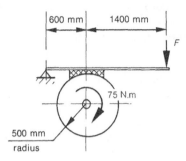

Fig. 10.7

10.9 A 35 kg cabinet, 0.75 m × 0.75 m × 1.8 m high, is pushed by a gradually increasing horizontal force applied 1.5 m above floor level. If the coefficient of friction between the cabinet and the floor is 0.4, determine if the cabinet will slide or tip.

10.3 *THE ANGLE OF FRICTION*

Let us now consider another approach to the analysis of friction on a horizontal plane. Referring to Figure 10.8, it is possible to combine the force of friction (F_f) and the normal reaction (F_n) into a single resultant force (F_r), representing a total reaction at the surface of contact to the action of weight (F_w) and applied force (F_p).

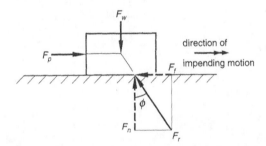

Fig. 10.8   *The friction force and the normal reaction combined into a single resultant force*

The resultant force (F_r) will be inclined at an angle to the normal direction. Using the rules for finding resultants of two mutually perpendicular forces yields:

$$F_r = \sqrt{F_f^2 + F_n^2}$$

and:
$$\tan \phi = \frac{F_f}{F_n}$$

But we know that at the moment of impending motion, i.e. when sliding motion is about to begin, the ratio of F_f to F_n is equal to the coefficient of static friction (μ). Therefore, for limiting friction:

$$\boxed{\tan \phi = \mu}$$

Under these circumstances, the angle ϕ, known as the **angle of friction**, is constant and depends only on the nature and condition of the surfaces in contact. The values of the angle of friction for each corresponding value of μ can easily be calculated and are summarised in Table 10.2.

Table 10.2   *The angle and coefficient of static friction*

μ	0.1	0.15	0.2	0.25	0.3	0.35	0.4	0.45	0.5
ϕ	5.71°	8.53°	11.31°	14.04°	16.70°	19.29°	21.80°	24.23°	26.57°

The use of the angle of friction enables problems involving sliding friction to be solved graphically as illustrated by the following example.

Example 10.4

A body of mass 3 kg rests on a horizontal surface, and the coefficient of friction between the two surfaces is 0.3. What horizontal force is required to start the body moving?

Solution

Weight of the body:

$$F_w = mg$$
$$= 3 \text{ kg} \times 9.81 \ \frac{\text{N}}{\text{kg}}$$
$$= 29.43 \text{ N}$$

Angle of friction:

$$\phi = \tan^{-1} 0.3$$
$$= 16.7°$$

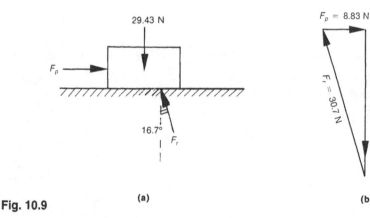

Fig. 10.9

The three forces in equilibrium are as shown in Figure 10.9(a). The answer is found by constructing a triangle of forces, Figure 10.9(b).

Therefore, the force required is:

$$F_p = 8.83 \text{ N}$$

10.4 *THE ANGLE OF REPOSE*

An easy way of observing the laws of friction and, in particular, of measuring the coefficient of friction between two surfaces is by means of a plane tilted slowly through an angle θ to the horizontal.

Take a book or a board and place on it a matchbox or similar object. Tilt the book slowly through a small angle θ. The weight of the box can now be resolved into components, one along $F_w \sin \theta$ and the other perpendicular to the surface of the book, $F_w \cos \theta$ (Fig. 10.10).

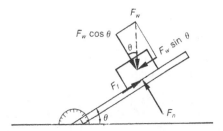

Fig. 10.10   *Friction on a tilted plane*

If the angle of tilting is small, the box will not slide down the book but will remain at rest, the force of friction being sufficient for equilibrium.

As the book is tilted more and more, the friction will reach its limiting value beyond which it cannot increase. At this moment, the box is on the point of slipping and all forces are exactly balanced:

$$F_n = F_w \cos \theta$$
$$F_f = F_w \sin \theta$$

Remembering that $\mu = \dfrac{F_f}{F_n}$, we write:

$$\mu = \frac{F_f}{F_n} = \frac{F_w \sin \theta}{F_w \cos \theta} = \tan \theta$$

$$\text{But} \quad \mu = \tan \phi$$
$$\therefore \ \tan \theta = \tan \phi$$
$$\text{or} \quad \theta = \phi$$

The value of the angle of inclination corresponding to impending motion is called the **angle of repose.** Clearly, the angle of repose is equal to the angle of static friction.

Understanding the concept of the angle of repose is important in the design of bins and hoppers for the storage and handling of granular materials, and for calculating the steepest angle to the horizontal which can be made by the inclined surface of a heap of loose material or an embankment.

Example 10.5

What is the steepest ramp on which a car can stand without slipping down if the coefficient of friction between the tyres and the ramp surface is 0.8?

Solution

$$\text{Angle of repose} = \text{angle of friction } \phi$$
$$= \tan^{-1} \mu$$
$$= \tan^{-1} 0.8$$
$$= 38.7°$$

 Problems

10.10 Determine the angle of friction corresponding to a coefficient of static friction of 0.6.

10.11 If the normal and frictional forces between two surfaces which are about to slip are 100 N and 35 N respectively, determine the coefficient of static friction, the angle of friction and the magnitude of the resultant force between the surfaces.

10.12 If the normal reaction between two surfaces is 120 N and the coefficient of friction is 0.25, determine the magnitude and direction of the total reaction for the case of limiting friction.

10.13 A body of 2 kg mass rests on a board which is gradually tilted until, at an angle of 27° to the horizontal, the body begins to move down the plane. Determine the coefficient of friction and the magnitude of the normal and frictional forces when the body begins to slip.

10.14 Solve problem 10.1 graphically.

10.15 Solve problem 10.2 graphically.

10.16 Solve problem 10.3 graphically.

10.17 Solve problem 10.4 graphically.

10.18 Solve problem 10.5 graphically.

10.19 Solve problem 10.6 graphically.

 Review questions

1. What is meant by *friction*?
2. Are frictional effects beneficial or harmful? State some examples.
3. What are the main two characteristics of sliding friction?
4. What is the ratio of the frictional and normal forces called?
5. What is the unit of the coefficient of static friction?
6. What is the effect of liquid lubricants on the magnitude of friction?
7. What is the difference between *static* and *kinetic* friction?
8. State the laws of dry sliding friction.
9. Define the *angle of friction*. How is it related to the coefficient of static friction?
10. What is meant by the *angle of repose*? How is it related to the angle of friction?

CHAPTER 11

Friction on inclined planes

One of the most common types of problems involving friction is that of starting or maintaining motion of a body on a sloping surface connecting two different levels. The importance of a good understanding of sliding friction on an inclined plane is not confined to raising or lowering a heavy object up or down a ramp. Devices such as screw jacks, power screws, cams and wedges operate essentially on the inclined plane principle. It is for this reason that we devote this chapter to the basic study of friction on an inclined plane.

Expected learning outcomes

After carefully studying the material presented in this chapter, working through all numerical examples, and successfully completing all practice problems, students should be able to:

1. draw a free-body diagram and analyse the equilibrium of forces for a body on an inclined plane;
2. solve problems involving impending or uniform motion on an inclined plane using the mathematical summation of force components;
3. solve problems involving impending or uniform motion on an inclined plane by graphical construction;
4. solve problems involving impending or uniform motion on an inclined plane by the sine rule method.

11.1 *FRICTION ON AN INCLINED PLANE*

In general, the mechanics of friction on an inclined plane can be divided into four possibilities:

1. There is no external push or pull applied to the body, apart from the force of gravity, and there is sufficient friction to prevent sliding. The body remains stationary on the incline. A car safely parked on a gentle slope is a good example. This situation occurs when the angle of inclination of the plane is less than the angle of repose (see Section 10.4).

2. An external force (F_p) is applied to the body, just sufficient to overcome static friction and to set the body in motion. When the body is on the verge of sliding motion, up or down the plane, the condition is referred to as **impending motion**, and the coefficient of static friction applies. All forces acting on the body are in static equilibrium and the law of friction applies. Therefore, we have the following three equations available for problem solving:

$$\Sigma F_x = 0, \quad \Sigma F_y = 0 \quad \text{and} \quad F_f = \mu F_n$$

3. The body is moving along the plane with uniform speed. In order to maintain continuous motion at constant speed, steady application of a somewhat smaller force (F_p) is required (see Section 10.1). Its magnitude depends on the value of the coefficient of kinetic friction. Otherwise, the body can be said to be in dynamic equilibrium, and the same three equations may be used for problem solving.*

4. The body is moving along the plane, assisted by an external force (F_p) against frictional resistance, but the forces are not in equilibrium. In this case the velocity is not constant and the problem can no longer be solved by the methods of statics alone. We will discuss this type of problem in Chapter 13.

A typical free-body diagram for cases 2 and 3 involves the following four forces: weight (F_w), applied push or pull (F_p), normal reaction force (F_n) and frictional force (F_f). See Figure 11.1(a) and (b). Note very carefully the direction of the frictional force, which must always be shown in opposition to that of the impending or actual motion.

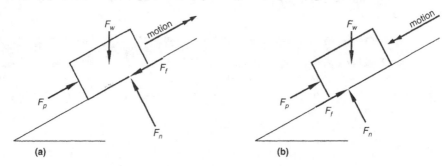

Fig. 11.1 (a) *Motion up* **(b)** *Motion down*

When a graphical solution is intended, it is useful to combine the normal reaction force (F_n) and the frictional force (F_f), which are related to each other by the angle of friction, and to represent them as a single reaction force (F_r), inclined to the normal direction by the angle of friction ϕ. It is extremely important to have the inclination of the reaction force (F_r) shown correctly with respect to the direction of motion. Examine Figure 11.2 very carefully, and compare it with the corresponding diagrams in Figure 11.1.

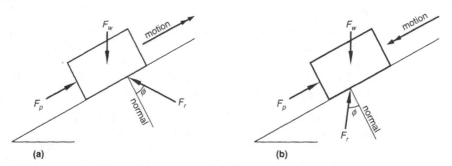

Fig. 11.2 (a) *Motion up* **(b)** *Motion down*

---

* In the worked examples and practice problems offered in this book, the distinction between coefficients of static friction and kinetic friction is not strongly emphasised. It is to be assumed that the value of the coefficient of friction given in any particular problem is the appropriate one for the situation, and its identity can be inferred from the context.

Problems involving friction and equilibrium of forces on an inclined plane can be solved by any one of three methods:

1. mathematically, by resolving all forces into components in two directions at 90° to each other, and then solving the equilibrium equations in each of the two directions;
2. graphically, by constructing a force triangle to scale, representing the equilibrium of forces acting on the body;
3. semigraphically, by sketching the triangle of forces and then solving it using the sine rule.

11.2 *MATHEMATICAL SOLUTION*

The mathematical method is best suited to problems in which the unknown applied force (F_p) is parallel to the plane.

Example 11.1

A 200 kg block rests on a 25° incline as shown in Figure 11.3(a). The coefficient of friction between the block and the plane is 0.2. Determine the magnitude of the force F_p, acting along the inclined plane, required to:

(a) start the block moving up the inclined plane
(b) prevent it slipping down the inclined plane

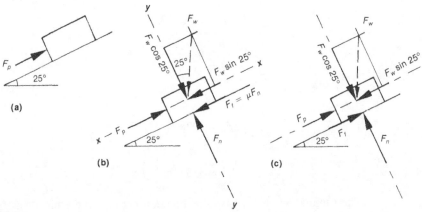

Fig. 11.3

Solution

(a) While motion is about to begin up the plane, the force of friction (F_f) will act down the plane in opposition to the impending motion, as shown in Figure 11.3(b).

The weight of the block can be resolved into components along and perpendicular to the plane:

$$(F_w)_x = F_w \sin 25°$$
$$= 200 \times 9.81 \times \sin 25°$$
$$= 829.2 \text{ N}$$
$$(F_w)_y = F_w \cos 25°$$
$$= 200 \times 9.81 \times \cos 25°$$
$$= 1778 \text{ N}$$

Summation of forces in the direction not containing the unknown force (F_p) yields:

$$\Sigma F_y = 0$$
$$F_n = F_w \cos 25°$$
$$\therefore F_n = 1778 \text{ N}$$

The law of friction can now be used to find the frictional force F_f:

$$F_f = \mu F_n$$
$$= 0.2 \times 1778$$
$$= 355.6 \text{ N}$$

Summation of forces in the other direction:

$$\Sigma F_x = 0$$
$$F_p = F_w \sin 25° + F_f$$
$$F_p = 829.2 + 355.6$$
$$\therefore F_p = 1184.8 \text{ N}$$

(b) The tendency for motion is down the plane, and the direction of friction will be reversed as shown in Figure 11.3(c). The solution follows the same basic steps:

$$(F_w)_x = F_w \sin 25°$$
$$= 829.2 \text{ N}$$
$$(F_w)_y = F_w \cos 25°$$
$$= 1778 \text{ N}$$

$\Sigma F_y = 0$:

$$\therefore F_n = 1778 \text{ N}$$

$F_f = \mu F_n$:

$$\therefore F_f = 0.2 \times 1778$$
$$= 355.6 \text{ N}$$

The summation of forces in the direction parallel to the plane has friction acting with the applied force, i.e. up the plane, to balance the component of weight acting along the plane.

$\Sigma F_x = 0$:

$$F_p + F_f = F_w \sin 25°$$
$$F_p + 355.6 = 829.2$$
$$\therefore F_p = 473.5 \text{ N}$$

By comparing the answers to parts (a) and (b), it can be seen that there is a range of magnitudes of force F_p between 473.5 N and 1184.8 N within which the block remains stationary on the plane. However, if the force is 1184.8 N or greater, the block will move up the plane. On the other hand, if the force is equal to or less than 473.5 N, the block will slip down the plane.

It is also worth noting that while the weight of the block is 1962 N, it can be pushed up the plane with a force of only 1184.8 N. It is suggested that the student should repeat this solution a few times using different values of the coefficient of friction and the angle of inclination to see what effect these have on the magnitude of the force required.

11.3 *GRAPHICAL SOLUTION*

The graphical method is relatively quick and simple. Its accuracy is quite acceptable for most practical purposes, provided the force triangle is drawn to a sufficiently large scale.

Example 11.2

A 100 kg block resting on a 30° incline is acted upon by a force F_p at 20° to the plane as shown in Figure 11.4(a). Determine the magnitude of the force required to start the motion up the plane if the coefficient of static friction is 0.2.

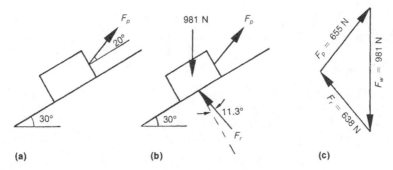

(a)   **(b)**   **(c)**

Fig. 11.4

Solution

The weight of the block is:

$$F_w = 100 \times 9.81$$
$$= 981 \text{ N}$$

The resultant of normal and frictional forces is inclined at the angle of friction from the normal direction, the angle of friction being:

$$\phi = \tan^{-1} \mu$$
$$= \tan^{-1} 0.2$$
$$= 11.3°$$

It is important to be careful when deciding which way the friction acts, in order to draw the correct direction of the resultant (see Fig. 11.4(b)).

A triangle of forces is then constructed (Fig. 11.4(c)) from which the magnitude of the applied force F_p is scaled off.

The answer, in this case, is 655 N.

11.4 SEMIGRAPHICAL SOLUTION

The semigraphical solution involves sketching the triangle of forces (not necessarily to scale) and then solving it by the sine rule method. It combines the best features of the other two methods, namely the helpful visual representation of the relationship between the forces as found in a graphical solution, with the precision of a mathematical method which can be carried out to any desired order of accuracy.

Example 11.3

Determine the magnitude of force F_p in the previous example by using the sine rule.

Solution

After sketching the triangle, the three angles within the triangle are evaluated by reference to the angles shown on the free-body diagram.

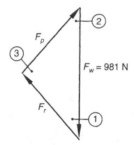

Fig. 11.5

In this case:

$$\text{Angle No. } 1 = 30° + 11.3°$$
$$= 41.3°$$
$$\text{Angle No. } 2 = 90° - (30° + 20°)$$
$$= 40°$$
$$\text{Angle No. } 3 = 180° - (41.3° + 40°)$$
$$= 98.7°$$

Force F_p can now be calculated quite simply using the sine rule:

$$\frac{981 \text{ N}}{\sin 98.7°} = \frac{F_p}{\sin 41.3°}$$
$$\therefore F_p = 655 \text{ N}$$

 Problems

11.1 A 150 kg box is resting on a 15° incline with a coefficient of friction of 0.35. What force acting along the plane is required to start the box moving up the plane?

11.2 What force parallel to the slipway is required to lift a 1.5 tonne boat up an incline of 20° if the coefficient of friction is 0.25?

11.3 A load of 20 kg resting on an inclined plane begins to slip when the plane is tilted to an angle of 25° to the horizontal. What force parallel to the plane will be necessary to keep the load from slipping down the plane if the angle of the slope is increased to 35°?

11.4 If the applied force in the previous problem is horizontal, what magnitude is required?

11.5 In problem 11.3, what horizontal force must be applied to the load to just start it moving up the plane inclined at 25°? at 35°?

11.6 What is the mass of the largest stone that can be dragged uphill at an angle of 20° by a tractor, capable of exerting a pull of 2 kN, if the coefficient of friction is 0.6?

11.7 A casting of mass 100 kg is pulled along a horizontal floor by a rope at 30° to the floor. The tension in the rope just sufficient to overcome friction is 300 N. Determine the coefficient of friction.

11.8 Determine the maximum and minimum values of the mass m between which static equilibrium will be maintained for the system shown in Figure 11.6. Take $\mu = 0.3$.

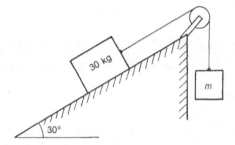

Fig. 11.6

11.9 Determine the range of values of mass m within which static equilibrium will be maintained for the system shown in Figure 11.7. Take $\mu = 0.25$.

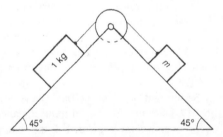

Fig. 11.7

Review questions

1. Briefly outline the different possibilities for rest or motion of a body on an inclined plane.
2. Sketch a typical free-body diagram of a body on an inclined plane. Explain the relationship between the direction of friction and that of impending motion.
3. Briefly outline the mathematical method of problem solving when friction on an inclined plane is involved.
4. Briefly outline the graphical method of problem solving when friction on an inclined plane is involved.
5. What is the advantage of the sine rule method?

CHAPTER 12

Screws and wedges

In this chapter we consider two practical applications of the inclined plane as an engineering component. First, we will direct our attention to the analysis of square-threaded screws, and then we will discuss some problems involving wedges. Both topics have a close relationship with the subject matter presented in the previous chapter, and both take into account the ever present phenomenon of friction.

Expected learning outcomes

After carefully studying the material presented in this chapter, working through all numerical examples, and successfully completing all practice problems, students should be able to:
1. analyse forces acting on the thread of a square-threaded screw;
2. solve problems involving the load–effort relation for a screw jack;
3. calculate forces involving simple arrangements of wedges.

12.1 *FRICTIONAL FORCES ON SCREW THREADS*

The analysis of forces acting on screw threads is very similar to that of a body sliding along an inclined plane. For our purposes here we limit our discussion to single-start square-threaded screws, which are most commonly used when large forces are applied along the axis of the screw. These screws are employed in screw jacks, clamping devices, presses and other mechanisms for the purpose of converting rotational effort exerted on the screw into an axial force.

A screw thread may be thought of as an inclined plane wrapped around a cylinder, at a mean diameter D (Fig. 12.1). A nut travels an axial distance L, called the **lead**, along the axis of the screw for every complete turn of the screw. The base length of the equivalent inclined plane, 'unwrapped' from one turn of the thread, is πD.

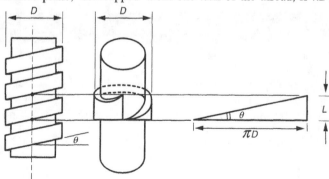

Fig. 12.1 *A screw thread*

The angle which the actual thread makes with a plane perpendicular to the axis of the screw is called the **helix angle**. The helix angle (θ) is the same as the angle of inclination of the equivalent inclined plane. It can be seen from Figure 12.1 that:

$$\tan \theta = \frac{L}{\pi D}$$

Example 12.1

What is the helix angle of a screw thread which has a lead of 10 mm and a mean diameter of 35 mm?

Solution

$$\tan \theta = \frac{L}{\pi D}$$
$$= \frac{10}{\pi \times 35}$$
$$= 0.090\,95$$
$$\therefore \theta = \tan^{-1} 0.090\,95$$
$$= 5.20°$$

Hence the helix angle is $\theta = 5.20°$.

It can be seen that in order for a screw of mean diameter D to exert a force F_L along its axis, there must be another force F_p acting on the nut at the mean radius of the thread. Graphical analysis of forces on the incline of the thread (Fig. 12.2) shows that:

$$F_p = F_L \tan (\phi + \theta)$$

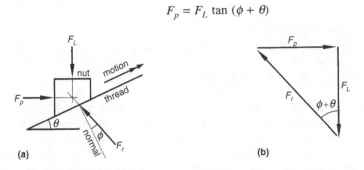

(a)   (b)

Fig. 12.2   (a) *Forces on the incline of the thread* **(b)** *Triangle of forces*

Since force F_p is acting at a radial distance $D/2$ from the axis of the screw, the turning moment M which must be applied is:

$$M = F_p \frac{D}{2}$$

and so the formula for the applied moment is:

$$M = \frac{F_L D}{2} \tan (\phi + \theta)$$

where ϕ is the angle of friction for the thread
     θ is the helix angle of the screw

Example 12.2

If the coefficient of friction for the screw in the previous example is 0.2, what is the moment required to exert an axial force of 540 N?

Solution

Angle of friction:

$$\phi = \tan^{-1} \mu$$
$$= \tan^{-1} 0.2$$
$$= 11.31°$$

Mean diameter:

$$D = 0.035 \text{ m}$$

Substitute into the formula for the applied moment:

$$M = \frac{F_L D}{2} \tan (\phi + \theta)$$
$$= \frac{540 \text{ N} \times 0.035 \text{ m}}{2} \tan (11.31° + 5.2°)$$
$$= 2.8 \text{ N.m}$$

Hence the required moment is 2.8 N.m.

12.2 *SELF-LOCKING SCREWS*

We recall (see Section 10.4, Angle of repose) that if the angle of friction is greater than the angle of inclination of an inclined plane, a body will not slide down the plane unless it is assisted by a reversed force, pushing it down the plane. It follows that if the angle of friction is greater than the helix angle of the thread, i.e. $\phi > \theta$, and the moment M is removed, the screw will remain self-locking, i.e. it will continue to support the load F_L by friction alone.

Example 12.3

Is the screw in the previous examples a self-locking screw?

Solution

The angle of friction $\phi = 11.31°$ and the helix angle $\theta = 5.2°$. Clearly, the angle of friction is greater than the helix angle. Therefore the screw is self-locking.

It can be shown that the reverse moment M', required to loosen a self-locking screw which is supporting a load F_L, is given by a very similar expression:

$$M' = \frac{F_L D}{2} \tan (\phi - \theta)$$

where ϕ is the angle of friction for the thread
      θ is the helix angle of the screw

Example 12.4

Determine the reverse moment that must be applied to the screw in the previous examples in order to loosen it under its load.

Solution

Substitute into the formula:

$$
\begin{aligned}
M' &= \frac{F_L D}{2} \tan (\phi - \theta) \\
&= \frac{540 \text{ N} \times 0.035 \text{ m}}{2} \tan (11.31° - 5.2°) \\
&= 1.01 \text{ N.m}
\end{aligned}
$$

Hence the reverse moment required is 1.01 N.m.

12.3 *THE SCREW JACK*

A simple screw jack, as shown in Figure 12.3, is a portable device for raising a heavy mass through a short distance. It consists of a vertical screw which is turned in a threaded base by an effort applied by hand to a horizontal lever arm attached to the screw cap.

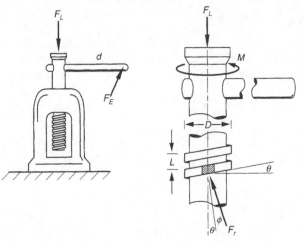

Fig. 12.3   *A screw jack*

The required turning moment M is produced by the effort force F_E applied at the end of the arm at a distance d from the centreline,* so that:

$$M = F_E d$$

Since the required moment is related to the load F_L by:

$$M = \frac{F_L D}{2} \tan (\phi + \theta)$$

we can combine these two expressions as follows:

$$F_E d = \frac{F_L D}{2} \tan (\phi + \theta)$$

Therefore, the effort required at the end of the arm of a simple screw jack can be calculated from:

$$F_E = \frac{F_L D}{2d} \tan (\phi + \theta)$$

Since the load on a screw jack is usually given as mass (m), the expression can be modified by substituting:

$$F_L = F_w$$
$$= mg$$

Hence:

$$\boxed{F_E = \frac{mgD}{2d} \tan (\phi + \theta)}$$

Example 12.5

Determine the horizontal force F_E, perpendicular to the 500 mm long handle of a simple screw jack, necessary to start lifting a 500 kg load. The square-threaded screw has a lead of 7 mm and a mean diameter of 80 mm. The coefficient of static friction for the screw is 0.15.

Solution
Angle of friction:

$$\phi = \tan^{-1} \mu$$
$$= \tan^{-1} 0.15$$
$$= 8.53°$$

---

* Only a very simple lever arm arrangement is considered here. Any difference in the ratio advantage due to a hand-operated crank-and-gear mechanism, found in many mechanical screw jacks, has to be taken into account where appropriate.

Helix angle:

$$\theta = \tan^{-1} \frac{L}{\pi D}$$

$$= \tan^{-1} \frac{7}{\pi \times 80}$$

$$= 1.60°$$

Therefore, the effort required at the end of the arm is:

$$F_E = \frac{mgD}{2d} \tan (\phi + \theta)$$

$$= \frac{500 \text{ kg} \times 9.81 \text{ N/kg} \times 0.08 \text{ m}}{2 \times 0.5 \text{ m}} \tan (8.53° + 1.60°)$$

$$= 70.1 \text{ N}$$

 ## *Problems*

12.1 Calculate the helix angle of a square thread if the mean diameter of the thread is 50 mm and the lead of the screw is 26 mm.

12.2 If the coefficient of friction for the screw in the previous problem is 0.18, what is the moment required to exert an axial force of 2 kN?

12.3 Determine the clamping force if the screw of a G-clamp is tightened with a turning moment $M = 3.5$ N.m. The screw has a mean diameter of 10 mm, a lead of 3 mm, and the coefficient of friction is 0.3.

12.4 Determine the moment M that must be applied to draw the end screws of a turnbuckle (Fig. 12.4) closer together if it has a square thread with a mean diameter of 12 mm and a lead of 2 mm. The axial tension force is 3 kN and the coefficient of friction is 0.25.

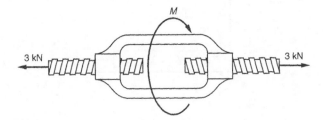

Fig. 12.4

12.5 Is the turnbuckle in the previous problem self-locking?

12.6 A G-clamp has a square-threaded screw with a mean diameter of 18 mm, a lead of 2.5 mm and a coefficient of friction of 0.22. If the clamping force is 900 N, determine the reverse moment that must be applied in order to loosen the screw.

12.7 A force F is applied to the handle of a square-threaded screw jack at a distance 660 mm from the centreline of the screw. The mean diameter of the screw is 50 mm and the lead is 4.8 mm. If the coefficient of friction is 0.1, determine the magnitude of the force F required to raise a load of 1.5 t.

12.8 Determine the mass which can be lifted by an effort of 50 N applied perpendicularly to the 370 mm handle of a screw jack, which has a square-threaded screw with a lead of 5 mm and a mean diameter of 60 mm. The coefficient of static friction for the screw is 0.2.

12.4 WEDGES

A **wedge** is a piece of hard material with two principal faces meeting at a sharply acute angle. The most common application of wedges is in making small adjustments to the position of heavy objects by applying to the wedge a force usually considerably smaller than the weight of the load. Wedges are also used as machine parts wherever small movements of a component in one direction is associated with large forces in a perpendicular direction. Because of the smallness of the wedge angle, and friction between all surfaces in contact, wedges usually have a self-locking characteristic, i.e. they will remain in place after being forced under the load.

The following example illustrates a method of solving problems involving wedges. You will recognise this method as a semigraphical approach, using the sine rule for solving the triangle of forces.

Example 12.6

A 3 tonne crate is raised by forcing a 3° wedge under it as shown in Figure 12.5. The coefficient of static friction is 0.27 at all surfaces in contact. Determine the minimum value of force F_p which must be applied to the wedge to move the crate upwards.

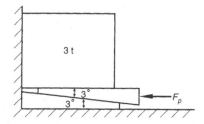

Fig. 12.5

Solution

There are two moving components in this problem: the crate and the wedge. Two separate free-body diagrams must be drawn and analysed. It is always wise to start with the component which carries the known force. In this case it is the crate, whose weight is:

$$F_w = mg$$
$$= 3000 \text{ kg} \times 9.81 \text{ N/kg}$$
$$= 29\,430 \text{ N}$$
$$= 29.43 \text{ kN}$$

The free-body diagram of the crate is drawn, as shown in Figure 12.6(a).

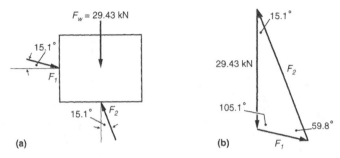

Fig. 12.6 (a) *Free-body diagram*   (b) *Triangle of forces*

Note carefully the directions of reaction forces at the surfaces subjected to friction. These are determined on the basis of the direction of impending motion of the crate in relation to the supporting surfaces, with which the crate makes contact. When the crate moves, it will move upwards relative to the wall, and to the right relative to the wedge.[*] Friction opposes both, and this determines the direction of the resultant forces F_1 and F_2. The angle at which these reaction forces are inclined to the normal direction at each surface is the angle of friction:

$$\phi = \tan^{-1} \mu$$
$$= \tan^{-1} 0.27$$
$$= 15.1°$$

The corresponding triangle of forces is then drawn as shown in Figure 12.6(b). We can now use the sine rule to solve for force F_2:

$$\frac{29.43}{\sin 59.8°} = \frac{F_2}{\sin 105.1°}$$

$$\therefore F_2 = 32.88 \text{ kN}$$

Now consider the wedge as a free body (Fig. 12.7(a)). It has the known reaction from force F_2 (= 32.88 kN) acting on it in the direction shown, as well as the unknown reaction force F_3. Once again, note very carefully the directions of these forces in relation to the wedge.

The triangle of forces is shown in Figure 12.7(b). Solving the triangle by the sine rule gives:

$$\frac{32.88}{\sin 71.9°} = \frac{F_p}{\sin 33.2°}$$

$$\therefore F_p = 18.9 \text{ kN}$$

---

[*] It is quite immaterial that in relation to a larger stationary frame of reference, it is the wedge that actually moves to the left while the crate is held stationary in the horizontal direction against the wall. We are concerned here with the relative motion of the crate with respect to the wedge, and that motion is from left to right.

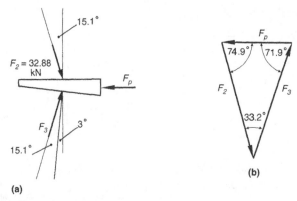

Fig. 12.7 **(a)** *The wedge as a free body* **(b)** *The triangle of forces*

 Problems

12.9 In a machine, the level of component A, which has a mass of 45 kg, is adjusted by a 5° wedge as shown in Figure 12.8. Assuming there is no frictional resistance in the roller supports, and the coefficient of friction between all sliding surfaces is 0.1, determine the force F_p required to raise the component.

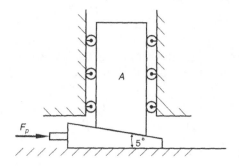

Fig. 12.8

12.10 In order to improve the effectiveness of the wedge in Example 12.6, some grease was used on the wedge, reducing the coefficient of friction on its surfaces from 0.27 to 0.15. Friction between the wall and the crate remained the same as before. Determine the new value of force F_p.

12.11 The mass of block A in Figure 12.9 is 100 kg and the mass of block B is 25 kg. If the coefficient of static friction between all surfaces is 0.35, determine the magnitude of force F_p required to just start moving block A to the right.

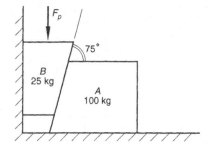

Fig. 12.9

12.12 Two wedge blocks are used to grip a flat bar specimen in a tensile testing machine as shown in Figure 12.10. If the angle θ of the wedges is 23°, and the coefficient of friction at surface A is 0.14, what is the coefficient of friction at B when the bar is about to slip? (*Hint*: Consider the wedge as a two-force body in equilibrium.)

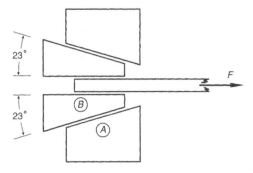

Fig. 12.10

12.13 If the wedge blocks in problem 12.12 are to be redesigned so that the minimum coefficient of friction at B is 0.47 while the coefficient at A remains 0.14, what is the maximum wedge angle that would not permit the specimen to slip under load?

Review questions

1. Define *helix angle* for a screw thread.
2. Explain how the helix angle can be calculated for a given thread.
3. What is a *self-locking screw*?
4. State the formula for calculating the effort required to lift a specified mass with a simple screw jack.
5. Briefly outline the method of analysing friction forces on wedges.

DYNAMICS

PART 5

LAWS OF MOTION ≡

If I have seen further, it is by standing on the shoulders of giants.

Sir Isaac Newton

CHAPTER 13

Linear motion

One of the primary concerns of mechanical engineering is with motion. Motion exists where there is a change in position or orientation of an object with reference to some other object or objects. Aeroplanes, trains and automobiles are part of everyday life. A reciprocating piston in an internal-combustion engine, a rotating flywheel, water flowing inside a pipe, and steam driving the rotor of a turbine are typical examples of different kinds of motion with which the science and practice of mechanical engineering are concerned.

Historically, mechanics is the oldest of the physical sciences and it provided the foundation for the growth and development of all other areas of physics. The study of motion in mechanics is called **dynamics**. Unlike statics, which goes back to Archimedes and the Greek philosophers, serious study of dynamics only began with the experimental work of Galileo (1564–1642), which led to the mathematical formulation of the fundamental laws of motion by Newton in his famous publication known as the *Principia.**

The part of dynamics that describes the geometry of motion in terms of two elementary concepts, position and time, is called **kinematics**. The part that relates the motion to the forces causing it is called **kinetics**.

---

* *Philosophiae Naturalis Principia Mathematica*, or *Mathematical Principles of Science*.

The problems of dynamics can be classified and studied according to the type of motion that exists. In this chapter, we discuss motion along a straight line, known as **rectilinear translation**. We shall simply refer to it as 'linear motion'. With some limitations our discussion also applies to motion along a curved path.

Expected learning outcomes

After carefully studying the material presented in this chapter, working through all numerical examples, and successfully completing all practice problems, students should be able to:

1. solve problems involving displacement, velocity and acceleration using the equations of linear motion;
2. solve problems involving acceleration of bodies in free fall;
3. state Newton's laws of motion;
4. solve problems involving acceleration against resistance and acceleration against gravity.

13.1 *DISPLACEMENT, VELOCITY AND ACCELERATION*

Let us now consider how linear motion of an object, such as a motor car, can be described in terms of its position and time. More specifically, motion is usually described in terms of linear displacement, linear velocity and linear acceleration—the concepts which are defined below.

Later in this and the following chapters, if it is clear from the context that motion is in a straight line, the word 'linear' will be omitted from the terms 'linear displacement', 'linear velocity' and 'linear acceleration'. Furthermore, provided that displacement, velocity and acceleration are measured along the path of the motion at every point, e.g. road distances and not 'as the crow flies', the definitions and the formulae that follow are also applicable to curved paths.

Linear displacement

Linear motion can generally be described as a change in the position of an object, or as the passage of an object from one place to another along its path. **Linear displacement**, *S*, of an object is a measure of its change of position along the path of its motion with respect to an arbitrary fixed point.

Displacement is a vector quantity, i.e. it has direction as well as magnitude. For our purposes, it is usually convenient to measure displacement from the initial position of the object in the direction of motion, which is assumed to be positive, in which case displacement is also a measure of the distance travelled by the object.*

The magnitude of linear displacement can be measured in metres, kilometres or millimetres, depending on the actual distance measured. However, when relating displacement to velocity, acceleration, and especially to force, work and power, it is usually necessary to convert to base units, i.e. metres, before any further calculations are performed.

---

* One should be aware that if reversal in the direction of motion occurs, e.g. in reciprocating or oscillatory motion, the displacement from a fixed origin as defined above is not equivalent to the total distance travelled by the object.

Linear velocity

Motion is a change in the position of an object which occurs in time. The time rate of change in the position of an object is known as its **linear velocity**, v. As such, velocity is a vector quantity, implying a magnitude and direction.

For motion which occurs in one direction only, as is the case in the majority of problems treated in this book, the directional sense of velocity coincides with that of displacement and is taken to be positive. If reversal in the direction of motion occurs, such as when an object is thrown upwards, reaches a maximum altitude and starts falling, problems can easily be solved by considering the different stages of motion separately.*

For linear motion in one direction, velocity at any point can best be understood as the distance travelled by an object per unit time in a specified direction along its linear path. The base unit of velocity is metre per second (m/s). However, the original unit of measurement commonly used to express vehicle speeds is kilometre per hour (km/h), which is not decimally related to the metre per second (1 km/h = 1000 m/3600 s = $\frac{1}{3.6}$ m/s). It is usually necessary to convert velocity from kilometres per hour to metres per second, particularly if calculations involve other related concepts, such as acceleration, power or momentum.

Note that speed is not a synonym for velocity, but a scalar quantity which expresses the magnitude of velocity only, without any reference to its direction. In your car, the speedometer reading tells you how fast you are travelling at any instant of time without any regard to the direction of your travel. Although in many situations the meaning is obvious, one should not regard the terms speed and velocity as equivalent.

If distances travelled in successive intervals of time are the same, the speed is said to be constant. Otherwise average speed can be calculated using total distance and the time taken to cover that distance.

Example 13.1

In the last Grand Prix race of the series on his way to the World Drivers' Championship title in 1980, Alan Jones of Australia established a new lap record for the 6 km road course at Watkins Glen, NY, by covering the distance in 1 min 43.8 s.

What was his speed in metres per second? What was it in kilometres per hour?

Solution

Distance: $\qquad\qquad\qquad S = 6$ km
$\qquad\qquad\qquad\qquad\quad = 6000$ m
Time: $\qquad\qquad\qquad\; t = 103.8$ s
Speed:

$$\frac{S}{t} = \frac{6000 \text{ m}}{103.8 \text{ s}}$$
$$= 57.8 \text{ m/s}$$

Converting to kilometres per hour:

$$\text{Speed} = 57.8 \text{ m/s} \times \frac{3600}{1000}$$
$$= 208.1 \text{ km/h}$$

---

* In more advanced problems involving oscillatory and curvilinear motion, it is usually necessary to treat velocities as vectors observing appropriate sign convention and the rules of vector mathematics.

Linear acceleration

If velocity is not constant but increases gradually at a uniform rate, an object is said to move with a uniformly accelerated motion.* The rate at which linear velocity changes with time is called **linear acceleration**, a.

If, over a period of time equal to t seconds, velocity of an object changes from its initial value v_0 to a final value v, it follows from the definition that acceleration a is the quotient of the increment of velocity $(v - v_0)$ and the time t:

$$a = \frac{v - v_0}{t}$$

It can readily be seen that the unit of acceleration must be the unit of velocity (m/s) divided by the unit of time (s), i.e. metre per second squared, m/s^2.

The relationship between initial and final velocities, time and acceleration for uniformly accelerated motion is usually stated as a formula in which final velocity is the subject:

$$v = v_0 + at$$

where v is final velocity in m/s
v_0 is initial velocity in m/s
a is acceleration in m/s^2
t is time taken in s

Example 13.2

A car starts from rest and accelerates at the rate of 1.2 m/s^2 for 15 s. Determine the velocity reached after 15 s.

Solution

Initial velocity: $v_0 = 0$
Time: $t = 15$ s
Acceleration: $a = 1.2$ m/s^2

Substitute into $v = v_0 + at$:

$$v = 0 + 1.2 \text{ m/s}^2 \times 15 \text{ s}$$
$$= 18 \text{ m/s}$$

Hence, velocity after 15 s is 18 m/s (64.8 km/h).

If, instead of increasing, velocity is gradually decreasing, the motion is said to be **uniformly decelerated**. Deceleration, or retardation, can be regarded as negative acceleration, i.e. acceleration acting in the direction opposite to velocity.

Example 13.3

If, after travelling for some distance at a constant velocity of 18 m/s, brakes are applied to the car producing a retardation of 2 m/s^2, determine the time taken to reduce its velocity to 10 m/s (36 km/h).

---

* The more general case of non-uniform acceleration is outside the scope of this book.

Solution

Initial velocity: $v_0 = 18$ m/s
Final velocity: $v = 10$ m/s
Acceleration: $a = -2$ m/s^2

Substitute into $v = v_0 + at$:

$$10 = 18 - 2t$$

$$\therefore t = \frac{10 - 18}{-2}$$

$$= 4 \text{ s}$$

Hence time taken is 4 s.

13.2 *EQUATIONS OF LINEAR MOTION*

In the case of uniformly accelerated linear motion, the distance travelled from the starting point is the product of time and average velocity:

$$S = t \times v_{av}$$

where simple arithmetic averaging of velocities gives:

$$v_{av} = \frac{v_0 + v}{2}$$

When these equations are combined, we have:

$$S = t\left(\frac{v_0 + v}{2}\right)$$

which is an additional independent equation to:

$$v = v_0 + at$$

Eliminating final velocity v from these equations yields:

$$S = v_0 t + \frac{at^2}{2}$$

Similarly, if time t is eliminated, we get:

$$2aS = v^2 - v_0^2$$

Given any three of the five variables, i.e. displacement, time, acceleration, and initial and final velocities, any problem involving uniformly accelerated linear motion can be solved using these equations.

Example 13.4

Find the total emergency stopping distance of a car and the total time taken from the point where the driver sights danger if the driver's reaction time before applying the brakes is 0.9 s, the initial velocity is 60 km/h and retardation due to the brakes is 7.5 m/s² (Fig.13.1).

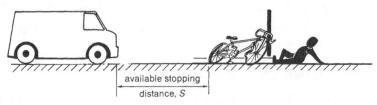

available stopping distance, S

Fig. 13.1

Solution

There are two stages in this problem:
(a) motion with uniform velocity before the brakes are applied, and
(b) uniformly decelerated motion that brings the car to rest.
Let us consider them separately.

(a) *Uniform motion*

Velocity (constant):
$$v = 60 \text{ km/h} \times \frac{1000}{3600}$$
$$= 16.67 \text{ m/s}$$

Time:
$$t = 0.9 \text{ s}$$

Displacement:
$$S = t \times v_{av}$$
$$= 0.9 \times 16.67 \text{ m/s}$$
$$= 15 \text{ m}$$

(b) *Decelerated motion*

Initial velocity:
$$v_0 = 16.67 \text{ m/s}$$
Final velocity:
$$v = 0$$
Acceleration:
$$a = -7.5 \text{ m/s}^2$$

Note: Retardation is negative acceleration.
Substitute into $v = v_0 + at$ to find time taken to bring the car to rest:

$$0 = 16.67 + (-7.5) \times t$$
$$\therefore t = 2.22 \text{ s}$$

Substitute into $S = t\,\dfrac{v_0 + v}{2}$ to find the displacement during the period of decelerated motion:

$$S = 2.22 \times \left(\frac{16.67 + 0}{2}\right)$$
$$= 18.5 \text{ m}$$

Combining the answers obtained in (a) and (b) yields:

$$\text{Total stopping distance} = 15 + 18.5$$
$$= 33.5 \text{ m}$$
$$\text{Total time taken} = 0.9 + 2.22$$
$$= 3.12 \text{ s}^*$$

13.3 *FREELY FALLING BODIES*

In ancient times, when people were ignorant of the effects of air resistance, an early Greek scientist–philosopher, Aristotle, had observed a leaf and a stone fall to the ground and had come to the general conclusion that a light body falls more slowly than a heavy one. Such was the appeal of the apparently irrefutable observation and the extraordinary influence of this outstanding thinker, that it took almost two thousand years and the perseverance of Galileo to dispel this erroneous principle.

Galileo suspected that if falling bodies were heavy enough to render air resistance negligible, Aristotle's idea was demonstrably wrong. He undertook a series of experiments which led him to the correct conclusions that the maximum speed attained by a falling body, in the absence of air resistance, is proportional to the time taken, and that the distance travelled from rest is proportional to the square of the time. This is consistent with the equations of linear motion if we take the initial velocity to be zero, i.e. $v_0 = 0$, and the acceleration due to gravity as a_g:

$$v = v_0 + at \quad \text{gives} \quad v = a_g t$$
$$\text{and} \quad S = v_0 t + \frac{at^2}{2} \quad \text{gives} \quad S = \frac{a_g t^2}{2}$$

Experiments conducted since the time of Galileo, combined with the principles expounded by Newton, set the value of gravitational acceleration at:

$$\boxed{a_g = 9.81 \text{ m/s}^2}$$

This is more or less constant anywhere on or near the surface of the Earth.[†]

---

* There must be a lesson in these results not only for the student of mechanics but also for the road safety conscious as well.

† The exact numerical value of gravitational acceleration is equal to the value of the gravitational constant and is, therefore, subject to local variations to the same extent (see Ch. 4). The relationship between gravitational acceleration and gravitational constant is illustrated in Example 13.8.

Example 13.5

A stone is dropped from the deck of a bridge and strikes the water below after 3.4 seconds of free fall. Neglecting air resistance, calculate the height of the bridge above the water and the velocity with which the stone strikes the water.

Solution

Since the initial velocity is zero, $v_0 = 0$, and $a_g = 9.81$ m/s^2, the distance of free fall is found from:

$$S = v_0 t + \frac{at^2}{2}$$
$$= 0 + \frac{9.81 \times 3.4^2}{2}$$
$$= 56.7 \text{ m}$$

The final velocity is found from:

$$v = v_0 + at$$
$$= 0 + 9.81 \times 3.4$$
$$= 33.4 \text{ m/s}$$

If a body such as a stone is projected vertically upwards with an initial velocity v_0, the acceleration due to gravity will be negative, i.e. it will be a retardation causing a decrease in velocity at the rate of 9.81 m/s^2.

Example 13.6

If a stone is thrown upwards from the edge of the bridge in the previous example with an initial velocity of 15.6 m/s, determine the maximum height reached above the deck of the bridge, the velocity with which it strikes the water below, and the total time taken.

Fig. 13.2

Solution

(a) Consider the upward motion first.

Initial velocity:	$v_0 = 15.6$ m/s
Acceleration:	$a_g = -9.81$ m/s^2
Final velocity at point A:	$v = 0$

Substitute into $2aS = v^2 - v_0{}^2$:

$$2 \times (-9.81) \times S = 0 - 15.6^2$$
$$\therefore S = 12.4 \text{ m}$$

Therefore the maximum height above the deck is 12.4 m.

Now using $v = v_0 + at$:

$$0 = 15.6 + (-9.81)t$$
$$\therefore t = 1.59 \text{ s}$$

Therefore, time taken to reach this height is 1.59 s.

(b) Consider free fall from the maximum height into the water below.
Measured from point A as the origin:

Displacement:
$$S = 56.7 \text{ m} + 12.4 \text{ m}$$
$$= 69.1 \text{ m}$$

Initial velocity: $\quad v_0 = 0$
Acceleration: $\quad a_g = 9.81 \text{ m/s}^2$

Substitute into $S = v_0 t + \dfrac{at^2}{2}$:

$$69.1 = 0 + \frac{9.81 t^2}{2}$$
$$\therefore t = \sqrt{\frac{69.1 \times 2}{9.81}}$$
$$= 3.75 \text{ s}$$

Therefore, time taken in free fall is 3.75 s.

Now using $v = v_0 + at$, velocity when striking the water is equal to:

$$v = 0 + 9.81 \times 3.75$$
$$= 36.8 \text{ m/s}$$

Combining the time taken for each part of this motion, total time from the moment the stone is thrown upwards to the moment it splashes into the water is:

$$t_{tot} = 1.59 \text{ s} + 3.75 \text{ s}$$
$$= 5.34 \text{ s}$$

It should be noted that since several alternative equations are available (see Section 13.2), there is usually more than one way in which a particular problem can be solved. The choice of a suitable sequence of calculations depends not only on the given information, but also on the selection of particular equations, which to some extent is a matter of personal preference.

Problems

13.1 When Kieren Perkins established a world record in the 1500 m freestyle swimming event at the 1994 Commonwealth Games in Canada, his time for the distance was 14 minutes and 41.66 seconds. What was his average speed?

13.2 What average speed should you travel at to cover the distance from Brisbane to Sydney, equal to 1020 km, in 12 h? Express your answer in kilometres per hour and in metres per second.

13.3 On a forward journey, a distance of 120 km was covered at an average speed of 100 km/h, and on the return journey at an average speed of 60 km/h. What was the average speed for the total journey?

13.4 Determine the acceleration required to increase the velocity of a motorcycle from 60 km/h to 100 km/h in 5 s.

13.5 A rocket is launched from rest with a constant upward acceleration of 18 m/s^2. Determine its velocity after 25 s.

13.6 Calculate the altitude reached by the rocket in the previous problem after 25 s.

13.7 A car travelling at 47 km/h accelerates to 97 km/h in a distance of 260 m. Calculate the acceleration and the time taken.

13.8 A train is moving at 56 km/h. If it is to pull up in 200 m, what must be the retardation and the time taken?

13.9 A launch approaches a wharf at a speed of 10.5 knots (= 5.4 m/s).* The engines are put in reverse when the launch is 10 m from the wharf, the retardation being 1.25 m/s^2. Will the launch come to rest before hitting the wharf? If not, with what velocity will it strike the wharf?

13.10 A train accelerates from rest at station *A* at a rate of 0.8 m/s^2 for 25 s, then travels at constant velocity for 77.5 s before it comes to rest at station *B* after a period of retardation lasting 20 s. What is the distance between stations *A* and *B,* and the average velocity of the train between the stations?

13.11 One train starts from rest and has an acceleration of 0.8 m/s^2. A second train has an initial velocity of 14.3 m/s and a retardation of 0.5 m/s^2. After how many seconds are they moving with the same velocity?

13.12 If the first train in the previous problem starts at the instant the second train passes it on parallel track and they move in the same direction, after how many seconds will the first train overtake the second train, and at what distance from the original point?

13.13 Determine the time taken and the final velocity when a hammer dropped accidentally from a height of 49 m on a construction site hits the ground.

13.14 A ball is kicked vertically upwards with a velocity of 35 m/s. Determine the greatest height and the time taken to reach it. Neglect air resistance.†

---

* A knot, originally defined as one nautical mile per hour, is a non-metric unit of speed, equal to 1.852 km/h, which is allowable in Australia for restricted application in marine and aerial navigation.

† In reality, motion of a relatively light object, such as a soccer ball, could be retarded quite significantly by air resistance. However, at this stage, the influence of air resistance will be ignored.

13.15 A rocket fired vertically burns its fuel for 25 s, which provides a thrust producing a constant upward acceleration of 8 m/s^2. Determine the altitude reached at the instant the rocket runs out of fuel, and the maximum altitude reached by the rocket. Determine also the total time, ground to ground, including free fall. Neglect air resistance.

13.4 *FORCE, MASS AND ACCELERATION*

The fundamental relationship between force, mass and acceleration is the central concept of dynamics. It is embodied in the system of mechanics summarised by Isaac Newton in his three **laws of motion**. Slightly reworded, these laws are as follows:

1. **First law.** If there is no unbalanced force acting on a body, the body will remain at rest or continue to move in a straight line with a constant linear velocity.
2. **Second law.** The acceleration of a body is proportional to the resultant force acting on it, inversely proportional to the mass of the body, and is in the direction of the force.
3. **Third law.** The forces of action and reaction are equal in magnitude and opposite in direction.

Taken in a different order, these laws take us through the basic principles of kinetics as follows.

The third law suggests that forces acting on a body always originate in other bodies, i.e. a push or pull experienced by an object is always a result of interaction with some other object. Single isolated forces do not exist.

The first law introduces the property of matter, often called **inertia**, which determines its resistance to a change in its state of rest or uniform motion. Mass of a body is then regarded as the quantitative measure of its **inertia**, which only a force can overcome.

The second law gives mathematical expression to the relationship between the mass of a body, the unbalanced force acting on it and the acceleration produced by the force:

$$F = m \times a$$

where F is the resultant of all forces acting on a body
m is the mass of the body
a is acceleration

It is very important to understand that the International System (SI) of units is a coherent system in which the product of any two unit quantities in the system is the unit of the resultant quantity. This applies to Newton's second law of motion for which the product of the SI units of mass and acceleration, i.e. kilogram and metre per second squared, is the SI unit of force, kg.m/s^2, which is given the special name **newton** (symbol N). Thus the second law provides the definition of the unit of force:

$$F = ma$$
$$1 \text{ N} = 1 \text{ kg} \times 1 \text{ m/s}^2$$

$$N = \frac{\text{kg.m}}{\text{s}^2}$$

Let us now consider a few examples of the applications of the second law of motion.

Example 13.7

Determine the net force required to give a body, of mass 300 kg, a horizontal acceleration of 2.5 m/s^2.

Solution

$$F = m \times a$$
$$= 300 \text{ kg} \times 2.5 \text{ m/s}^2$$
$$= 750 \text{ kg.m/s}^2$$
$$= 750 \text{ N}$$

Example 13.8

Determine the acceleration of a body of mass 25 kg due entirely to its own weight.

Solution

The weight of the body, i.e. the force of gravity acting on it, is:

$$F_w = mg$$
$$= 25 \text{ kg} \times 9.81 \ \frac{\text{N}}{\text{kg}}$$
$$= 245.3 \text{ N}$$

Therefore, from $F = ma$, acceleration due to gravity is:

$$a = \frac{F}{m}$$
$$= \frac{245.3 \text{ N}}{25 \text{ kg}}$$
$$= \frac{245.3 \text{ kg.m/s}^2}{25 \text{ kg}}$$
$$= 9.81 \text{ m/s}^2$$

Obviously, this only proves what we already knew about acceleration of a body under the action of gravity only, i.e. freely falling bodies. Furthermore, it is apparent that acceleration due to gravity is independent of mass.

Example 13.9

Determine the acceleration of a body sliding down a smooth surface inclined to the horizontal at 35°.

Solution

It is necessary to consider components of weight acting along and at right angles to the surface. Let the weight be $F_w = mg$. The perpendicular component of weight is balanced by the normal reaction at the surface, as shown in Figure 13.3.

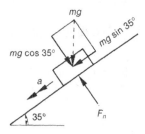

Fig. 13.3

The parallel component, equal to $mg \sin 35°$, is not balanced by any other force, assuming no frictional resistance, i.e. a smooth surface. Therefore, from $F = ma$:

$$a = \frac{F}{m}$$
$$= \frac{mg \sin 35°}{m}$$
$$= 9.81 \times \sin 35°$$
$$= 5.63 \text{ m/s}^2$$

The acceleration is in the direction of the net unbalanced force, i.e. parallel to the plane as shown.

Example 13.10

Determine the force required to accelerate a vehicle, of mass 1.5 t, from rest to 60 km/h in 12 s.

Solution

The acceleration required is found from $v = v_0 + at$, where $v = 60$ km/h $= 16.67$ m/s:

$$16.67 = 0 + a \times 12$$
$$\therefore a = 1.389 \text{ m/s}^2$$

Applying Newton's law:

$$F = ma$$
$$= 1500 \text{ kg} \times 1.389 \text{ m/s}^2$$
$$= 2083 \text{ N}$$
$$= 2.08 \text{ kN}$$

13.5 *ACCELERATION AGAINST RESISTANCE*

It has already been stated that, in the equation $F = ma$, the force is the net accelerating force, i.e. the resultant of all forces applied to the body.* The resultant unbalanced force, i.e. net accelerating force, is the difference between the applied push or pull F_p and the resistance force F_r.

$$F = F_p - F_r$$

The resistance is usually due to friction of some kind, which may include sliding friction, friction in bearings and air friction.

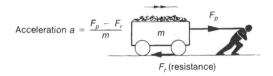

$$\text{Acceleration } a = \frac{F_p - F_r}{m}$$

F_r (resistance)

Fig. 13.4   *Using Newton's law to find acceleration against resistance*

Acceleration produced by the resultant force is found from Newton's law (Fig. 13.4):

$$a = \frac{F}{m}$$
$$= \frac{F_p - F_r}{m}$$

It should be clear from this equation that if the applied force F_p is equal to the resistance force F_r, there will be no acceleration, i.e. the motion, if any, will continue at constant velocity. Therefore, a vehicle moving on a level road at a constant speed requires a force equal to the tractive resistance to maintain uniform motion. Any force in excess of tractive resistance will accelerate the vehicle. On the other hand, a tractive effort which is less than tractive resistance will result in retardation.

Example 13.11

A train of total mass 120 t is travelling at 60 km/h on level track. The tractive resistance is 80 N/t. Calculate the tractive effort required to accelerate the train to 100 km/h in 30 s.

---

* For the sake of clearer understanding of the relationship between force and acceleration given by Newton's second law, we purposely avoided the expression 'inertia force' in this book. This term is sometimes used to denote a fictitious quantity equal and opposite to the resultant of the otherwise unbalanced system of actual forces acting on an accelerating body. It is also equal to the negative product of mass and acceleration of the body.

An 'inertia force' is not a real force. It does not represent any actual interaction that could be described as a push or a pull, and it does not obey Newton's third law. Its only merit, as a useful mathematical device, lies in the principle that any accelerated body can be treated as if it is in equilibrium under the action of the real forces and the fictitious force, leading to a technique of solving some problems in dynamics by the methods of statics.

However, if our purpose here is to learn about the phenomena of dynamics, introducing an artificial state of equilibrium, where in reality none exists, does not seem to be justified.

Solution
Acceleration must be calculated using $v = v_0 + at$, where $v = 100$ km/h $= 27.78$ m/s, and $v_0 = 60$ km/h $= 16.67$ m/s:

$$v = v_0 + at$$
$$27.78 = 16.67 + a \times 30$$
$$\therefore a = 0.37 \text{ m/s}^2$$

The net accelerating force required to accelerate the mass is:

$$F = ma$$
$$= 120\,000 \text{ kg} \times 0.37 \text{ m/s}^2$$
$$= 44\,400 \text{ N}$$
$$= 44.4 \text{ kN}$$

Tractive resistance, which is often stated as resistance per unit mass of the vehicle, is equal to:

$$F_r = 80 \; \frac{\text{N}}{\text{t}} \times 120 \text{ t}$$
$$= 9600 \text{ N}$$
$$= 9.6 \text{ kN}$$

Therefore, from the balance of forces:

$$F = F_p - F_r$$
$$44.4 \text{ kN} = F_p - 9.6 \text{ kN}$$
$$\therefore F_p = 44.4 + 9.6$$
$$= 54 \text{ kN}$$

The required tractive effort F_p is 54 kN.

Example 13.12
Determine the time taken and the distance travelled by the train in the previous example when it is brought to rest from 100 km/h by a braking force of 72.7 kN.

Solution
If tractive resistance F_r is unchanged:

$$F_r = 9.6 \text{ kN}$$

In this case, the resistance will assist the braking effort F_b in slowing the train down. Therefore the total decelerating force will be:

$$F = F_r + F_b$$
$$= 9.6 + 72.7$$
$$= 82.3 \text{ kN}$$

Deceleration is therefore equal to:

$$a = \frac{F}{m}$$

$$= \frac{82\,300 \text{ N}}{120\,000 \text{ kg}}$$

$$= 0.686 \text{ m/s}^2$$

Remembering that deceleration is negative acceleration, we can substitute into $2aS = v^2 - v_0^2$ and solve for distance travelled during the period of retardation:

$$2aS = v^2 - v_0^2$$
$$2 \times (-0.686)S = 0 - 27.78^2$$
$$\therefore S = 562 \text{ m}$$

The time taken can be found from:

$$S = t\left(\frac{v + v_0}{2}\right)$$

$$562 = t\left(\frac{0 + 27.78}{2}\right)$$
$$\therefore t = 40.5 \text{ s}$$

13.6 *ACCELERATION AGAINST GRAVITY*

When motion occurs in a vertical direction, such as that of a lift, the force required to produce acceleration is influenced by the force of gravity, i.e. by the weight, F_w, of the object.

The resultant unbalanced force in this case is the difference between the applied force F_p, which is often provided by the pull in a cable, and the force of gravity F_w (Fig. 13.5).

$$F = F_p - F_w$$

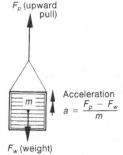

Fig. 13.5 *Using Newton's law to find acceleration against gravity*

Gravity in this case can be regarded as the resistance to upward acceleration. The acceleration produced by the resultant force is found from Newton's law:

$$a = \frac{F}{m}$$

$$= \frac{F_p - F_w}{m} \text{, where } F_w = mg$$

It follows that if the applied pull F_p is equal to the weight F_w, there will be no acceleration, i.e. the object will remain in a state of rest or, if moving, will continue to move with constant velocity.

A force F_p greater than the weight will produce an upward acceleration, while a force less than the weight will allow downward acceleration.

Example 13.13

A loaded lift has a total mass of 1500 kg. Determine the force in the cables when:
(a) the acceleration of 2 m/s² is upwards
(b) the acceleration of 2 m/s² is downwards

Solution

Net accelerating force is:

$$F = ma$$
$$= 1500 \text{ kg} \times 2 \text{ m/s}^2$$
$$= 3000 \text{ N}$$
$$= 3 \text{ kN}$$

The force of gravity (weight of the lift) is:

$$F_w = mg$$
$$= 1500 \text{ kg} \times 9.81 \ \frac{\text{N}}{\text{kg}}$$
$$= 14\,715 \text{ N}$$
$$= 14.7 \text{ kN}$$

(a) When acceleration is upwards:

$$F = F_p - F_w$$
$$3 \text{ kN} = F_p - 14.7 \text{ kN}$$
$$\therefore F_p = 17.7 \text{ kN}$$

Hence the force F_p in the cable is 17.7 kN.

(b) When acceleration is downwards, weight is greater than the force in the cable:

$$F = F_w - F_p$$
$$3 \text{ kN} = 14.7 \text{ kN} - F_p$$
$$\therefore F_p = 11.7 \text{ kN}$$

Hence the force F_p in the cable is 11.7 kN.

Problems

13.16 Calculate the acceleration of a car of mass 1.2 t on a level road if the net accelerating force is 6 kN.

13.17 A boy and his bicycle have a combined mass of 45 kg. When travelling at 20 km/h on a level road, he ceases to pedal and finds his speed reduced to 16 km/h in 10 s. What is the total force resisting motion?

13.18 A planing-machine table has a mass of 450 kg and must be accelerated from rest to a velocity of 0.35 m/s in 100 mm. Determine the net accelerating force required.

13.19 When a ship of mass 6000 t is launched, it leaves the slip with a velocity of 2.5 m/s and is then stopped within a distance of 150 m from the slip. What is the magnitude of the force that brings it to rest?

13.20 A trolley of mass 100 kg is to be pulled up a smooth incline at 10° to the horizontal with an acceleration of 2.5 m/s².
What force parallel to the incline is required if friction at the wheels is negligible?

13.21 A locomotive exerts a pull of 30 kN on a train whose total mass is 200 t. If total tractive resistance is 8 kN on level track, how long will it take to reach a velocity of 50 km/h, starting from rest?

13.22 The train in the previous problem reaches an upward slope of 4° to the horizontal. Determine the pull that must be exerted by the locomotive to maintain a steady velocity of 50 km/h. Assume that tractive resistance remains the same.

13.23 A train of mass 500 t is drawn by a locomotive of mass 50 t, which exerts a tractive effort of 75 kN, while tractive resistance is 60 N/t. Determine the time and distance required to reach a velocity of 80 km/h from rest, on level track.

13.24 A lift cage, together with its load, has a mass of 2000 kg and is raised vertically by a cable. Determine the pull exerted by the cable if the lift reaches a velocity of 5 m/s after rising 5 m from rest.

13.25 A lift has a mass of 1000 kg. Calculate the force in the lifting cable when the lift is:
(a) moving at constant velocity
(b) moving upwards with an acceleration of 1.6 m/s²
(c) moving upwards with a retardation of 1.4 m/s²
(d) moving downwards with an acceleration of 1.0 m/s²

13.26 A hoist lowers a load of 3 t at a uniform velocity of 2.4 m/s by means of a wire rope. Calculate the force in the rope required to bring the load to rest in 3 m.

13.27 A man in a lift holds a spring balance graduated to read in newtons, with a 2 kg mass hanging from it. What will the readings of the spring balance be when the lift:
(a) accelerates upwards at 2 m/s²?
(b) moves upwards with a constant velocity?
(c) moves upwards with a retardation of 2 m/s²?

13.7 *ACCELERATION AGAINST DRY SLIDING FRICTION*

We have already met with some problems involving acceleration of vehicles against tractive resistance. The underlying cause of resistance to continuous motion is usually found in some combination of different forms of friction. In this section, we are about to examine typical cases in which dry sliding friction between a body and a supporting surface is responsible for resistance to motion.

While studying these examples, observe a common sequence of steps in the mathematical method of dealing with such problems. This sequence can be summarised as follows:

Step 1
Draw a free-body diagram resolving all forces into components along the plane (*x*-direction) and perpendicular to the plane (*y*-direction). Note that the force of friction F_f must always be shown in the direction opposite to that of motion.

Step 2
Since there is no motion perpendicular to the plane, write the equation of equilibrium of forces in the *y*-direction, $\Sigma F_y = 0$. Solve for the normal reaction F_n.

Step 3
Using the law of dry sliding friction, $F_f = \mu F_n$, calculate the frictional resistance force F_f.

Step 4
Since there is motion along the plane, write the expression for the net accelerating force *F*, i.e. add all forces assisting motion and subtract all forces opposing motion in the *x*-direction.

Step 5
Substitute into Newton's law formula, $F = ma$, and then calculate the required unknown, which may be either the acceleration or the applied force F_p.

Example 13.14

A block of mass 26.3 kg is sliding along a horizontal supporting surface with an acceleration of 1.7 m/s² when pulled by a horizontal applied force F_p. If the coefficient of kinetic friction between the block and the surface is 0.35, what is the magnitude of the applied force?

Solution

Draw the free-body diagram as shown in Figure 13.6.

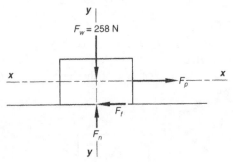

Fig. 13.6

In this case all forces are either parallel or perpendicular to the plane. Therefore, there is no need to resolve them into components.

The weight of the block is:

$$F_w = 26.3 \times 9.81$$
$$= 258 \text{ N}$$

There is no motion perpendicular to the plane. Therefore, the forces must be balanced in the *y*-direction:

$$F_n = F_w$$

Normal reaction:

$$F_n = 258 \text{ N}$$

Using the law of friction:

$$F_f = \mu F_n$$
$$= 0.35 \times 258$$
$$= 90.3 \text{ N}$$

Therefore frictional resistance force F_f is 90.3 N.

The net accelerating force F in the x-direction is equal to the unknown applied force F_p less resistance F_f:

$$F = F_p - F_f$$
$$\therefore F = F_p - 90.3$$

Substitute this into Newton's law formula:

$$F = ma$$
$$F_p - 90.3 = 26.3 \times 1.7$$
$$\therefore F_p = 135 \text{ N}$$

Hence the applied force F_p is 135 N.

Example 13.15

The block shown in Figure 13.7 has a mass of 5 kg. The coefficient of friction between the block and the surface is 0.32. The force F_p of 18 N is applied at 25° to the horizontal as shown. Calculate the acceleration of the block, and hence the distance moved in 6 s, starting from rest.

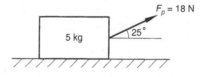

Fig. 13.7

Solution

In this case, it is necessary to resolve the applied force into components:

$$F_{px} = 18 \cos 25°$$
$$= 16.31 \text{ N}$$
$$F_{py} = 18 \sin 25°$$
$$= 7.61 \text{ N}$$

The free-body diagram for the block is shown in Figure 13.8.

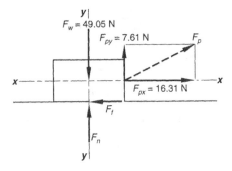

Fig. 13.8

The weight of the block is:

$$F_w = 5 \times 9.81$$
$$= 49.05 \text{ N}$$

There is no motion perpendicular to the plane. Therefore, the forces must be balanced in the y-direction. The vertical component of the applied force is included in this equilibrium equation:

$$F_n + F_{py} = F_w$$
$$F_n + 7.61 = 49.05$$
$$\therefore F_n = 41.44 \text{ N}$$

Hence, the normal reaction F_n is 41.44 N.

Using the law of friction:

$$F_f = \mu F_n$$
$$= 0.32 \times 41.44$$
$$= 13.26 \text{ N}$$

Hence, the frictional resistance force F_f is 13.26 N.

The net accelerating force F in the x-direction is equal to the horizontal component of the unknown applied force F_{px} less frictional resistance F_f:

$$F = F_{px} - F_f$$
$$= 16.31 - 13.26$$
$$= 3.05 \text{ N}$$

Substitute this into Newton's law formula:

$$F = ma$$
$$3.05 = 5a$$
$$\therefore a = 0.61 \text{ m/s}^2$$

Hence, the acceleration along the plane is 0.61 m/s^2.

The distance moved in 6 s is now found using the appropriate equation of linear motion:

$$S = v_0t + \frac{at^2}{2}$$

$$= 0 + \frac{0.61 \times 6^2}{2}$$

$$= 11 \text{ m}$$

Let us apply the same method to another similar problem, this time involving frictional resistance on an inclined plane.

Example 13.16

A box of mass 110 kg is accelerating downwards along a plane inclined at 35° to the horizontal against a restraining force of 185 N acting up along the plane. The coefficient of friction is 0.43. Calculate the acceleration of the box, and hence the time taken for it to slide 3 m along the plane.

Solution

The weight of the box is:

$$F_w = 110 \times 9.81$$
$$= 1079 \text{ N}$$

In this case, it is necessary to resolve the weight into components:

$$F_{wx} = 1079 \sin 35°$$
$$= 618.9 \text{ N along the plane}$$
$$F_{wy} = 1079 \cos 35°$$
$$= 883.9 \text{ N perpendicular to the plane}$$

The free-body diagram for the box is drawn as shown in Figure 13.9:

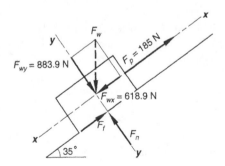

Fig. 13.9

Note carefully the direction of frictional resistance force F_f which is shown in opposition to motion, i.e. up the plane. There is no motion perpendicular to the plane. Therefore, the forces must be balanced in the y-direction. Only that component of weight which is perpendicular to the plane is relevant to this equilibrium equation:

$$F_n = F_{wy}$$
$$= 883.9 \text{ N}$$

Thus, the normal reaction F_n is 883.9 N.

Using the law of friction:

$$F_f = \mu F_n$$
$$= 0.43 \times 883.9 \text{ N}$$

Therefore, frictional resistance force F_f is 380.0 N.

The net accelerating force F is equal to the algebraic sum of all forces acting in the x-direction along the plane. These include the applied force F_p, the force of frictional resistance F_f, and the x-component of the weight F_{wx}. Forces acting in the direction of acceleration are taken to be positive. Forces acting in the opposite direction are taken to be negative.

$$F = F_{wx} - F_p - F_f$$
$$= 618.9 - 185 - 380.0$$
$$= 53.9 \text{ N}$$

Therefore the net accelerating force F is 53.9 N. Substitute this into Newton's law formula:

$$F = ma$$
$$53.9 = 110a$$
$$\therefore a = 0.49 \text{ m/s}^2$$

Hence, the acceleration of the box along the plane is 0.49 m/s².

The time taken for the box to slide 3 m along the plane is now found by substituting values into the following equation of linear motion:

$$S = v_0 t + \frac{at^2}{2}$$

$$3 = 0 + \frac{0.49t^2}{2}$$

$$\therefore t = 3.5 \text{ s}$$

Hence, the time taken is 3.5 s.

 # Problems

13.28 A block of mass 2.5 kg is sliding along a horizontal supporting surface with an acceleration of 2.9 m/s² when pulled by a horizontal applied force F_p.

   If the coefficient of kinetic friction between the block and the surface is 0.52, what is the magnitude of the applied force?

13.29 In an automated materials-handling operation, a metal component of mass 2.3 kg is pushed by a constant horizontal force F_p of 3.84 N along a horizontal table-top surface before falling into a chute. The coefficient of friction between the component and the supporting surface is 0.16.

If the motion starts from rest, how long does it take the component to move a horizontal distance of 800 mm, and what is the velocity at the end of this period of time?

13.30 A box of mass 127 kg is pulled along a horizontal surface by a force of 350 N, which is inclined at 30° to the horizontal. The coefficient of friction between the block and the surface is 0.2. Calculate the acceleration of the box.

13.31 A block of mass 66 kg is to be accelerated upwards along a plane inclined at 10° to the horizontal by force F_p, which is parallel to the plane and has a magnitude of 340 N. The coefficient of friction is 0.3.

Calculate the acceleration of the block, and hence the time taken for it to slide 4.4 m upwards along the plane, starting from rest.

13.32 A box of mass 110 kg is placed at the top of an inclined plane, which makes an angle of 35° to the horizontal, and is then released. If the coefficient of static friction between the box and the plane is 0.57, prove that the box will start sliding down the plane, unassisted by any external push.

If the coefficient of kinetic friction is 0.43, calculate the acceleration of the box, and hence the time taken for it to slide 3 m along the plane. Compare this problem with Example 13.16.

13.33 Determine the magnitude of the horizontal force F_p required to push a block of mass 30 kg, up along an inclined plane, with a constant acceleration of 0.8 m/s² (Fig. 13.10). The plane makes an angle of 25° to the horizontal, and the coefficient of kinetic friction between the plane and the block is 0.21. (*Hint*: The solution involves setting up simultaneous equations in terms of *x* and *y* components of the unknown force, i.e. $F_p \cos 25°$ and $F_p \sin 25°$, and then solving them for F_p.)

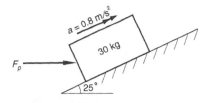

Fig. 13.10

Review questions

1. What is meant by:
   (a) *dynamics?*   (b) *kinematics?*   (c) *kinetics?*
2. Define *linear displacement, velocity* and *acceleration*, and state their symbols and the units used.
3. What is implied by the term *uniformly accelerated motion?*
4. List four equations of linear motion.
5. What is the acceleration of a body in free fall?
6. State Newton's laws of motion.
7. What is the relationship between the force acting on a body, the mass of the body, and the acceleration produced by the force?
8. What is the relationship between the newton and the kilogram?
9. Briefly outline the steps for solving problems involving acceleration against friction on inclined planes.

CHAPTER 14

Rotational motion

Up to this point we have considered linear motion, which could be described as motion in which the body moves from one place to another. Let us now discuss the case of **rotation**, or rotary motion, during which a body turns around a fixed axis in such a way that every particle of the body, except the axis, travels along a circular path. In rotary motion, the body as a whole does not move from one place to another. Common examples of rotation are flywheels, pulleys, shafts, gears, and turbine rotors.

You may find that this chapter is in some ways very similar to the previous chapter dealing with linear motion. this is because there is a definite analogy between linear and rotational motion, so that previously discussed concepts of linear displacement, velocity and acceleration have their rotational equivalents, which can be described by relations similar to equations and laws of linear motion.

Expected learning outcomes

After carefully studying the material presented in this chapter, working through all numerical examples, and successfully completing all practice problems, students should be able to:
1. solve problems involving angular displacement, velocity and acceleration using the equations of rotational motion;
2. state the relationship between torque and angular acceleration;
3. calculate the mass moment of inertia for objects consisting of disc or cylindrical elements;
4. solve problems involving angular acceleration against resistance.

14.1 *ANGULAR DISPLACEMENT, VELOCITY AND ACCELERATION*

Let us now consider how rotational motion of an object, such as a flywheel, can be described in terms of its orientation in space and time. More specifically, rotation is usually described in terms of angular displacement, angular velocity and angular acceleration—the concepts which are defined below.

Angular displacement

When a body undergoes rotational motion its orientation in space changes. **Angular displacement** of a rotating object is a measure of its change of orientation with respect to a fixed radius as an arbitrary datum. This is usually denoted by the Greek letter θ (theta).

Angular displacement is a rotational quantity, which has direction as well as magnitude. For our purposes, it is usually convenient to measure angular displacement from the initial position of the object, in the direction of rotation which is assumed to be positive, in which case angular displacement is simply the angle through which the object turns.

The magnitude of angular displacement can be measured in revolutions, degrees or radians. Revolutions are the most convenient units for measuring angular displacement of mechanical components such as shafts and flywheels. However, when relating angular displacement to torque, work and power, it is necessary to convert to base units, i.e. radians, before any further calculations are performed. Remember that one revolution contains 360 degrees or 2π radians.

$$1 \text{ revolution} = 360° = 2\pi \text{ rad}$$

Angular velocity

The rate at which a body changes its angular position is called its **angular velocity,** usually denoted by the symbol ω.[*] Angular velocity is also a quantity, with a magnitude and direction.

For rotational motion in a particular direction, the directional sense of angular velocity coincides with that of displacement and can be taken to be positive. The magnitude of angular velocity at any instant of time can be described as the angle turned through per unit time. the base unit of angular velocity is radian per second (rad/s). However, the most common practical unit of measurement used to express rotational speeds of mechanical components is the revolution per minute (rpm), which is not decimally related to the radian per second. It is normally required to convert angular velocity from revolutions per minute to radians per second,[†] particularly if calculations involve other related concepts, such as angular acceleration, energy or power.

If the angle turned through in each successive interval of time is the same, the angular velocity is said to be constant. Otherwise, average angular velocity can be calculated using total angular displacement and the time in which it occurs.

Example 14.1
A cam in a mechanism makes 500 revolutions in 2 minutes 37 seconds. What is its average angular velocity in revolutions per minute and in radians per second?

Solution
Total angular displacement: $\theta = 500$ revolutions
Time: $t = 157$ s
    $= 2.617$ min

---

[*] Omega — the last letter of the Greek alphabet.
[†] To convert revolutions per minute to radians per second, multiply speed in rpm by $2\pi/60$. For example:

$$955 \text{ rpm} = 955 \times \frac{2\pi}{60}$$
$$= 100 \text{ rad/s}$$

Average angular velocity:

$$\omega = \frac{\theta}{t}$$

$$= \frac{500 \text{ revolutions}}{2.617 \text{ min}}$$

$$= 191 \text{ rpm}$$

Also equal to:

$$\omega = \frac{500 \times 2\pi \text{ rad}}{157 \text{ s}}$$

$$= 20 \text{ rad/s}$$

Angular acceleration

If angular velocity is not constant, but is increasing gradually at a uniform rate, an object is said to be rotating with a **uniformly accelerated motion**.[*] The rate at which angular velocity is changing with time is called **angular acceleration, α.**[†]

If, over a period of time equal to t seconds, angular velocity of an object changes from its initial value ω_0 to a final value ω, it follows from the definition that angular acceleration is the quotient of the increment of angular velocity $(\omega - \omega_0)$ and the time t:

$$\alpha = \frac{\omega - \omega_0}{t}$$

It can easily be seen that the unit of angular acceleration must be the unit of angular velocity, rad/s, divided by the unit of time, s, i.e. radian per second squared, rad/s^2.

The relationship between the initial and final angular velocities, time, and angular acceleration for uniformly accelerated rotational motion is usually stated as a formula in which final angular velocity is the subject:

$$\omega = \omega_0 + \alpha t$$

where ω is the final angular velocity in rad/s
   ω_0 is the initial angular velocity in rad/s
   α is the angular acceleration in rad/s^2
   t is the time taken in s

Note the similarity between this relationship and that for linear velocity and acceleration, $v = v_0 + at$.

Example 14.2

A flywheel starts from rest and is accelerated at the rate of 2.4 rad/s^2 for 30 s. Determine the angular velocity reached after 30 s.

Solution

Initial angular velocity: $\omega_0 = 0$
Time: $t = 30$ s
Angular acceleration: $\alpha = 2.4$ rad/s^2

---

[*] The more general case of non-uniform angular acceleration is not covered in this book.
[†] This is alpha — the first letter of the Greek alphabet.

Substitute into $\omega = \omega_0 + \alpha t$:

$$\omega = 0 + 2.4 \text{ rad/s}^2 \times 30 \text{ s}$$
$$= 72 \text{ rad/s}$$

Hence, angular velocity after 30 s is 72 rad/s (= 687.5 rpm).

If, instead of increasing, angular velocity is gradually decreasing, the rotation is said to be **uniformly decelerated**. Angular deceleration or retardation is regarded as negative acceleration, i.e. angular acceleration acting in the direction opposite to angular velocity.

Example 14.3

If, after rotating for some time at a constant angular velocity of 72 rad/s, brakes are applied to the flywheel producing a retardation of 4 rad/s², determine the time taken to reduce its angular velocity to 40 rad/s (382 rpm).

Solution

Initial angular velocity: $\omega_0 = 72 \text{ rad/s}$
Final angular velocity: $\omega = 40 \text{ rad/s}$
Angular acceleration: $\alpha = -4 \text{ rad/s}^2$

Substitute into $\omega = \omega_0 + \alpha t$:

$$40 = 72 - 4t$$

Hence, time taken:

$$t = \frac{72 - 40}{4}$$
$$= 8 \text{ s}$$

14.2 *EQUATIONS OF ROTATIONAL MOTION*

In the case of uniformly accelerated rotational motion, the angular displacement from the initial position is the product of time and the average velocity:

$$\theta = t \times \omega_{av}$$

where simple arithmetic averaging of angular velocities gives:

$$\omega_{av} = \frac{\omega_0 + \omega}{2}$$

When these equations are combined, we have:

$$\theta = t\left(\frac{\omega_0 + \omega}{2}\right)$$

which is an additional independent equation to:

$$\omega = \omega_0 + \alpha t$$

Eliminating final velocity ω from these equations yields:

$$\theta = \omega_0 t + \frac{\alpha t^2}{2}$$

Similarly, if time t is eliminated, we get:

$$2\alpha\theta = \omega^2 - \omega_0^2$$

Note very carefully how these equations compare with those for linear motion in Section 13.2.

Any problem involving kinematics of rotational motion can be solved using these equations. However, great care must be exercised in the use of appropriate units. It is usually better to make all necessary unit conversions before substituting into the appropriate equations.

Example 14.4

A flywheel turns at a constant angular velocity of 150 revolutions per minute for 45 seconds before a brake is used to bring it to rest with a retardation of 0.5 radians per second squared.

Determine the total time and the total angular displacement of the wheel.

Solution

There are two stages in this problem: motion with uniform angular velocity before the brake is applied, and uniformly decelerated motion that brings the flywheel to rest. Each stage can be considered separately and the results combined.

(a) *Uniform motion*

Angular velocity (constant):
$$\omega = 150 \text{ rpm}$$
$$= 150 \times \frac{2\pi}{60}$$
$$= 15.71 \text{ rad/s}$$

Time: $t = 45 \text{ s}$

Angular displacement:
$$\theta = t \times \omega$$
$$= 45 \text{ s} \times 15.71 \text{ rad/s}$$
$$= 706.9 \text{ rad}$$

This is equal to 112.5 revolutions.

(b) *Decelerated motion*

Initial angular velocity: $\omega_0 = 15.71$ rad/s
Final angular velocity: $\omega = 0$
Angular acceleration: $\alpha = -0.5$ rad/s^2 (Note the negative sign.)

Substitute into $\omega = \omega_0 + \alpha t$ to find the time taken to bring the flywheel to rest:

$$0 = 15.71 - 0.5t$$
$$\therefore\ t = 31.4\ \text{s}$$

Substitute into $\theta = t\left(\dfrac{\omega_0 + \omega}{2}\right)$ to find the angular displacement during the period of decelerated motion:

$$\theta = 31.4\left(\frac{15.71 + 0}{2}\right)$$
$$= 246.6\ \text{rad}$$

This is also equal to 39.3 revolutions.

Combining the answers obtained in (a) and (b) yields:

Total angular displacement = 706.9 + 246.6 rad   *or*   = 112.5 + 39.3 revolutions
= 953.5 rad   = 151.8 revolutions
Total time taken = 45 s + 31.4 s
= 76.4 s

 # Problems

14.1 (a) Convert 78 revolutions to radians.
(b) Convert 3500 radians to revolutions.
(c) Convert 657 rpm to rad/s.
(d) Convert 243 rad/s to rpm.

14.2 A flywheel, running at constant angular velocity, completes 12 revolutions in 5 seconds. Calculate its angular velocity in revolutions per minute and in radians per second.

14.3 What is the angular speed of each of the three hands of a clock expressed in revolutions per minute and in radians per second?

14.4 A flywheel with an initial angular velocity of 35 rad/s is accelerated at 12 rad/s^2 to a final angular velocity of 275 rad/s. Determine the time taken, and the angular displacement in radians during that time.

14.5 An electric motor starting from rest takes 24 seconds to reach its normal operating speed of 1440 rpm. What is its angular acceleration in rad/s^2? Calculate also the total number of revolutions made by the motor during this period.

14.6 A rotor of a steam turbine revolving at 8000 rpm slows down to 2000 rpm in 15.7 s after steam supply has been adjusted. Determine the angular deceleration, and the number of revolutions made by the rotor in that time.

14.7 The output shaft of an electric motor accelerates from rest, at the rate of 10 rad/s^2, to its final angular velocity in 15.2 s. What is the final angular velocity in rad/s and rpm? Determine also the total number of revolutions made by the shaft in that period.

14.8 A heavy flywheel, rotating initially at 1500 rpm was allowed to slow down to 1120 rpm in 10.2 min due to air and bearing friction, and was then brought to rest from this speed by the application of a brake which caused an angular deceleration of 1.15 rad/s^2. Calculate the total time in minutes and the total number of revolutions made in coming to rest from the initial speed of 1500 rpm.

14.9 During a test on a large boring mill, its circular worktable, revolving in a horizontal plane, was observed to accelerate from rest at the rate of 0.2 rad/s^2 for 18 s, then rotate at constant speed for 90 s before coming to rest after a period of retardation lasting 24 s. Determine the total number of revolutions made by the worktable during the test.

14.3 *TORQUE*

The analogy between rotation and linear motion is not limited to the relationships between displacement, velocity and acceleration. It can also be extended to the study of kinetics of rotation, i.e. the causes of rotational motion.

As with motion in a straight line, rotation, once started, tends to continue with constant angular velocity unless an outside influence acts to increase or decrease the velocity. In rectilinear motion, force is the external agent which acts to change the conditions of rest or uniform linear motion of a body. In rotational kinetics, **torque** T is the analogue of force, which influences change in the state of rest or rotational motion of a component capable of rotating about its axis.

Let us start by describing torque as a 'pure turning effort'.* As such, torque is somewhat similar in its effect to the turning moment of a force couple, and is measured in the same units, i.e. newton metres (N.m), or their multiples or submultiples such as kilonewton metres (kN.m). In fact, if all action and reaction forces acting on the component are properly identified and taken into account, torque is always a result of a force couple, or a cumulative effect of several force couples.

Thus the concept of torque is closely related to that of moment of a force about a point, as discussed earlier in this book (see Ch. 6). However, we should not take the terms 'torque' and 'moment' to be completely synonymous, as this could lead to much misunderstanding. There are several features which distinguish torque from moment of a force. Let us consider them:

1. The term 'torque' is usually used to describe a sustained turning effort applied to mechanical components, such as gears, shafts and flywheels, in situations which involve a period of continuous rotation. On the other hand, moments often refer to static forces, such as those acting on a beam resting on its supports, or to forces that produce only a small amount of rotation, through a small angle and of discontinuous nature, such as the movement of a foot pedal about its hinge.

2. Torque is always specified in relation to the geometrical axis of a mechanical component about which the component physically rotates, or can be expected to rotate, whereas moment of a force can be calculated quite meaningfully about any

---

* The word 'torque' derives its origin from a Latin word for twisting, or twisted necklace as worn by the ancient Gauls.

convenient reference point, often in circumstances which do not involve any actual rotation about the chosen reference point.[*]

3. Unlike moment of a force, torque does not have to be readily identifiable with any particular single force located at a fixed distance from the axis of rotation. For example, within an electric motor, there are distributed magnetic forces acting on the rotor, but to an outside observer, the effort produced by the output shaft appears simply as a pure turning effort.

In spite of what has already been said about the differences between torque and moment, it should be pointed out that there is always a close relationship between torque and the forces acting on various rotating components, such as gears, sprockets and pulleys, as the following example will illustrate.

Example 14.5

A gearbox consists of two spur gears in mesh. The larger gear has a 200 mm pitch-circle diameter, and the pinion, i.e. the smaller gear, has a 50 mm pitch-circle diameter.[†] (*Note*: Mating gears have their pitch circles tangent.) The input shaft transmits a torque through the pinion of 160 N.m. Determine the tangential force between the two gears, and the value of output torque.

Solution

Input torque $T_{in} = 160$ N.m. This produces a force between the two gears, such that the moment of the force about the centreline of the pinion is equal to the transmitted torque:

$$\text{Force} \times \text{radius} = \text{torque}$$
$$F \times 0.025 \text{ m} = 160 \text{ N.m}$$
$$\therefore F = \frac{160 \text{ N.m}}{0.025 \text{ m}}$$
$$= 6400 \text{ N}$$
$$= 6.4 \text{ kN}$$

But the same force, or, to be precise, its equal and opposite reaction force, is also applied to the larger gear at a distance of 0.1 m from its centreline, producing a turning moment of:

$$M = F \times d$$
$$= 6400 \text{ N} \times 0.1 \text{ m}$$
$$= 640 \text{ N.m}$$

This is transmitted as torque through the output shaft:

i.e. $\quad T_{out} = 640$ N.m

---

[*] When dealing with systems in which forces and motions do not all lie in the same plane, a further theoretical distinction can be made, namely that torque relative to a given axis is a scalar quantity, whereas moment of a force is a vector.

[†] The **pitch-circle diameter** is the diameter of the circle which by a pure rolling action would transmit the same motion as the actual gear wheel.

14.4 *TORQUE AND ROTATIONAL MOTION*

It has already been stated that rotation of a body tends to continue with constant angular velocity unless an outside unbalanced torque acts to increase or decrease the velocity. Therefore it can be concluded that every physical body possesses a property of rotational inertia, which determines its resistance to a change in its state of rest or uniform rotational motion, i.e. its resistance to angular acceleration. The correct technical term to describe this property is **mass moment of inertia**, often abbreviated to **moment of inertia**, with the symbol I.[*]

Mathematically, mass moment of inertia of a body with respect to its axis of rotation is a function of mass distribution in the body relative to the axis. The most common elementary shape for a rotating object is a disc (flywheel) or a cylinder (shaft),[†] for which the mass moment of inertia about the axis is given by the following expression:

$$I = \frac{mr^2}{2}$$

where m is mass in kg
   r is radius in m

It follows that the SI unit of mass moment of inertia is kg.m^2.

Example 14.6

Calculate the mass moment of inertia of a flywheel in the form of a 650 mm diameter disc, 70 mm thick, having a mass of 185 kg.

Solution

$$\text{Moment of inertia } I = \frac{mr^2}{2}$$
$$= \frac{185 \text{ kg} \times (0.325 \text{ m})^2}{2}$$
$$= 9.77 \text{ kg.m}^2$$

Later in this chapter we will examine in more detail the procedure for calculating mass moments of inertia for objects of composite shape, consisting of several disc or cylindrical elements, made from materials of specified density.

In the meantime, let us return to the relationship between torque and rotational motion. For a rigid body rotating about a fixed axis, the laws of motion have the same form as those for rectilinear motion, with torque (T) replacing force (F), mass moment of inertia (I) replacing mass (m), and angular acceleration (α) replacing linear acceleration (a).

The relationship between mass moment of inertia of a rotating body, the torque acting on it and the angular acceleration produced by the torque is identical to that between

---

[*] Alternatively, J is sometimes used as a symbol for mass moment of inertia. Both are acceptable symbols according to the International Standards Organisation.

[†] The two are of fundamentally the same geometrical shape. A disc is a relatively thin, flat, round object of cylindrical shape, and conversely, a cylinder can be seen as a disc of relatively small diameter elongated in the direction of its axis.

mass, force and acceleration in linear motion. If we substitute rotational terms, instead of linear terms, into Newton's second law equation, $F = ma$, we have:

$$\boxed{T = I \times \alpha}$$

where T is net unbalanced torque, N.m
I is mass moment of inertia, kg.m^2
α is angular acceleration, rad/s^2

Dimensional homogeneity of this equation is not immediately apparent, but can be demonstrated as follows:

$$T = I \times \alpha$$
$$\text{N.m} = \text{kg.m}^2 \text{ rad/s}^2$$

After the metre cancels out and the radian is left out because it is a dimensionless ratio, we have:

$$\text{N} = \text{kg.m/s}^2$$

which is homogeneous by definition of the newton.

Example 14.7
Determine the net torque required to give a flywheel with a mass moment of inertia of 0.75 kg.m^2 an angular acceleration of 16 rad/s^2.

Solution
$$T = I \times \alpha$$
$$= 0.75 \text{ kg.m}^2 \times 16 \text{ rad/s}^2$$
$$= 12 \text{ N.m}$$

Example 14.8
Determine the torque required to accelerate a turbine rotor, undergoing a dynamic balancing test, from rest to a speed of 15 000 rpm in 80 s, if the mass moment of inertia of the rotor is 11.5 kg.m^2.

Solution
The angular acceleration required is found from $\omega = \omega_0 + \alpha t$,

where $\omega = 15\,000 \times \dfrac{2\pi}{60} = 1571$ rad/s:

$$\omega = \omega_0 + \alpha t$$
$$1571 = 0 + \alpha \times 80$$
$$\therefore \alpha = 19.63 \text{ rad/s}^2$$

Therefore, torque required:

$$T = I \times \alpha$$
$$= 11.5 \text{ kg.m}^2 \times 19.63 \text{ rad/s}^2$$
$$= 225.8 \text{ N.m}$$

Example 14.9

Determine the angular acceleration of a flywheel in the form of a disc 400 mm in diameter and having a mass of 60 kg, if the applied torque is 24 N.m.

Solution

The mass moment of inertia of the flywheel is:

$$I = \frac{mr^2}{2}$$
$$= \frac{60 \text{ kg} \times (0.2 \text{ m})^2}{2}$$
$$= 1.2 \text{ kg.m}^2$$

Substituting into $T = I \times \alpha$ yields:

$$24 \text{ N.m} = 1.2 \text{ kg.m}^2 \times \alpha$$

Hence the angular acceleration is:

$$\alpha = \frac{24 \text{ N.m}}{1.2 \text{ kg.m}^2}$$
$$= 20 \text{ rad/s}^2$$

14.5 *ACCELERATION AGAINST RESISTANCE*

It must be understood that in the equation $T = I \times \alpha$, the torque T is net accelerating torque, i.e. the resultant of all torques applied to the body. The resultant unbalanced torque, i.e. net accelerating torque, is the difference between the applied turning effort T_a and the resistance torque T_r (Fig. 14.1):

$$T = T_a - T_r$$

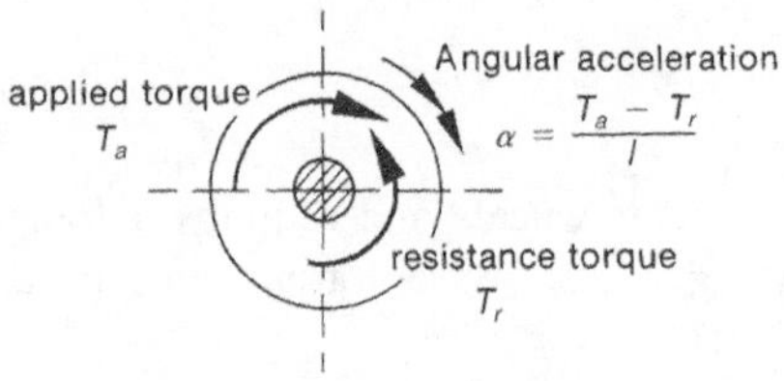

Fig. 14.1   *Angular acceleration against resistance*

The resistance is usually due to friction in the bearings, axle friction etc. Acceleration produced by the resultant torque is found from $T = I \times \alpha$:

$$\alpha = \frac{T}{I} = \frac{T_a - T_r}{I}$$

It follows from this equation that if the applied torque T_a is equal to the resistance torque T_r, there will be no acceleration, i.e. the rotation, if any, will continue at constant angular velocity. Thus, a flywheel rotating at a constant speed requires a torque equal to the friction torque to maintain uniform rotation. Any torque in excess of friction torque will accelerate the wheel. On the other hand, an applied torque which is less than friction torque will result in retardation.

Example 14.10

A flywheel of mass moment of inertia equal to 53 kg.m^2 is rotating at 300 rpm. The frictional resistance is 16 N.m. Calculate the torque that must be applied in order to accelerate the wheel to 500 rpm in 15 s.

Solution

Angular velocity can be calculated using $\omega = \omega_0 + \alpha t$, where:

$$\omega = 500 \times \frac{2\pi}{60}$$
$$= 52.36 \text{ rad/s}$$
$$\omega_0 = 300 \times \frac{2\pi}{60}$$
$$= 31.42 \text{ rad/s}$$

Substituting:
$$52.36 = 31.42 + \alpha \times 15$$
$$\therefore \ \alpha = 1.4 \text{ rad/s}^2$$

The net accelerating torque required to accelerate the wheel is:

$$T = I \times \alpha$$
$$= 53 \text{ kg.m}^2 \times 1.4 \text{ rad/s}^2$$
$$= 74 \text{ N.m}$$

Therefore, the total torque that must be applied is the sum of the accelerating torque and resistance torque:

$$T_a = 74 \text{ N.m} + 16 \text{ N.m}$$
$$= 90 \text{ N.m}$$

 # Problems

14.10 A car has 750 mm diameter wheels and requires a force of 0.5 kN between the road surface and each driving wheel. What torque must be applied to each of the driving wheels to supply this force?

14.11 A pull in a bicycle chain of 250 N applied to a sprocket of diameter 80 mm provides the necessary torque to the rear wheel. If the wheel is 600 mm in diameter, what is the force between the wheel and the ground?

14.12 A 300 mm diameter pulley is subjected to belt tensions equal to 500 N and 200 N. Determine the torque acting on the pulley.

14.13 Two shafts are connected by a coupling having six bolts equally spaced around a 300 mm diameter circle, as shown in Figure 14.2. Assuming that the load is shared equally between the bolts, and neglecting friction, determine the tangential force transmitted by each bolt if the torque transmitted is 5000 N.m.

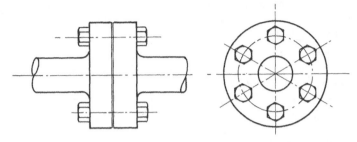

Fig. 14.2

14.14 Calculate the angular acceleration of a cylindrical drum mounted on a shaft through its geometrical axis and subjected to a net accelerating torque of 60 N.m. The mass moment of inertia of the shaft and drum assembly is 12 kg.m^2.

14.15 A spin-dryer with a load of washing rotates at 200 rpm until the power is turned off. It takes 2.5 min for it to come to rest. Given that the mass moment of inertia of the loaded spin-dryer is 1.72 kg.m^2, calculate the friction torque responsible for bringing the dryer to rest.

14.16 A wheel of a car has a moment of inertia of 1.6 kg.m^2. When tested for dynamic balance, it is accelerated from rest to a speed of 1500 rpm by a net torque of 20.6 N.m. Calculate the time taken and the number of turns made by the wheel to reach the final speed.

14.17 A flywheel in the form of a disc of uniform thickness, 500 mm in diameter and mass 150 kg, has frictional resistance of 5 N.m and an applied torque of 25 N.m acting on it. If it starts from rest, determine the angular velocity reached after 1 min.

14.18 The flywheel in problem 14.17 rotates at a constant speed of 3000 rpm. It is to be brought to rest in 20 s by applying a force F to the brake lever as shown in Figure 14.3.

If the coefficient of friction between the shoe and the wheel is 0.55, calculate the force F required. Assume that frictional resistance remains at 5 N.m.

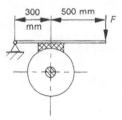

Fig. 14.3

14.19 A cast-iron pulley, 1.1 m in diameter, has a mass moment of inertia of 42.5 kg.m^2, and average resistance torque of 39 N.m. When accelerating from rest, the pulley is subjected to constant belt tensions of 800 N and 420 N on tight and slack sides of the belt respectively. Determine how long it takes to accelerate the pulley to 1375 rpm.

14.6 *MASS MOMENT OF INERTIA OF RIGID BODIES*

Mass moment of inertia (I) of a rigid body about its axis of rotation is defined mathematically as the sum of the products of the mass elements of the body and the squares of their perpendicular distances from the axis.

Expressions for the mass moments of inertia of the most common rotating solids, such as a disc, a solid cylinder and a hollow cylinder, as well as for a mass concentrated at a fixed distance from the centre of rotation, have been derived by integration and are given in Table 14.1.

Table 14.1   *Mass moments of inertia of rigid bodies about the axis of rotation*

Rotating body	Mass moment of inertia, I (kg.m^2)
Point mass at radius r	mr^2
Thin shell or hoop about its axis	mr^2
Solid cylinder about its axis	$\dfrac{mr^2}{2}$
Hollow cylinder about its axis	$\dfrac{m}{2}(r_o^{\,2} + r_i^{\,2})$

m = total mass
r = radius

When the mass of the body is not given, it must be calculated as a product of volume and density before standard formulae from Table 14.1 can be used.[*] Thus it can be concluded that mass moment of inertia of a rotating component depends not only on its geometrical shape and size, but on the density of the material from which it is made. The density of a material is usually given the symbol ρ.

---

[*] **Density** of a substance is the mass of unit volume of that substance, expressed in such units as kilograms per cubic metre. (See also Section 26.2.)

Example 14.11

Determine the mass moment of inertia of a flywheel in the form of a disc, 50 mm wide × 300 mm diameter, if the material is steel (density ρ of steel is 7800 kg/m³).

Solution

Volume of material:

$$V = l \times \frac{\pi D^2}{4}$$

$$= 0.05 \times \frac{\pi 0.3^2}{4}$$

$$= 0.003\,53 \text{ m}^3$$

Mass of flywheel:

$$V \times \rho = 0.003\,53 \times 7800$$

$$= 27.57 \text{ kg}$$

Moment of inertia (disc):

$$I = m\frac{r^2}{2}$$

$$= 27.57 \times \frac{0.15^2}{2}$$

$$= 0.3101 \text{ kg.m}^2$$

$$\doteqdot 0.310 \text{ kg.m}^2$$

For any element of mass in a composite body, which is not concentric with the common axis of rotation, the moment of inertia of the element about the common axis has to be calculated using the formula known as the **transfer formula**:

$$\boxed{I = I_c + md^2}$$

where I is the moment of inertia of the element about the common axis

I_c is the moment of inertia of the element about its own centroidal axis, using the appropriate expression from Table 14.1

m is the mass of the element

d is the perpendicular distance of the element from the common axis

Example 14.12

A disc of diameter 80 mm and mass 1.96 kg is attached to a flywheel so that it rotates at a distance of 90 mm from the axis of rotation of the wheel, as shown in Figure 14.4. Calculate the contribution this disc makes to the total moment of inertia of the flywheel.

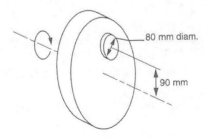

Fig. 14.4

Solution
Centroidal moment of inertia of the disc:

$$I_c = \frac{mr^2}{2}$$
$$= \frac{1.96 \text{ kg} \times (0.04 \text{ m})^2}{2}$$
$$= 0.001\,57 \text{ kg.m}^2$$

Transfer term:

$$md^2 = 1.96 \text{ kg} \times (0.09 \text{ m})^2$$
$$= 0.015\,88 \text{ kg.m}^2$$

Moment of inertia of the disc about the common axis of rotation:

$$I = I_c + md^2$$
$$= 0.001\,57 + 0.015\,88$$
$$= 0.017\,45 \text{ kg.m}^2$$
$$\doteqdot 0.0175 \text{ kg.m}^2$$

If any mass is removed from a solid body, e.g. by drilling a hole, the mass removed and its corresponding moment of inertia are regarded as negative, as will be seen from part (c) of the following example.

Example 14.13

Determine the mass moment of inertia of the flywheel shown in Figure 14.5 if the material is steel.

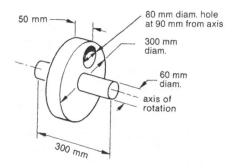

50 mm

80 mm diam. hole
at 90 mm from axis

300 mm
diam.

60 mm
diam.

axis of
rotation

300 mm

Fig. 14.5

Solution
There are three components to be considered: the disc, shaft and hole.
(a) *Disc*
The mass moment of inertia has already been calculated (see Example 14.11 above).

$$I_{\text{disc}} = 0.3101 \text{ kg.m}^2$$

(b) *Shaft*
The length of the shaft not included as part of the disc is 250 mm.

Volume:

$$V = l\frac{\pi D^2}{4}$$

$$= 0.25 \times \frac{\pi 0.06^2}{4}$$
$$= 0.000\,707 \text{ m}^3$$

Mass:

$$V \times \rho = 0.000\,707 \times 7800$$
$$= 5.51 \text{ kg}$$

Mass moment of inertia:

$$I = m\frac{r^2}{2}$$

$$= 5.51 \times \frac{0.03^2}{2}$$
$$= 0.0025 \text{ kg.m}^2$$

(c) *Hole*

This can be regarded as a solid disc removed, therefore all values here are regarded as negative.

Volume:

$$V = l\frac{\pi D^2}{4}$$

$$= 0.05 \times \frac{\pi 0.08^2}{4}$$
$$= 0.000\,251 \text{ m}^3 \text{ (negative)}$$

Mass:

$$V \times \rho = 0.000\,251 \times 7800$$
$$= 1.96 \text{ kg (negative)}$$

Centroidal mass moment of inertia:

$$I_c = m\frac{r^2}{2}$$

$$= 1.96 \times \frac{0.04^2}{2}$$
$$= 0.001\,57 \text{ (negative)}$$

Transfer term:

$$md^2 = 1.96 \times 0.09^2$$
$$= 0.0159 \text{ (negative)}$$

Mass moment of inertia about common axis of rotation:

$$I = I_c + md^2$$
$$= 0.001\,57 + 0.0159$$
$$= 0.0175 \text{ kg.m}^2 \text{ (negative)}$$

Therefore the total mass moment of inertia of the flywheel is:

$$I = \Sigma(I)$$
$$= 0.3101 + 0.0025 - 0.0175$$
$$= 0.2951 \text{ kg.m}^2$$

For convenience, this solution can also be tabulated as follows:

Table 14.2 *Solution to Example 14.13 in tabular form*

Element	m	d	I_c	md^2	I
disc	27.57	—	0.3101	—	0.3101
shaft	5.51	—	0.0025	—	0.0025
hole	−1.96	0.09	−0.0016	−0.0159	−0.0175

$$\Sigma(I) = 0.2951 \text{ kg.m}^2$$

 # Problems

14.20 For each of the following steel shafts (density of steel is 7800 kg/m³), determine the mass moment of inertia:
  (a) 750 mm long, 60 mm diameter, solid shaft
  (b) 750 mm long, 75 mm outside diameter, 45 mm inside diameter, hollow shaft

14.21 Determine the mass moment of inertia of the aluminium pulley shown in Figure 14.6. Density of aluminium is 2560 kg/m³.

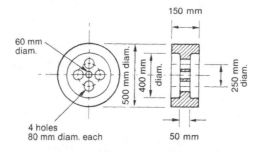

Fig. 14.6

 ## Review questions

1. Define *angular displacement, angular velocity* and *angular acceleration,* and state the symbols and units used.
2. What is implied by the term *uniformly accelerated rotation*?
3. List four equations of rotational motion.
4. Define *torque* and state the SI unit of torque.
5. State the relationship between torque and angular acceleration.
6. State the unit of mass moment of inertia.

CHAPTER 15

Circular motion

In this chapter we will consider motion of bodies, such as motor vehicles, in a circular path. This type of motion is closely related to rotation in so far as the body moves at the end of a radius which rotates about the centre of the circular path. At the same time, circular motion is also related to linear motion because the circumference of the circle along which the body moves is a line, albeit a curved one.

We will start by considering the relation between rotation and circular motion, and then discuss the concept and effects of a special kind of force called **centripetal force**, which acts to keep the body moving in a circular path.

Expected learning outcomes

After carefully studying the material presented in this chapter, working through all numerical examples, and successfully completing all practice problems, students should be able to:
1. convert angular displacement, velocity and acceleration to corresponding circumferential terms, and vice versa;
2. calculate centripetal acceleration and relate it to centripetal force;
3. solve problems involving the stability of vehicles on curved roads, the banking of roads and railway tracks, and forces that are due to out-of-balance masses.

15.1 *RELATION BETWEEN ROTATION AND CIRCULAR MOTION*

It is useful to note the difference between the rotation of an object about an axis and the linear motion of a point travelling in a circular path. In rotary motion, the object simply spins, usually around its own geometrical axis. In circular motion, a point or an object moves along a circumference of a circle at a fixed distance r from the centre of its circular path, such as a car driven around a curve of a certain radius.

There is, however, a relationship between the angular terms that describe the rotary motion, i.e. angular displacement, velocity and acceleration, and their linear counterparts measured along the circumference. This relationship stems from the definition of the radian as the angle subtended at the centre of a circle by an arc equal in length to the radius. If we consider circular motion of a point P on a rotating disc at a radius r from the centre, as in Figure 15.1, it can be seen that:

$$S = r \times \theta$$

where S is the linear displacement along the circumference
r is the radius, in the same units as displacement
θ is the corresponding angular displacement in radians

Fig. 15.1 *Circular motion*

Dividing the equation by time, i.e. $\dfrac{S}{t} = r \times \dfrac{\theta}{t}$, produces the equation relating instantaneous angular velocity, ω, to linear velocity along the circular path, v:

$$v = r \times \omega$$

Now, dividing this new equation by time, i.e. $\dfrac{v}{t} = r \times \dfrac{\omega}{t}$, it follows that the angular and linear accelerations are similarly related:

$$a = r \times \alpha$$

Example 15.1

On the former British passenger liner *Queen Mary* there were four gearboxes transmitting power from turbines to a propeller shaft rotating at 180 revolutions per minute at normal cruising speed. Each large driven gear was approximately 4 metres in pitch-circle diameter.

For a point on the pitch circle of the gear, determine the linear velocity, and the distance travelled along the circumference for each revolution.

Solution

Radius of circular motion: $r = 2$ m

Angular velocity:

$$\omega = 180 \times \frac{2\pi}{60}$$
$$= 18.85 \text{ rad/s}$$

Angular displacement:

$$\theta = 1 \text{ revolution}$$
$$= 2\pi \text{ rad}$$

Substitute these values into the appropriate formulae, and solve.

Linear circumferential velocity:

$$v = r \times \omega$$
$$= 2 \text{ m} \times 18.85 \text{ rad/s}$$
$$= 37.7 \text{ m/s}^*$$

Linear displacement per revolution:

$$S = r \times \theta$$
$$= 2 \text{ m} \times 2\pi \text{ rad}$$
$$= 12.57 \text{ m}$$

The comparison between linear motion of a vehicle and the rotation of its wheels also depends on the above-mentioned relationships. We know that when a wheel rolls on a road surface, its axis is actually moving forward relative to the road. One can also visualise this situation from the driver's point of view in terms of a 'fixed' axis of rotation, in relation to which the road surface is 'moving backwards' under the wheel. In either case, the point of contact between the wheel and the road has momentarily a linear velocity v and an angular velocity ω, which are related by $v = r \times \omega$.

Example 15.2

Determine the speed of a car in kilometres per hour when its wheels, which are 650 mm in diameter, are rotating at 440 rpm (Fig. 15.2).

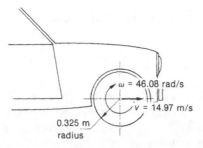

$\omega = 46.08$ rad/s

$v = 14.97$ m/s

0.325 m
radius

Fig. 15.2

Solution
Angular velocity:

$$\omega = 440 \times \frac{2\pi}{60}$$
$$= 46.08 \text{ rad/s}$$

---

* Note that since the radian is dimensionless (see Ch. 3, Section 3.4), it may be used to compose units involving angular measure, and left out when transition to linear units is involved, as above, without disturbing dimensional homogeneity of an equation.

Radius of rotation:

$$r = \frac{0.65 \text{ m}}{2}$$
$$= 0.325 \text{ m}$$

Linear velocity:

$$v = r \times \omega$$
$$= 0.325 \text{ m} \times 46.08 \text{ rad/s}$$
$$= 14.97 \text{ m/s}$$

Converting to kilometres per hour:

$$v = 14.97 \times \frac{3600}{1000}$$
$$= 53.9 \text{ km/h}$$

15.2 *CENTRIPETAL ACCELERATION*

When we consider uniformly accelerated linear motion, two defining equations could be said to apply. The change of velocity Δv is given by:[*]

$$\Delta v = v_2 - v_1$$

and acceleration, as the measure of change of velocity over time, is given by:

$$a = \frac{\Delta v}{t}$$

These equations work in a purely algebraic sense, when all we do is substitute numerical values and calculate, provided the motion under investigation is linear. This is clearly so when motion is in a straight line (rectilinear motion). It is also true in the case of circular motion for the relationships between velocity and acceleration measured along the circumferential direction of motion, i.e. in the direction tangential to the circular path.

Acceleration measured along the circumference is usually called **tangential acceleration**, and when necessary can be distinguished from any other kind by the subscript t, as in a_t. It should be obvious that our discussion of circular motion in the preceding section of this chapter referred only to tangential acceleration, without specifically identifying it as such.

Now, let us examine the case of circular motion a little more closely. To make things simpler, let us assume initially that a body under investigation moves in a circle of radius r with constant speed, i.e. it is in uniform circular motion. It is clear, then, that since the magnitude of the velocity remains the same, then $v_2 = v_1 = v$, and the value of tangential acceleration a_t must be equal to zero. Is this the end of the story? Apparently not! We observe that, as the body moves in a circular path from position 1 to position 2 as shown

---

[*] Δ is 'delta', the fourth letter of the Greek alphabet. It is used as a symbol in mathematics and science to signify a difference or a change in some quantity.

in Figure 15.3(a), the *direction* of its velocity is constantly changing. Therefore, it must be concluded that there is some change in velocity, even if it is only one of direction, and when there is a change in velocity there must be some kind of acceleration involved.

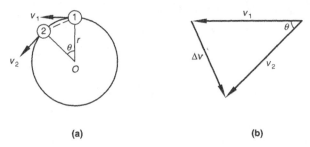

(a)   **(b)**

Fig. 15.3   **(a)** *Change in velocity for uniform circular motion*   **(b)** *Triangle of velocity vectors*

Velocity, like force, is a vector quantity. This means that it has magnitude and direction, and for this reason, addition of velocities that involve changes of direction must always be done vectorially.* The velocity equation given above can be transposed into $v_2 = v_1 + \Delta v$, and can be represented graphically as a triangle of velocity vectors, as shown in Figure 15.3(b). This is just a graphical way of saying that velocity at point 1 (v_1) plus the amount of directional change (Δv) that occurs between points 1 and 2 results in a new velocity at point 2 (v_2). This is so, even if the magnitudes of the two velocities remain the same.

The velocity triangle in Figure 15.3(b) is similar to the physical triangle O–1–2 in Figure 15.3(a), as both have two equal sides and the same included angle θ. Therefore, it follows that:

$$\frac{\Delta v}{v} = \frac{\text{length of chord 1–2}}{r}$$

When the angle θ is very small, the length of the chord 1–2 is very nearly equal to the length of the arc 1–2, i.e. to the amount of linear displacement S of the body along the circumference, which takes place over the correspondingly short interval of time t. Therefore:

$$\frac{\Delta v}{v} = \frac{S}{r}$$

We also know that when the velocity along the path is uniform, then:

$$S = vt$$

Hence: $$\frac{\Delta v}{v} = \frac{vt}{r}$$

---

* For our purposes here, the rules for vectorial addition of velocities are exactly the same as those described previously for the graphical addition of forces (see Section 4.6).

and:
$$\Delta v = \frac{v^2 t}{r}$$

Since acceleration is defined as the change of velocity over time, substitution yields:

$$\text{Acceleration} = \frac{\Delta v}{t} = \frac{v^2 t}{rt} = \frac{v^2}{r}$$

Since we are dealing with motion in which the *magnitude* of the velocity does not change, this acceleration must be perpendicular to the direction of the velocity. This can also be seen from the triangle of velocities since the angle θ is very small. In fact, this acceleration is directed along the radius towards the centre of the circular path. Therefore, it is called a **normal**, **radial** or **centripetal acceleration**, a_c.[*]

In this derivation, as the time interval t and the corresponding angle θ approach zero, i.e. they are taken to be extremely small, the approximation of chord length to arc length S becomes more accurate, and the conclusion reached comes to represent the exact instantaneous value of centripetal acceleration. Furthermore, it can be demonstrated by more rigorous analysis that, since centripetal acceleration a_c is an instantaneous quantity, the expression for a_c holds true for variable conditions as well as for uniform conditions of circular motion.[†]

In summary, it can be said that at each instant, when a body is moving with velocity v along a circular path of radius r, it experiences a centripetal acceleration, i.e. an acceleration directed radially towards the centre of the circle, which is given by:

$$\boxed{a_c = \frac{v^2}{r}}$$

where a_c is the centripetal acceleration in m/s^2
        r is the radius of the circular path in m
        v is the linear velocity along the circumference in m/s

Example 15.3
Calculate the centripetal acceleration of a car which is travelling at 108 km/h around a curve of radius 200 m.

Solution
Convert velocity to metres per second:

$$v = 108 \text{ km/h}$$
$$= \frac{108}{3.6} \text{ m/s}$$
$$= 30 \text{ m/s}$$

---

[*] The word 'centripetal' comes from Latin *centrum*, meaning 'centre', and *petere*, meaning 'to seek'.

[†] If the speed of a body changes as it travels along its circular path, the total change in its velocity vector is due to the combined effect of centripetal acceleration (change of direction) and tangential acceleration (change of speed). However, detailed analysis of such situations is outside the scope of this book.

Substitution gives:

$$a_c = \frac{v^2}{r}$$
$$= \frac{30^2}{200}$$
$$= 4.5 \text{ m/s}^2$$

15.3 *CENTRIPETAL FORCE*

According to Newton's laws of motion, every moving body would persevere in its state of uniform motion in a straight line unless it is compelled to change that state by an external unbalanced force impressed on it. In the case of circular motion, the path is not a straight line, and there is a change of direction of velocity, as we have already discussed. Therefore, there must exist an unbalanced force acting on the body, which is the cause of this change.

The rate of change of velocity in this case has been identified as centripetal acceleration a_c. We shall therefore refer to the force which is responsible for this acceleration as the **centripetal force** F_c. Applying Newton's second law results in the following expression, which can be used for calculating the centripetal force acting on a body of mass m:

$$\boxed{F_c = ma_c = \frac{mv^2}{r}}$$

Example 15.4
What is the magnitude of the centripetal force which causes the car in Example 15.3 to travel in a circular path if the mass of the car is 1.2 t?

Solution
Mass in kilograms:

$$m = 1.2 \times 1000 \text{ kg}$$
$$= 1200 \text{ kg}$$

Hence, centripetal force:

$$F_c = ma_c$$
$$= 1200 \text{ kg} \times 4.5 \text{ m/s}^2$$
$$= 5400 \text{ N}$$
$$= 5.4 \text{ kN}$$

Now let us explore the nature of the force we decided to call centripetal force F_c.

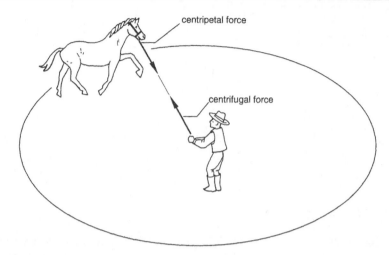

centripetal force

centrifugal force

Fig. 15.4 *Centripetal and centrifugal forces*

When a racehorse, limited by the length of its tether, is exercised in a small circle, it would most probably prefer to gallop in a straight line. However, it is prevented from doing so by the pull of the rope held by its trainer. Every time the horse tries to extend the range of its movement, the rope tightens and the horse experiences a centripetal pull telling it to stay on the circular track. At the same time the trainer experiences an equal but opposite pull of the rope on his hands. This can be described as the centrifugal reaction to the centripetal pull (Fig. 15.4).[*]

There is one very important lesson to be learned from this rather imperfect illustration. The centripetal pull is experienced by the horse, i.e. by the moving object, and is directed radially towards the centre of the circular path. The centrifugal pull is experienced by the trainer, i.e. by the object which provides the controlling force, and is directed radially away from the centre.

A horse is an intelligent animal and soon learns to behave in the manner expected of it without being prompted all the time by a physical force. However, in the world of inanimate objects a constant centripetal force must be applied continuously in order to maintain circular motion. For instance, if you make a simple model aeroplane fly in a horizontal circle at the end of a string which is attached to the aeroplane at one end and held in your hand at the other, the string will pull the plane into the circular path all the time. Otherwise the plane would move off in a straight line. While the string is continuously pulling the model plane towards you as the centre of the circle, and the radial distance between you and the plane remains the same, it is also true that there is a constant equal and opposite centrifugal reaction on your hand.

In the case of the car in Examples 15.3 and 15.4, the moving object is the car, and the centripetal force experienced by it is generated by road surface friction acting on its turned front wheels. It is because of the existence of this constant centripetal force, directed towards the centre of curvature of the road, that the car is impelled to follow the circular path. It can also be said that a centrifugal reaction force is experienced by the road surface.

---

[*] The word 'centrifugal' comes from *centrum*, meaning 'centre', and *fugere*, meaning 'to fly from'.

15.4 *CENTRIPETAL FORCE AND ANGULAR VELOCITY*

Occasionally we have to solve problems in which circular motion of a mass is closely related to the rotation of some physical component, such as a shaft, a flywheel or a washing-machine drum, whose speed of rotation is given in revolutions per minute or in radians per second.

Remembering that circumferential velocity v of a point P moving in a circle of radius r about the centre point O (see Fig. 15.1) is related to the angular velocity ω of the radial line OP by $v = r\omega$, we can easily convert angular velocity ω of the rotating component to the corresponding linear velocity v of the mass along its circular path.

Alternatively, the formula for centripetal force can be modified in terms of angular velocity ω. We can determine, by substitution, that centripetal acceleration of a mass located at point P is:

$$a_c = \frac{v^2}{r} = \frac{(\omega r)^2}{r} = \omega^2 r$$

Hence centripetal force acting on the mass is given by:

$$\boxed{F_c = m\omega^2 r}$$

Example 15.5

Calculate the centripetal force acting on a 150 g balancing mass attached to a car wheel, at a radius of 200 mm, when the wheel rotates at 955 rpm.

Solution

Angular velocity:

$$\omega = 955 \times \frac{2\pi}{60}$$
$$= 100 \text{ rad/s}$$

Hence centripetal force can be calculated using the modified formula:

$$F_c = m\omega^2 r$$
$$= 0.15 \times 100^2 \times 0.2$$
$$= 300 \text{ N}$$

 ## Problems

15.1 A mass at the end of a 700 mm pendulum moves through an arc of 20°. Determine the angular displacement in radians and the corresponding linear displacement along the arc.

15.2 If, at a particular instant, the velocity of the mass in problem 15.1 is 0.3 m/s and deceleration is 0.15 m/s², determine the corresponding angular velocity and deceleration in rad/s and rad/s² respectively.

15.3 A fan impeller is 800 mm in diameter and rotates at a constant angular velocity of 600 rpm. Determine the tip speed, i.e. the circumferential velocity, of its blades.

15.4 A pump is to be driven through a V-belt drive at a speed of 580 rpm by an electric motor rotating at 1450 rpm. The driver pulley is 100 mm in diameter.
Determine the linear velocity of the belts and the diameter of the driven pulley.

15.5 A car accelerates from rest at the rate of 1.2 m/s². Determine the angular acceleration, angular velocity and angular displacement of its wheels after 15 s if the wheel diameter is 650 mm.

15.6 Given that the linear speed of a conveyor belt is 1.5 m/s, and belt thickness is 12 mm, determine the centripetal acceleration of a point on the outer surface of the belt when it is moving around a 780 mm diameter pulley.

15.7 Determine the centripetal acceleration of a child riding a wooden horse on a merry-go-round if the radial distance from the centre of rotation is 3.5 m and the platform takes 12 s to complete each revolution.

15.8 Calculate the centripetal force acting on a 1.2 t vehicle when it is travelling at 60 km/h around a bend of 90 m radius.

15.9 A centrifuge for the separation of sugar crystals in molasses must exert a force of 6 newtons per gram mass. If the filter drum has a radius of 350 mm, determine the required speed in revolutions per minute.

15.10 The radius of curvature of a humpback bridge at its uppermost point is 90 m. When a car travelling over the bridge reaches this point, at what speed, in kilometres per hour, will the wheels leave the road surface? (*Hint*: Centripetal force is provided by gravity.)

15.5 *STABILITY OF VEHICLES ON LEVEL CURVED ROADS*

Like any other force, centripetal force is always caused by some form of interaction between physical objects. One of the most common forms of interaction between two surfaces in contact is due to dry sliding friction, such as that which occurs between the road surface and the tyres of a moving vehicle.

In order to make a vehicle travel in a circular path around a horizontal curve on a level road, it is necessary to have a centripetal force of sufficient magnitude act on the vehicle through the points of contact between the wheels and the road surface. The magnitude of the required force is dependent on the mass m of the vehicle, the radius r of road curvature, and the circumferential speed v, as explained in Section 15.3.

Since this force is generated by friction, its maximum possible value is limited by the coefficient of static friction μ between the road and the tyres. It can be seen that frictional force is also dependent on the mass of the vehicle, since on a horizontal surface the normal reaction F_n is equal to the weight of the vehicle $F_w = mg$. Therefore:

$$F_c = F_r = \mu F_n = \mu mg$$

We know from experience that, for a given set of road and tyre conditions, the ability of a vehicle to successfully negotiate a curve on a level road appears to depend primarily on the speed of travel. Two rather undesirable alternatives are not uncommon, namely the vehicle skidding or overturning.

Let us now examine the criteria that determine the critical conditions beyond which velocity of the vehicle may not be increased without one of these disastrous alternatives taking place.

Example 15.6

A car of mass 1.35 t is travelling around a horizontal curve of radius 115 m. If the coefficient of static friction between the road surface and the tyres is 0.5, determine the minimum speed at which the vehicle may skid.

Solution

The maximum centripetal force that can be provided by frictional contact is:

$$F_c = \mu mg$$
$$= 0.5 \times 1350 \text{ kg} \times 9.81 \text{ N/kg}$$
$$= 6622 \text{ N}$$

Substitute into the formula for centripetal force:

$$F_c = \frac{mv^2}{r}$$
$$6622 = \frac{1350 \times v^2}{115}$$
$$\therefore v = 23.75 \text{ m/s}$$
$$= 85.5 \text{ km/h}$$

Friction has the uncanny ability to progressively build up resistance to sliding motion to match the physical demands of the situation. Therefore, if the car in the example is travelling at slower speeds below 85.5 km/h, just sufficient friction would be generated between the wheels and the road surface to provide the necessary centripetal force, corresponding exactly to the lower speed of travel, to enable the car to go around the curve without sliding. However, friction cannot increase indefinitely. In our example, it cannot increase beyond its limiting value of 6622 N, and any attempt to travel faster than 85.5 km/h would result in a skid, i.e. the vehicle would continue in a straight line instead of moving in the circular path.

However, everything we have said so far does not discount the possibility of the vehicle overturning as it attempts to negotiate the curve. The analysis of the tendency for a vehicle to overturn requires a different criterion to be investigated. Here again, experience teaches us that stability against overturning has a lot to do with the speed of the vehicle travelling in a curved path, but it also depends to a large extent on the dimensions of the vehicle itself. In particular, it depends on the relationship between the wheel base, i.e. the width between the wheels of the vehicle, and the height of its **centre of mass** above the ground.[*]

Example 15.7

If the car in Example 15.6 has its centre of mass 0.65 m above the ground, and the wheel base is 1.5 m wide, determine the minimum speed at which the vehicle may overturn.

---

[*] The centre of mass is also known as the **centre of inertia** or the **centre of gravity**, since both the linear inertia of the entire mass and the force of gravity (weight) acting on it can be considered to be located at this point.

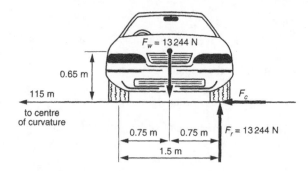

Fig. 15.5

Solution

At the critical instant when the car is about to overturn, its entire weight rests on the outer wheels. Therefore, while the reaction on the inner wheels is zero, the total vertical reaction force at the outer wheels is equal to the weight of the car.

$$F_r = F_w$$
$$= mg$$
$$= 1350 \text{ kg} \times 9.81 \text{ N/kg}$$
$$= 13\ 244 \text{ N}$$

Together with weight F_w, which is acting through the centre of mass, the reaction force F_r constitutes a couple which has a perpendicular distance between the pair of its forces equal to one-half of the wheel-base width. The moment due to this couple is:

$$M = F\frac{w}{2}$$
$$= 13\ 244 \text{ N} \times \frac{1.5 \text{ m}}{2}$$
$$= 9933 \text{ N.m}$$

In order to make the car follow the circular path, a horizontal centripetal force F_c must exist, acting through the points of effective contact between the tyres and the road surface. The moment of the centripetal force about the centre of mass depends on the height of the centre of mass above the road surface:

$$M = F_c h$$
$$= F_c \times 0.65$$

The critical condition occurs when the two moments are just balanced against each other:

$$F_c \times 0.65 = 9933$$
$$\therefore F_c = \frac{9933 \text{ N.m}}{0.65 \text{ m}}$$
$$= 15\ 282 \text{ N}$$

Substitute into the formula for centripetal force:

$$F_c = \frac{mv^2}{r}$$

$$15\,282 = \frac{1350 \times v^2}{115}$$

$$\therefore v = 36.08 \text{ m/s}$$
$$= 130 \text{ km/h}$$

Therefore we may conclude that, provided there is adequate friction to generate the required centripetal force without sliding,[*] the car will start overturning, i.e. rotating about its centre of mass, as soon as its velocity exceeds 130 km/h.

In order to determine the maximum safe speed for a given set of conditions, compare the two separate, relatively independent criteria obtained above, one for skidding ($v = 85.5$ km/h, Example 15.6) and the other for overturning ($v = 130$ km/h, Example 15.7). It is not hard to see that in this case, if the car is travelling under the specified conditions of road surface friction and radius of curvature, skidding may occur at a speed lower than the speed which is necessary to cause overturning. Therefore, the maximum limit of safety is the lower of the two speeds, 85.5 km/h.

It is not difficult to translate the method of solution used above into two separate expressions which would enable the limiting criteria for vehicle stability to be found by direct substitution of given data, followed by simple calculation.

Limit of safe velocity in km/h to avoid skidding:

$$v = 3.6\sqrt{\mu g r}$$

Limit of safe velocity in km/h to avoid overturning:

$$v = 3.6\sqrt{\frac{wgr}{2h}}$$

It is useful to note that, since all forces acting on a vehicle in this situation happen to be proportional to the mass of the vehicle, the mass is eventually cancelled out from the calculations, rendering the criteria for safe velocity independent of mass.

15.6 *BANKING OF ROADS AND TRACKS*

In order to counteract, at least partially, the undesirable tendency for skidding and overturning when vehicles round a curve at high speeds, it is common practice to bank highways, railway tracks and velodrome surfaces by giving them a transverse slope suitable for the estimated normal speed of travel.

---

[*] The flanged wheels on railway locomotives and carriages make it practically impossible for a train to skid off the rails. However, overturning may occur at excessive speeds, as for any vehicle moving in a curved path.

When a vehicle travels around a banked curve at the exact speed v for which the curve is designed, there is no side thrust on the wheels, the total reaction being perpendicular to the inclined surface, as illustrated in Figure 15.6(a). The centripetal force F_c is in effect the horizontal resultant of two forces acting on the vehicle, namely the weight F_w and the normal reaction F_n. This resultant, which is directed towards the centre of road curvature, can be found by constructing a triangle of forces as shown in Figure 15.6(b).

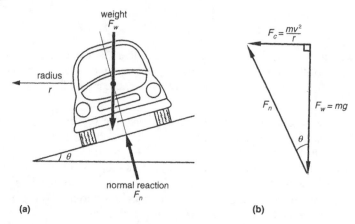

Fig. 15.6   (a) *Travelling on a banked curve*   **(b)** *Triangle of forces*

Since the angle θ within the triangle of forces is equal to the angle of inclination of the road surface to the horizontal, we can also write:

$$\tan \theta = \frac{F_c}{F_w} = \frac{mv^2/r}{mg}$$

After the mass m is cancelled out, we obtain the following expression which can be used to calculate the correct angle of inclination for the expected normal speed of travel:

$$\boxed{\tan \theta = \frac{v^2}{gr}}$$

Example 15.8

A highway curve with a radius of 355 m is to be banked at an angle to suit an estimated normal speed of 110 km/h. What is the required angle of inclination?

Solution

The speed in metres per second is $v = 30.56$ m/s.
The tangent of the required angle is calculated from:

$$\tan \theta = \frac{v^2}{gr}$$

$$= \frac{30.56^2}{9.81 \times 355}$$

$$= 0.2681$$

Hence, the required angle of banking to suit the specified speed is:

$$\theta = \tan^{-1} 0.2681$$
$$= 15°$$

The expression given above can be transposed into another form for calculating the correct speed of travel for a curve banked at a known angle θ:

$$v = \sqrt{gr \tan \theta}$$

Example 15.9

What is the most comfortable speed for a car when travelling on a 300 m radius highway curve banked at 12°?

Solution

For this road, the exact speed of travel, without any tendency for skidding or overturning, is found from:

$$v = \sqrt{gr \tan \theta}$$
$$= \sqrt{9.81 \times 300 \times \tan 12°}$$
$$= 25 \text{ m/s}$$
$$= 90 \text{ km/h}$$

In railroad engineering, the difference in height between the outer and inner rails on a banked railway curve is usually referred to as the **superelevation.** Over 60 per cent of the world's railroads are built to a standard gauge width of 1.435 metres.[*]

If we let w represent the width between the rail centres, as shown in Figure 15.7, the superelevation e will be given by:

$$e = w \sin \theta$$

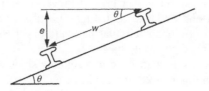

Fig. 15.7   *The superelevation (e) of a railway track*

Example 15.10

Calculate the superelevation for a standard gauge railway track, on a curve of radius 1 km, if there is to be no side thrust on the wheels travelling at 150 km/h.

Solution

The speed in metres per second is $v = 41.67$ m/s.

---

[*] Originally defined as 4 feet $8\frac{1}{2}$ inches between rail centres, and used on all major railway lines in Australia, Great Britain, USA and Europe.

The angle of banking corresponding to this speed is calculated from:

$$\tan \theta = \frac{v^2}{gr}$$

$$= \frac{41.67^2}{9.81 \times 1000}$$

$$= 0.177$$

$$\therefore \ \theta = \tan^{-1} 0.177$$

$$= 10°$$

Hence, the required superelevation is:

$$e = w \sin \theta$$

$$= 1.435 \times \sin 10°$$

$$= 0.25 \text{ m}$$

In practice, from the point of view of passenger comfort, the exact amount of superelevation on railway tracks is not particularly critical, provided the amount used is attained very gradually and is strictly maintained throughout the length of the curve.

The term 'superelevation' can also be used to describe the amount by which the outer edge of a highway of known width is elevated above its inner edge. However, unlike the standard gauge railway track with its definite and unchanging distance between the two rails, highways do not necessarily have a standard uniform width. Therefore, it is more common to state the **transverse inclination** of a road surface, expressed as either an angle from the horizontal or a percentage gradient.

15.7 *CENTRIFUGAL FORCE DUE TO OUT-OF-BALANCE MASSES*

So far, considerable emphasis has been placed on the force we call centripetal force, i.e. the force applied to a moving mass which causes the mass to travel in a circular path. However, we should also remember that whenever a centripetal force exists, there must always be an equal and opposite centrifugal reaction force, which is directed radially outwards from the centre of the circular path and experienced by another component in the interacting system of physical objects.

Therefore, if a mass m is attached to a rotating shaft at a distance r from the axis of rotation, as shown in Figure 15.8, there will be an out-of-balance centrifugal force acting on the shaft away from its axis. Such an out-of-balance effect can be severely detrimental to the efficient running of rotating machinery, resulting in excessive vibrations and wear.

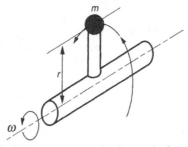

Fig. 15.8   *An out-of-balance mass attached to a shaft*

Example 15.11

Calculate the out-of-balance centrifugal force acting on a shaft which is rotating at 700 rpm and carries a 0.6 kg mass at a radial distance of 250 mm from its axis.

Solution

In the case of a single mass, the centrifugal reaction force is exactly equal in magnitude to the centripetal force acting on the out-of-balance mass.

$$\text{Angular velocity} = 700 \times \frac{2\pi}{60}$$
$$= 73.3 \text{ rad/s}$$

Hence, the centrifugal force is:

$$F_c = m\omega^2 r$$
$$= 0.6 \times 73.3^2 \times 0.25$$
$$= 806 \text{ N}$$

When several masses are made to revolve in one plane about a common axis of rotation, the resultant pull on the axis is found by vectorial addition of their individual centrifugal effects, as illustrated by the following example.

Example 15.12

A turntable revolving about its centre at 95.5 rpm has a metal object of mass 5 kg attached at a radius of 0.3 m, and another object of mass 7 kg attached at a radius of 0.5 m. The angle between the objects is 90°, as shown in Figure 15.9(a). Determine the magnitude and direction of the out-of-balance centrifugal force.

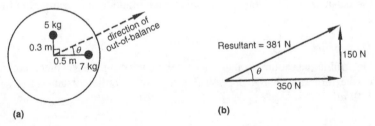

(a)

(b)

Fig. 15.9

Solution

$$\text{Angular velocity} = 95.5 \times \frac{2\pi}{60}$$
$$= 10.0 \text{ rad/s}$$

Hence, the individual centrifugal forces are as follows.
(i) For the 5 kg mass:

$$F_c = m\omega^2 r$$
$$= 5 \times 10.0^2 \times 0.3$$
$$= 150 \text{ N}$$

(ii) For the 7 kg mass:

$$F_c = m\omega^2 r$$
$$= 7 \times 10.0^2 \times 0.5$$
$$= 350 \text{ N}$$

With the two forces acting at 90° to each other, the resultant centrifugal force is found by constructing the right-angled triangle to scale (Fig. 15.9(b)), or solving it by Pythagoras' rule.

Magnitude:

$$\sqrt{150^2 + 350^2} = 381 \text{ N}$$

Direction:

$$\tan \theta = \frac{150}{350}$$
$$= 0.4286$$
$$\therefore \ \theta = \tan^{-1} 0.4286$$
$$= 23.2°$$

Therefore, the resultant out-of-balance force of 381 N makes an angle of 23.2° with the radial position of the 7 kg mass.

It is possible to balance any arrangement of out-of-balance masses revolving in the same plane by a single balancing mass, provided it is positioned in the direction exactly opposite to that of the resultant out-of-balance force. The size of the balancing mass and its distance from the axis of rotation must be selected so that the centrifugal force due to the balancing mass would be equal in magnitude to the resultant out-of-balance force.

Example 15.13

Determine the required magnitude of the balancing mass which is to be placed at the rim of the turntable in the previous example at a radial distance from its centre equal to 1.19 m.

Solution

The resultant out-of-balance force is 381 N. Therefore, for the balancing mass:

$$F_c = m\omega^2 r$$
$$381 = m \times 10.0^2 \times 1.19$$
$$\therefore \ m = 3.2 \text{ kg}$$

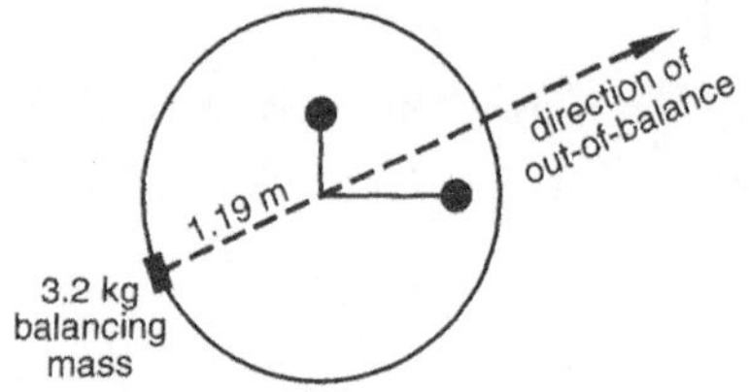

Fig. 15.10

Problems

15.11 A truck of mass 3.5 t is travelling around a level road that has a horizontal curvature of radius 200 m. The coefficient of static friction between the road surface and the wheels is 0.75. Calculate the minimum vehicle speed, in kilometres per hour, at which skidding will occur.

15.12 A car of mass 1.35 t and with a wheel-base width of 1.5 m has a load of 375 kg placed on a roof-rack so that the combined centre of mass is 0.9 m above the ground.
   Calculate the minimum speed at which the car may overturn when travelling around a horizontal curve of radius 115 m on a level road.

15.13 A loaded panel van has a total mass of 2.3 t and a wheel-base width of 1.45 m, and its centre of mass is 1.24 m above the ground. Assuming the coefficient of static friction between the tyres and the road is 0.71, calculate the maximum limit of safe speed on a level road that has a horizontal radius of curvature of 95 m.
   What is likely to occur if the vehicle enters the curve at 85 km/h?

15.14 A circular path of a velodrome is to be designed with a horizontal radius of 30 m and for an estimated speed of 64 km/h. What is the angle at which the track should be banked in order to eliminate any tendency for skidding upwards or downwards at this speed?

15.15 A standard gauge railway line, 1435 mm wide between rail centres, is laid with the outer rail superelevated by 100 mm on a curve of radius 2.2 km. What is the ideal train speed for this curve?

15.16 In a machine, a horizontal arm is driven by a vertical spindle which makes 500 rpm. A 2 kg block is attached to the free end of the arm and revolves in a horizontal circle, 0.58 m from the axis of the spindle. Determine the centripetal force acting on the block and the centrifugal force on the vertical spindle, clearly stating their directions with respect to the axis of rotation.

15.17 The out-of-balance of a car wheel is equivalent to a mass of 25 kg with an eccentricity (i.e. radial distance from the axis) of 0.72 mm. The overall diameter of the tyre is 600 mm. Calculate the out-of-balance centrifugal force when the car is travelling at 115 km/h.

15.18 In order to balance the wheel in Problem 15.17, a balancing mass is to be attached to the rim of the wheel at 180 mm from the axis. What should the balancing mass be?

15.19 A circular plate rotates on a vertical spindle through its centre, making 4 revolutions per second. It supports two masses of 10 kg and 15 kg at radii of 0.75 m and 0.6 m respectively. The angle between the masses is 90°. Determine the resultant out-of-balance centrifugal force acting on the spindle due to the revolving masses.

15.20 If it is necessary to balance the two masses in the previous problem by placing a single balancing mass at a radial distance of 1.5 m from the spindle, what should the balancing mass be?

Review questions

1. State the relations between rotation and circular motion.
2. Explain the concept of *centripetal acceleration* and state the formula for calculating its value.
3. Define *centripetal force* and state the formula for calculating its value.
4. Give two examples of instability of vehicles on level curved roads.
5. Explain the reason for the banking of roads and railway tracks.
6. Define the term *superelevation*.
7. Explain the difference between *centripetal force* and *centrifugal force*.

PART 6

WORK AND ENERGY

≡

Now, to have access to the science of motion, one does not need to scale the heights of mathematics first. On the contrary, nature itself reveals its beauty in full splendour, and even a person with meagre abilities can see a multitude of things that have hitherto been hidden from the greatest minds.

Robert Mayer
on the law of conservation of energy

CHAPTER 16

Work and power

This chapter is concerned with the concepts of work and power as they apply to the linear and rotational motion of moving objects and mechanical components.

Expected learning outcomes

After carefully studying the material presented in this chapter, working through all numerical examples, and successfully completing all practice problems, students should be able to:
1. distinguish between the concepts and units of mechanical work and power;
2. calculate the work done by a constant force;
3. calculate average and instantaneous power associated with linear and rotational motion;
4. calculate the work done to accelerate a body in linear or rotational motion;
5. calculate the work done in deforming a coil spring from its free length.

16.1 *MECHANICAL WORK*

Mechanical work is defined in terms of linear or rotational effort, when the result produced by the effort is the movement or rotation of a body.

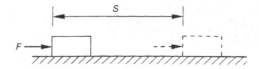

Fig. 16.1

In linear motion, if a force F is applied to a body which moves in a straight line a distance S in the direction of the force (Fig. 16.1), then the work W done by the force on the body is said to be the product of the force and the distance:

$$W = F \times S$$

The unit of work follows from this definition. If the units of force and distance are newtons and metres respectively, then the unit of work is the newton metre. Work is a new physical quantity different from both force and distance. It is therefore convenient to give the unit of work a special name. The SI unit of work equal to 1 N.m is called the **joule**, denoted by J.

Example 16.1

Determine the work done by a force of 50 N moving a distance of 3 m in the direction of the force.

Solution

Work done:

$$W = F \times S$$
$$= 50 \text{ N} \times 3 \text{ m}$$
$$= 150 \text{ J}$$

Example 16.2

A hoist lifts a load of 1.5 t through a vertical distance of 20 m. Determine the amount of work done against gravity.

Solution

The work is done against gravity, i.e. against the weight of the load.

The weight of the load is:

$$F_w = mg$$
$$= 1500 \text{ kg} \times 9.81 \ \frac{\text{N}}{\text{kg}}$$
$$= 14\,715 \text{ N}$$

The work done against this force is:

$$W = F \times S$$
$$= 14\,715 \text{ N} \times 20 \text{ m}$$
$$= 294\,300 \text{ J}$$
$$= 294.3 \text{ kJ}$$

In rotational terms, torque T is analogous to force, i.e. torque is a turning effort. Similarly, angular displacement θ is analogous to linear displacement. This analogy can be used to arrive at the expression for work done by torque on a rotating object:

$$W = T \times \theta$$

Since torque is measured in newton metres and angular displacement in radians, which are dimensionless, it can be seen that work done by torque can also be measured in joules.

Example 16.3

A flywheel makes 200 revolutions while the torque applied to it is 35 N.m. Determine the work done.

Solution

Angular displacement must be expressed in radians:

$$\theta = 200 \text{ revolutions}$$
$$= 200 \times 2\pi$$
$$= 1257 \text{ rad}$$

Work done:

$$W = T \times \theta$$
$$= 35 \times 1257$$
$$= 43\,980 \text{ J}$$
$$= 44 \text{ kJ}$$

16.2 *POWER*

When work is being done continuously over a period of time, the time rate of doing work is called **power**, P.

$$P = \frac{W}{t}$$

where P is average power
W is work done
t is time taken to do the work

The unit of power follows from the definition and is equal to one joule of work per second of time, i.e. J/s. The name given to the unit of power is the **watt**,[*] denoted by W.

---

[*] The SI unit was named after the Scottish engineer James Watt who, in the late 18th century, according to a historical anecdote, established another unit of power, called **horsepower**, after actual experiments with strong dray horses. The horsepower, equal to 746 watts, is about 50 per cent more than the rate that an average horse can sustain for a working day.

(One has to be careful to avoid confusion between W as a symbol for work done and W as a unit symbol for power.)

Example 16.4

If it takes 27 s to lift the load in Example 16.2, what is the average power required?

Solution

$$\text{Power} = \frac{W}{t}$$
$$= \frac{294.3 \text{ kJ}}{27 \text{ s}}$$
$$= 10.9 \text{ kJ/s}$$
$$= 10.9 \text{ kW}$$

When work is done by a force moving with a constant linear velocity v, we can substitute $S = vt$ into the expression for power:

$$P = \frac{W}{t} = \frac{F \times S}{t} = \frac{Fvt}{t} = Fv$$

Hence, for power in linear motion of a force:

$$\boxed{P = Fv}$$

This relationship also applies if v is interpreted as average velocity.

Example 16.5

A train moving at 63 km/h requires 40 kN of tractive effort at this speed. Determine the driving power.

Solution

$$v = 63 \text{ km/h}$$
$$= 17.5 \text{ m/s}$$

Power:

$$P = Fv$$
$$= 40 \text{ kN} \times 17.5 \text{ m/s}$$
$$= 700 \text{ kN.m/s}$$
$$= 700 \text{ kW}$$

When work is done by a torque applied through a rotating member such as a shaft turning with a constant angular velocity $\omega = \dfrac{\theta}{t}$, substitution yields:

$$P = \frac{W}{t} = \frac{T \times \theta}{t} = \frac{T\omega t}{t} = T\omega$$

Therefore, the power produced by a torque in rotational motion is:

$$P = T\omega$$

where ω is constant or average angular velocity in radians per second.

Example 16.6

An output shaft of an electric motor rotates at 1450 rpm and produces a torque of 81 N.m. What is the shaft power?

Solution

$$\omega = 1450 \text{ rpm}$$
$$= \frac{2\pi \times 1450}{60}$$
$$= 151.8 \text{ rad/s}$$
$$P = T\omega$$
$$= 81 \text{ N.m} \times 151.8 \text{ rad/s}$$
$$= 12\,300 \text{ W}$$
$$= 12.3 \text{ kW}$$

 Problems

16.1 A locomotive applies a tractive effort of 120 kN over a distance of 2 km. Calculate the work done.

16.2 A crane lifts a crate of mass 2.5 t through a height of 12 m. Calculate the work done.

16.3 A box of mass 80 kg is hauled along a horizontal floor by a force, parallel to the floor, for a distance of 25 m. If the coefficient of sliding friction between the floor and the box is 0.4, calculate the amount of work done.

16.4 A torque of 60 N.m is applied to a flywheel to turn it through 150 revolutions at constant speed. What is the work done?

16.5 A radar antenna of mass 3500 kg turns on rollers mounted on a horizontal circular track of diameter 2 m. If the frictional resistance of the rollers is 0.5 kN/t, calculate the work required to turn the antenna through one complete revolution.

16.6 Determine the power developed by an engine when it does 97.5 MJ of work in 25 min.

16.7 If a lift having a total mass of 1.6 t travels a vertical distance from the ground to the tenth floor in 17.5 s and the distance between floors is 3.5 m, calculate the average power required.

16.8 A train of total mass 650 t is hauled by a locomotive along a level track at a constant speed of 60 km/h. If the tractive resistance is 85 newtons per tonne mass of the train, calculate the power developed at this speed.

16.9 The power output from an engine is measured by means of a belt brake arrangement as shown in Figure 16.2. The brake-drum diameter is 800 mm, over which a belt supports two masses of 55 kg each. One side of the belt is attached to a spring balance which reads 375 N. The engine speed is 450 rpm. Calculate the power dissipated in friction between the belt and the drum.

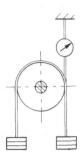

Fig. 16.2

16.10 A motor vehicle is travelling at 100 km/h against total air and frictional resistance of 950 N. Neglecting any transmission losses, estimate the torque which the engine must develop if its rotational speed is 2200 rpm.

16.3 *WORK AND ACCELERATION*

When a constant force or torque is available to accelerate a body in linear or rotational motion, the work done can be related to the acceleration produced.

Work required to accelerate a mass

If a body of mass m is acted upon by a net accelerating force F over a distance S, the work done is $W = FS$. At the same time:

$$S = v_{av} \times t$$

$$= \left(\frac{v + v_0}{2}\right)t$$

and:

$$F = ma$$

$$= m\left(\frac{v - v_0}{t}\right)$$

Substituting into the work equation:

$$W = F \times S$$

$$= m\left(\frac{v - v_0}{t}\right) \times \left(\frac{v + v_0}{2}\right)t$$

$$= \frac{m}{2}(v - v_0) \times (v + v_0)$$

$$= \frac{m}{2}(v^2 - v_0^2)$$

Therefore the work done in accelerating a mass m from an initial velocity v_0 to a final velocity v is given by:

$$W = \frac{m}{2}(v^2 - v_0^2)$$

It will be seen later that this equation has a special significance with respect to the quantity called kinetic energy. At this stage, however, we will use it simply to calculate the work required to change the velocity of a body.

Example 16.7

Determine the force, work and average power required to accelerate a car of mass 1.2 t from rest to 72 km/h in 16 s.

Solution

Final velocity:

$$v = 72 \text{ km/h}$$
$$= 20 \text{ m/s}$$

Acceleration:

$$a = \frac{v - v_0}{t}$$
$$= \frac{20 - 0}{16}$$
$$= 1.25 \text{ m/s}^2$$

Force required:

$$F = ma$$
$$= 1200 \text{ kg} \times 1.25 \text{ m/s}^2$$
$$= 1500 \text{ N}$$
$$= 1.5 \text{ kN}$$

Work required:

$$W = \frac{m}{2}(v^2 - v_0{}^2)$$
$$= \frac{1200}{2}(20^2 - 0)$$
$$= 240\,000 \text{ J}$$
$$= 240 \text{ kJ}$$

Alternatively, distance travelled can be determined:

$$S = \left(\frac{v + v_0}{2}\right)t$$
$$= \left(\frac{20 + 0}{2}\right) \times 16$$
$$= 160 \text{ m}$$

Hence:

$$W = F \times S$$
$$= 1.5 \text{ kN} \times 160 \text{ m}$$
$$= 240 \text{ kJ}$$

Average power can now be calculated:

$$P = \frac{W}{t}$$
$$= \frac{240 \text{ kJ}}{16 \text{ s}}$$
$$= 15 \text{ kW}$$

It should be understood that these results refer to the force, work and power associated with the acceleration of the mass only. Work and power due to other causes, such as frictional resistance, must be allowed for separately if required.

Work required for rotational acceleration

By analogy with linear motion, if a body of mass moment of inertia I is acted upon by a net accelerating torque T over an angular displacement θ, the work done is $W = T \times \theta$. At the same time:

$$\theta = \omega_{av} \times t$$
$$= \left(\frac{\omega + \omega_0}{2} \right) t$$

and:

$$T = I \times \alpha$$
$$= I \left(\frac{\omega - \omega_0}{t} \right)$$

Substituting into the work equation:

$$W = T \times \theta$$
$$= I \left(\frac{\omega - \omega_0}{t} \right) \left(\frac{\omega + \omega_0}{2} \right) t$$
$$= \frac{I}{2} (\omega - \omega_0)(\omega + \omega_0)$$
$$= \frac{I}{2} (\omega^2 - \omega_0^2)$$

Therefore the work done in accelerating a body of mass moment of inertia I from initial angular velocity ω_0 to a final angular velocity ω is given by:

$$W = \frac{I}{2} (\omega^2 - \omega_0^2)$$

Example 16.8

Determine the torque, work and average power when a flywheel of mass moment of inertia of 53 kg.m^2 is accelerated from 700 rpm to 1500 rpm in 24 s.

Solution

Initial angular velocity:

$$\omega_0 = 700 \text{ rpm}$$
$$= 73.3 \text{ rad/s}$$

Final angular velocity:

$$\omega = 1500 \text{ rpm}$$
$$= 157.1 \text{ rad/s}$$

Angular acceleration:

$$\alpha = \frac{\omega - \omega_0}{t}$$
$$= \frac{157.1 - 73.3}{24}$$
$$= 3.49 \text{ rad/s}^2$$

Torque required:

$$T = I\alpha$$
$$= 53 \text{ kg.m}^2 \times 3.49 \text{ rad/s}^2$$
$$= 185 \text{ N.m}$$

Work done:

$$W = \frac{I}{2}(\omega^2 - \omega_0^2)$$
$$= \frac{53}{2}(157.1^2 - 73.3^2)$$
$$= 511\,400 \text{ J}$$
$$= 511.4 \text{ kJ}$$

Alternatively, angular displacement can be determined:

$$\theta = \left(\frac{\omega + \omega_0}{2}\right)t$$
$$= \left(\frac{157.1 + 73.3}{2}\right) \times 24$$
$$= 2764 \text{ rad}$$

Hence:

$$W = T \times \theta$$
$$= 185 \text{ N.m} \times 2764 \text{ rad}$$
$$= 511\,400 \text{ J}$$
$$= 511.4 \text{ kJ}$$

Average power can now be calculated:

$$P = \frac{W}{t}$$
$$= \frac{511.4 \text{ kJ}}{24 \text{ s}}$$
$$= 21.3 \text{ kW}$$

16.4 *WORK DONE IN DEFORMING COIL SPRINGS*

In many practical situations, mechanical work is done by a force whose magnitude does not remain constant over the entire period during which work is being done. The full mathematical treatment of the effects of variable forces can be quite complex and usually requires some knowledge of differential and integral calculus. For this reason, it is outside the scope of this book.

However, it is possible for us to consider the relatively simple but useful case of work done in deforming a coil spring, which involves a variable force changing at a uniform rate. Let us start by considering the undeformed length of a coil spring, usually referred to as its **free length**. It is found that within the normal operating limits of the material, the force required to produce axial compression or elongation of the spring is directly proportional to the amount of deformation from the free length. We can express this as a formula:

$$\boxed{F = kx}$$

where x is the amount of elongation or compression
  F is the force corresponding to the amount of deformation
  k is the constant of proportionality

The constant of proportionality k is known as the **spring modulus**, or simply as the **spring constant**, and represents the force required to deform the spring a unit distance. The most common practical unit for expressing spring constants is the newton per millimetre (N/mm). However, as will be seen below, it is sometimes necessary to convert from this practical unit into the more fundamental newton per metre (N/m) to ensure dimensional homogeneity of calculations performed.

Example 16.9

Determine the initial, final and average force when a coil spring is stretched 12 mm from its free length. The spring constant is 200 N/mm.

Solution

The initial force is zero, as the term 'free length' implies. Therefore, $F_0 = 0$. The final force, i.e. at the end of stretching, is:

$$F = kx$$
$$= 200 \text{ N/mm} \times 12 \text{ mm}$$
$$= 2400 \text{ N}$$

The average force is:

$$F_{av} = \frac{F_0 + F}{2}$$
$$= \frac{0 + 2400}{2}$$
$$= 1200 \text{ N}$$

The work done by a variable force is the product of the average value of the force and the displacement, which is equal to the amount of deformation produced.

Example 16.10

Determine the amount of work done by the variable force when stretching the spring in the previous example.

Solution

Work done:

$$W = F_{av}x$$
$$= 1200 \text{ N} \times 0.012 \text{ m}$$
$$= 14.4 \text{ J}$$

Let us now combine the expressions used above into a single equation for calculating the work done in deforming a spring. It was established previously that $F_0 = 0$ and $F = kx$. Therefore, the average force is:

$$F_{av} = \frac{F_0 + F}{2}$$
$$= \frac{0 + kx}{2}$$
$$= \frac{kx}{2}$$

Substitution yields:

$$W = F_{av}x$$
$$= \frac{kx}{2} x$$
$$= \frac{kx^2}{2}$$

Therefore, work done in deforming a coil spring from its free length is:

$$\boxed{W = \frac{kx^2}{2}}$$

where k is the spring constant in N/m

      x is the amount of elongation or compression in m

Example 16.11

Use the above formula to check the answer obtained in the previous example.

Solution

Using base units, $k = 200\,000$ N/m and $x = 0.012$ m. Substitute:

$$W = \frac{kx^2}{2}$$
$$= \frac{200\,000 \times 0.012^2}{2}$$
$$= 14.4 \text{ J}$$

 # Problems

16.11 A motor car of mass 1 t is travelling at 40 km/h. Determine the average power required to accelerate the car to 80 km/h over a distance of 300 m. Neglect friction and air resistance.

16.12 If total resistance to motion of the car in the previous problem is 365 N, determine the average power required to accelerate the car, as specified above, taking into account the resistance force.

16.13 A flywheel, having a mass moment of inertia of 65 kg.m^2, is to be accelerated from rest to a speed of 750 rpm in 30 s. If bearing friction is neglected, calculate the work that must be done to accomplish this.

16.14 If it is found that the average power actually required to accelerate the flywheel described in the previous problem is 9 kW, what is the magnitude of friction torque in the bearings?

16.15 A train of mass 200 t, travelling at 60 km/h, is brought to rest in a distance of 600 m by a constant braking force. If total tractive resistance is 18 kN, what is the value of the braking force and the average power dissipated by the brakes?

16.16 A flywheel revolving at 450 rpm has its speed uniformly reduced to 150 rpm in 3 min. The wheel, which is a disc of uniform thickness, has a mass of 880 kg and is 1200 mm in diameter. The bearing friction is 5 N.m. Calculate the braking torque applied and the average power dissipated by the brake.

16.17 A block of mass 245 kg is placed on a platform supported by six springs, with each spring taking an equal share of the total load. The vertical deflection of the platform when the load is placed on it is observed to be 16 mm. Assuming that the weight of the platform itself is negligible, calculate the spring constant for each of the six springs.

16.18 For a coil spring with a modulus of 50 N/mm, calculate the amount of work required to stretch it by 20 mm from free length, and the final force necessary to maintain it in the stretched condition.

Review questions

1. Define *mechanical work*.
2. State the relations between:
   (a) force and work for linear motion
   (b) torque and work for rotational motion
3. Define *power*.
4. State the relations between:
   (a) force and power for linear motion
   (b) torque and power for rotational motion
5. Explain how the work done in accelerating a body can be related to its change of velocity:
   (a) for linear motion
   (b) for rotational motion
6. Explain what is meant by the term *spring modulus*.
7. Show how the work of deforming a coil spring can be evaluated.

CHAPTER 17

Mechanical energy

The concept of mechanical energy is one of the most useful in engineering science. However, the colloquial meaning of the word 'energy' is very different from its precise scientific meaning. The idea of energy, by itself, is very abstract and historically was a matter of great confusion among many of the ablest scientific minds until well into the 19th century.

The essential unity of the concept of energy in its different forms, e.g. mechanical, thermal, electrical, chemical, was not clear until the development of the steam engine prompted men such as James Watt, James Joule and Robert Mayer to explore the relations between different forms of energy in action, such as work and heat, and to proclaim the law of conservation of energy.

In this chapter, we will not include non-mechanical forms of energy, such as heat, in our discussion because they do not fit our description of mechanical systems of bodies at rest or in motion. Therefore, our definition of energy, at this stage, is restricted to mechanical energy only.

Expected learning outcomes

After carefully studying the material presented in this chapter, working through all numerical examples, and successfully completing all practice problems, students should be able to:
1. define *mechanical energy* and state the SI unit of energy;
2. calculate gravitational potential energy;
3. calculate kinetic energy of linear and rotational motion;
4. calculate strain energy stored in a coil spring.

17.1 *MECHANICAL ENERGY*

In the previous chapter, we saw how doing mechanical work can alter the condition of a physical object in terms of:

1. lifting it to a higher elevation against the force of gravity;
2. accelerating the object against the inertia of its mass; or
3. stretching or compressing a spring against the ability of its material to resist such deformation.

It took many centuries of scientific thought to recognise that the work done to lift or to accelerate a body, or to stretch a piece of elastic material, does not disappear without a trace, but is in fact stored within the body by reason of:

1. its increased elevation above some datum level;

2. its higher linear or rotational velocity; or
3. the amount of elastic deformation produced.

Furthermore, mechanical energy is, at least in theory, fully retrievable in the form of work if the process is reversed. Consequently, mechanical energy is usually defined as a physical quantity stored in a material body by virtue of its position, its motion or its strained condition, and is a measure of the capacity of the body to do work.* If energy is stored work, its units must correspond to the units of work. Therefore, the SI unit of energy is the joule.

By its very nature, stored energy is always latent, i.e. not directly observable or measurable. However, its quantity can be indirectly evaluated using measurements of other related physical quantities, such as mass, height, velocity and elongation. On the other hand, the effects of energy in action are always immediately obvious, e.g. a drop hammer driving a pile into the ground, or a clock mechanism driven by a spring. The amount of work done can be taken as a measure of the amount of energy released.

It is true that in most practical situations friction appears to detract from the quantity of work available when stored mechanical energy is released. Some energy is inevitably converted into heat and dissipated to the surroundings. However, if properly allowed for, frictional effects can be included in the overall energy accounts of mechanical systems, as we will see in subsequent chapters of this book.

Initially, let us begin by ignoring friction and discussing three forms of mechanical energy called gravitational potential energy, kinetic energy and elastic strain energy, and their relation to the work done in the absence of frictional losses.

17.2 *POTENTIAL ENERGY*

The **potential energy** of a body is the energy that a body possesses due to its position in the gravitational field.

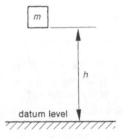

Fig. 17.1 *A raised mass has potential energy*

As an example of this, consider a block of mass m raised a distance h above the ground (Fig. 17.1). The force of gravity acting on the block is $F_w = mg$, and the work done in lifting the block is $W = Fh = mgh$. This work is stored in the body as potential energy with respect to the ground as the datum. In symbols, potential energy can be stated as follows:

$$PE = mgh$$

---

* The word 'energy' is derived from the Greek *energeia, en* meaning 'in' and *ergon* meaning 'work'.

Example 17.1

Calculate the potential energy of a drop hammer which has a mass of 1 t and is raised 1.5 m above the pile head before being allowed to drop freely in order to drive it into the ground.

Solution

Potential energy of the hammer relative to the pile is:

$$PE = mgh$$

$$= 100 \text{ kg} \times 9.81 \ \frac{\text{N}}{\text{kg}} \times 1.5 \text{ m}$$

$$= 14\,715 \text{ J}$$

It is interesting to note that a rudimentary idea of potential energy goes back to Galileo who recognised that when a load is lifted with a pulley system, the force applied multiplied by the distance through which that force must be applied, i.e. the work done, remains constant even though the force and distance may vary.

17.3 *KINETIC ENERGY*

In the previous chapter, we found that in order to accelerate a body in linear motion, work must be done on that body. As a result of doing W units of work, the velocity changes, so that:

$$W = \frac{m}{2}(v^2 - v_0{}^2) = \frac{mv^2}{2} - \frac{mv_0{}^2}{2}$$

The quantity $\frac{mv^2}{2}$ was first recognised in the 17th century and was then called *vis viva*, or 'living force'. In the 19th century, it was finally accepted that *vis viva* was not a force but a form of mechanical energy now called 'kinetic energy'.

The **kinetic energy** of a body is the energy which it possesses due to its velocity. Kinetic energy is proportional to the mass and the square of the velocity:

$$\boxed{KE = \frac{mv^2}{2}}$$

Example 17.2

Calculate the kinetic energy of a vehicle of mass 1720 kg, moving with a velocity of 80 km/h.

Solution

Velocity:

$$v = 80 \text{ km/h}$$

$$= 22.2 \text{ m/s}$$

Kinetic energy:

$$KE = \frac{mv^2}{2}$$
$$= \frac{1720 \text{ kg} \times (22.2 \text{ m/s})^2}{2}$$
$$= 424\,700 \text{ J}$$
$$= 424.7 \text{ kJ}$$

Kinetic energy of rotating bodies can also be calculated using rotational analogues of the linear terms mass and velocity, namely mass moment of inertia I and angular velocity ω. Therefore, for a rotating body:

$$KE = \frac{I\omega^2}{2}$$

Example 17.3

Calculate the kinetic energy of a flywheel with mass moment of inertia 61 kg.m^2, and rotating at 250 rpm.

Solution
Angular velocity:

$$\omega = 250 \text{ rpm}$$
$$= 26.18 \text{ rad/s}$$

Hence, kinetic energy stored in the flywheel at this speed is given by:

$$KE = \frac{I\omega^2}{2}$$
$$= \frac{61 \times 26.18^2}{2}$$
$$= 20\,900 \text{ J}$$
$$= 20.9 \text{ kJ}$$

17.4 STRAIN ENERGY

The third form of mechanical energy is the energy of elastic deformation, such as that stored in a stretched or compressed coil spring. This form of energy is called **strain energy**.

Strain energy is the result of work having been done in stretching or compressing the spring against the stiffness of its material to reach its 'strained' condition. Like other kinds of mechanical energy, strain energy is potentially recoverable when the spring is allowed to return to its original free length.

We know that the amount of work required to stretch or compress a simple coil spring from its free length is given by:

$$W = \frac{kx^2}{2}$$

This amount of work is then stored within the material of the spring and becomes its strain energy:

$$SE = \frac{kx^2}{2}$$

where SE is the strain energy, in J, of a coil spring*
  k is the spring modulus in N/m
  x is the amount of elongation or compression in m

Example 17.4

Given that a bumper spring of stiffness 20 N/mm is compressed by 150 mm from its free state, what is the strain energy at the end of compression?

Solution

Note that in order to obtain correct units for energy, it is necessary to use base units when using the formula for strain energy stored in the spring at the end of compression:

$$SE = \frac{kx^2}{2}$$
$$= \frac{20\,000 \text{ N/m} \times (0.15 \text{ m})^2}{2}$$
$$= 225 \text{ J}$$

 Problems

17.1 Determine the potential energy of a lift cage of mass 2.3 t when its position is 35 m above ground level.

17.2 Determine the amount of work to be done if the lift cage in problem 17.1 is to be elevated a further 10 m.

17.3 Calculate the height to which a drop hammer of mass 800 kg must be lifted in order to store 10.6 kJ of potential energy.

17.4 Calculate the kinetic energy of a 200 t train moving at 85 km/h.

17.5 Determine the change in the kinetic energy of the train in the previous problem if its speed is reduced from 85 km/h to 60 km/h.

17.6 Calculate the kinetic energy of a car of mass 1.2 t at 60 km/h, 80 km/h and 100 km/h.
    By comparing the energy stored at these speeds, determine the work done in increasing the speed by 20 km/h:
(a) from 60 km/h to 80 km/h
(b) from 80 km/h to 100 km/h

17.7 A car of mass 950 kg is initially travelling at 60 km/h. If the speed increases so that its kinetic energy increases by 165 kJ, determine the final speed of the car.

---

* The abbreviation SE is used here to stand for 'strain energy'. However, within the limited context of this and subsequent chapters, we may think of it as specifically signifying 'spring energy', i.e. energy stored in a coil spring. The more general aspects of strain, elasticity and stiffness will be discussed in some detail in Chapter 25.

17.8 Calculate the kinetic energy of a flywheel with mass moment of inertia 13.5 kg.m^2 and rotating at 1200 rpm.

17.9 Estimate the mass moment of inertia of a flywheel if a change in kinetic energy corresponding to a change in velocity from 500 rpm to 900 rpm is equal to 324 kJ.

17.10 A steel disc with a 300 mm diameter and 50 mm thick is rotated at 400 rpm. Calculate the change in kinetic energy if the speed is doubled. Take the density of steel as 7800 kg/m^3.

17.11 Calculate the required spring modulus for a buffer spring which must absorb 144 J of energy within the allowable amount of compression not exceeding 240 mm. What force will be required to produce the maximum allowable amount of deformation in the spring?

17.12 Calculate the amount of strain energy stored in a coil spring of 44 N/mm modulus when it is stretched the first 58 mm from its free length. Calculate the total elongation from the free length if an additional amount of 74 J of work was done to further stretch the spring.

Review questions

1. Define *mechanical energy* and state the SI unit of energy.
2. Define and state the formula for potential energy.
3. Define and state the formula for kinetic energy for:
   (a) linear motion
   (b) rotation
4. Define and state the formula for strain energy in a coil spring.

CHAPTER 18

Conservation of energy

Now we are ready to examine one of the most fundamental laws of nature, known as the **principle of conservation of energy**. It simply states that energy can be neither created nor destroyed, but only transformed from one form to another.

Expected learning outcomes

After carefully studying the material presented in this chapter, working through all numerical examples, and successfully completing all practice problems, students should be able to:
1. state the *principle of conservation of energy*;
2. perform calculations involving conversion from one form of mechanical energy into another, for example:
   (a) potential energy into kinetic energy
   (b) kinetic energy into potential energy
   (c) potential energy into strain energy
   (d) kinetic energy into strain energy

18.1 *TRANSFORMATION OF POTENTIAL AND KINETIC ENERGY*

Let us now examine the conservative relation between the most common forms of mechanical energy: potential and kinetic energy.

If we let PE_1 and KE_1 be potential and kinetic energy amounts stored in a body in its *initial* state, and PE_2 and KE_2 be potential and kinetic energy stored in a body in its *final* state, the conservation of energy principle can be stated as follows:

$$PE_1 + KE_1 = PE_2 + KE_2$$

This equation is true only if no external work is done on the body and no loss of mechanical energy occurs due to friction.

Example 18.1

An object of mass 3 kg is dropped from a height of 12 m. Using the conservation of energy principle, calculate the velocity with which it strikes the ground (Fig. 18.1).

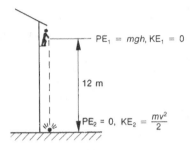

Fig. 18.1

Solution

Before the object is dropped, its kinetic energy KE_1 is zero and its potential energy relative to the ground is:

$$PE_1 = mgh$$
$$= 3 \times 9.81 \times 12$$
$$= 353.2 \text{ J}$$

When it strikes the ground, its height above ground is zero, therefore $PE_2 = 0$. Its kinetic energy is:

$$KE_2 = \frac{mv^2}{2}$$
$$= \frac{3v^2}{2}$$
$$= 1.5v^2$$

Substitute into $PE_1 + KE_1 = PE_2 + KE_2$:

$$353.2 + 0 = 0 + 1.5v^2$$
$$\therefore v = 15.3 \text{ m/s}$$

Example 18.2

A car of a roller coaster has a mass of 500 kg and is released from a height of 18 m at the top of the first incline (Fig. 18.2). Calculate its velocity at the lowest point and also at the top of the second incline which is 5 m below the top of the first incline.

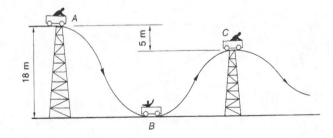

Fig. 18.2

Solution

Calculate potential and kinetic energy at points A, B and C.

$$PE_A = mgh$$
$$= 500 \times 9.81 \times 18$$
$$= 88\,290 \text{ J}$$

$$KE_A = \frac{mv^2}{2}$$
$$= 0 \text{ (since } v = 0)$$
$$PE_B = 0 \text{ (since } h = 0)$$

$$KE_B = \frac{mv^2}{2}$$
$$= \frac{500}{2}v_B^2$$
$$= 250v_B^2$$

$$PE_C = mgh$$
$$= 500 \times 9.81 \times 13$$
$$= 63\,765 \text{ J}$$

$$KE_C = \frac{mv^2}{2}$$
$$= \frac{500}{2}v_C^2$$
$$= 250v_C^2$$

Equating total energy, PE + KE, for all points gives:

$$88\,290 + 0 = 0 + 250v_B^2 = 63\,765 + 250v_C^2$$

from which $v_B = 18.8$ m/s and $v_C = 9.9$ m/s.

In situations involving rotating parts, kinetic energy of rotation must be taken into account when calculating the overall energy balance.

Example 18.3

A metal cylinder of diameter 100 mm and mass 30 kg is allowed to roll freely down an incline as shown in Figure 18.3. If it starts from rest at the top of the incline, what will be its linear and rotational speeds at the bottom of the incline?

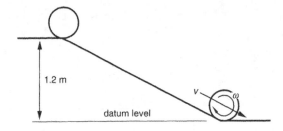

1.2 m

datum level

Fig. 18.3

Solution

When the cylinder begins to roll from the top of the incline, its initial potential energy is:

$$PE_1 = mgh$$
$$= 30 \times 9.81 \times 1.2$$
$$= 353.2 \text{ J}$$

and its initial kinetic energy is zero because it starts from rest.

$$KE_1 = 0$$

When the cylinder reaches the bottom of the incline, its final potential energy is zero:

$$PE_2 = 0$$

The final kinetic energy of the cylinder at the bottom of the incline consists of two parts, due to a combination of linear motion of its centre of mass down the incline, and the rotation about its own axis.

Kinetic energy associated with linear velocity v is given by:

$$KE_2 \text{ (linear)} = \frac{mv^2}{2}$$
$$= \frac{30v^2}{2}$$

Kinetic energy of rotation depends on the mass moment of inertia, which, for a cylindrical object, is found from:

$$I = \frac{mr^2}{2}$$
$$= \frac{30 \times 0.05^2}{2}$$
$$= 0.0375 \text{ kg.m}^2$$

Hence:

$$KE_2 \text{ (rotation)} = \frac{I\omega^2}{2}$$
$$= \frac{0.0375\omega^2}{2}$$

Now the conservation of energy principle can be stated as follows:

$$PE_1 + KE_1 = PE_2 + KE_2 \text{ (linear)} + KE_2 \text{ (rotation)}$$
$$\therefore 353.2 + 0 = 0 + \frac{30v^2}{2} + \frac{0.0375\omega^2}{2}$$

We must recognise that the centre of mass of the cylinder moves down the incline with linear velocity which is related to the angular velocity of its rotation about its axis (see Section 15.1):

$$v = r\omega$$
$$= 0.05\omega$$

Substitution of $v = 0.05\omega$ into the above conservation of energy equation yields:

$$353.2 + 0 = 0 + \frac{30(0.05\omega)^2}{2} + \frac{0.0375\omega^2}{2}$$

Solving this equation for angular velocity gives:

$$\omega = 79.24 \text{ rad/s}$$
$$= 12.6 \text{ revolutions per second}$$

Hence linear velocity:

$$v = 0.05 \times 79.24$$
$$= 3.96 \text{ m/s}$$

18.2 *TRANSFORMATION OF ENERGY INVOLVING SPRINGS*

When interaction of mechanical components involves an elastic member, such as a coil spring, the conservation of energy equation must be extended to allow for the elastic strain energy of the compressed or extended spring where appropriate:

$$\boxed{PE_1 + KE_1 + SE_1 = PE_2 + KE_2 + SE_2}$$

Consider the following two examples of kinetic or potential energy being converted into the energy stored in a spring.

Example 18.4

A locomotive of mass 25 t is to be brought to rest in 125 mm from a velocity of 1.8 km/h by two buffer springs (Fig. 18.4). What must the spring modulus be for each of the two springs? Assume the engine is disengaged.

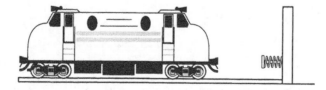

Fig. 18.4

Solution

Velocity before impact is:

$$v = 1.8 \text{ km/h}$$
$$= 0.5 \text{ m/s}$$

The energy levels of the locomotive and spring just before impact are:

$$PE_1 = 0$$

$$KE_1 = \frac{mv^2}{2}$$

$$= \frac{25\,000 \times 0.5^2}{2}$$

$$= 3125 \text{ J}$$

$$SE_1 = 0$$

After the impact, when the locomotive is brought to rest, the springs will be compressed by 0.125 m, and the energy levels will be as follows:

$$PE_2 = 0$$
$$KE_2 = 0$$

$$SE_2 = 2 \text{ springs} \times \frac{k \times 0.125^2}{2}$$

$$= 0.015\,63k$$

Substitute into the conservation of energy equation:

$$PE_1 + KE_1 + SE_1 = PE_2 + KE_2 + SE_2$$
$$0 + 3125 + 0 = 0 + 0 + 0.015\,63k$$

Hence the required spring modulus for each spring is found to be:

$$k = 200 \times 10^3 \text{ N/m}$$
$$= 200 \text{ N/mm}$$

Example 18.5

A block of mass 15 kg falls 350 mm and strikes a coil spring of modulus 10 N/mm (Fig. 18.5). What will be the maximum compression of the spring?

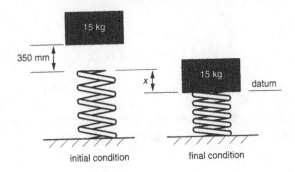

Fig. 18.5

Solution

When the mass comes momentarily to rest and before rebound occurs, the spring will experience its maximum compression x. The lowest position of the mass, corresponding to the maximum compression of the spring, will be $(0.35 + x)$ metres below its initial level. If the lowest position is taken as the datum level, potential energies of the mass will be:

$$PE_1 = mgh$$
$$= 15 \times 9.81 \times (0.35 + x)$$
and $\quad PE_2 = 0$

Kinetic energy is zero in both cases, since the block comes momentarily to rest:

$$\therefore KE_1 = KE_2 = 0$$

The initial strain energy in the spring is zero:

$$SE_1 = 0$$

and final strain energy of the compressed spring is given by:

$$SE_2 = \frac{kx^2}{2}$$
$$= \frac{10\,000x^2}{2}$$
$$= 5000x^2$$

Substitute into the conservation of energy equation and solve for the unknown x:

$$PE_1 + KE_1 + SE_1 = PE_2 + KE_2 + SE_2$$
$$15 \times 9.81 \times (0.35 + x) + 0 + 0 = 0 + 0 + 5000x^2$$
$$51.50 + 147.2x = 5000x^2$$

Rearrange into the usual form of a quadratic equation and solve:

$$5000x^2 - 147.2x - 51.50 = 0$$
$$x = \frac{147.2 \pm \sqrt{147.2^2 - 4 \times 5000 \times (-51.5)}}{2 \times 5000}$$
$$\therefore x_1 = 0.1173 \text{ m}, \, x_2 = -0.0878 \text{ m}$$

From the two alternative mathematical answers obtained when solving the quadratic equation, we must choose the positive one as representing the greatest amount of deformation suffered by the spring at the end of compression.[*] Thus the spring will be compressed by the maximum amount of 117 mm.

---

[*] The alternative negative answer is also meaningful in its own way. In fact it describes the amount by which the spring would be stretched after rebound if the block were to become attached to the spring, as by a hook, during the initial impact. However, since the negative answer does not represent the maximum amount of compression, and does not constitute a direct answer to the question asked, it should be discarded.

18.3 *CONSERVATION OF MECHANICAL ENERGY*

It is now time to summarise and emphasise what has been learned so far about the principle of conservation of energy in its relation to physical objects at rest and in motion, and to systems of interacting mechanical components such as railway trucks and buffer springs.

One of the most fundamental facts in dynamics is that whenever a body or a system of mechanical components is allowed to move, the change of position or configuration results in redistribution of total energy involving transformations between potential energy, kinetic energy and strain energy, sometimes in some combination of all three. In dynamics, as a general rule, we are not concerned with other forms of energy such as chemical, electrical or nuclear energy.

In addition to changes in the three forms of mechanical energy, friction usually causes a portion of the initial amount of energy to be converted into heat and 'lost' by dissipation into the surroundings, no longer available in mechanical form. However, in many idealised cases in engineering mechanics, friction is assumed to be negligible so that the only kinds of energy considered are the three forms of mechanical energy, namely potential, kinetic and strain energy.

In every scientific experiment ever conducted to verify conservation of mechanical forms of energy, in which the effects of friction were eliminated as far as practically possible, it was found that, provided there was no work done on the system by any external force,[*] the sum of its potential, kinetic and strain energies always remained constant.

The mathematical expression of this principle, as it applies to physical objects and to systems of interacting mechanical components, is the one we have been using to solve problems throughout this chapter:

$$PE_1 + KE_1 + SE_1 = PE_2 + KE_2 + SE_2$$

Just remember its three important limitations:

1. Only mechanical forms of energy are considered.
2. Losses of energy due to friction are negligible.
3. There is no work input from any external force.

In its most general formulation, the law of conservation of energy takes into account all possible forms of energy, such as thermal energy of steam, chemical energy of foods and of fuels, electrical energy used for heating, lighting and electrical appliances, and atomic energy hidden within all fundamental particles of matter.[†]

While our primary concern here is with mechanical forms of energy, one should not lose sight of the fact that mechanical energy is only one link in many chains of useful transformations of energy employed in engineering. For example, potential energy of water in a dam is converted into pressure and kinetic energy of water flowing in a pipe leading to a turbine, which is then converted into work output, which in turn is used to

---

[*] The force of gravity is not regarded here to be an external force because its influence has already been embodied within the potential energy term.

[†] In nuclear physics, according to Einstein's theory of relativity, matter is itself regarded as a form of energy, and the mass of an atomic particle as a measure of its energy content. Therefore, when energy is said to be 'created' during a nuclear reaction, i.e. released by annihilating matter, the otherwise separate laws of conservation of mass and energy merge with each other. However, this is well outside the scope of everyday mechanical engineering, and therefore of this book.

drive an electric generator. The electric energy is delivered through a distribution network to its user devices, such as pumps, fans, electric trains, refrigerators, machine tools and heaters; these transform the electric energy into other forms, many of which are again mechanical in nature.

In summary, the law of conservation of energy states that energy can never be created or destroyed — it can only be transformed from one kind into another.

 # Problems

In the following problems assume that friction and other resistances are negligible.

18.1 A drop hammer is raised to a height of 3 m above its striking point and released. Calculate its velocity at the instant of striking.

18.2 A ball is kicked vertically upwards with a velocity of 35 m/s. Determine the greatest height reached. (Compare with problem 13.14.)

18.3 A swinging pendulum of an Izod impact testing machine falls freely through a vertical height of 950 mm and then strikes a specimen at the lowest point of its travel.
Determine the kinetic energy and the velocity of the 7 kg pendulum on impact.

18.4 A 40 kg boy has a velocity of 5.5 m/s as he passes the lowest position on a swing. If the effective length of the rope is 4.5 m, determine the angle it makes with the vertical when the boy reaches the maximum height.

18.5 If the handbrakes of a car, parked on a 15° incline, fail and the car rolls freely 100 m down the slope before hitting a tree, determine the velocity on impact.

18.6 A semitrailer with disabled brakes is forced to use a safety ramp to avoid an accident. If the ramp is 100 m long at 20° to the horizontal, and the vehicle has a velocity of 25 km/h when it reaches the top of the ramp, what was its initial velocity?

18.7 A box slides down a smooth chute with an initial velocity of 2.9 m/s. If the chute is 10 m long and inclined at 35° to the horizontal, determine the velocity at the bottom of the chute.

18.8 Two masses of 5 kg and 3 kg are connected by a fine string passing over a smooth pulley. Determine the velocity with which the masses will be moving after each mass has moved 1 m from rest. (*Hint*: The total energy of the system is the sum of the energies of its component parts.)

18.9 A small projectile of mass 28 g is fired vertically upwards from a spring-operated gun. Just before the trigger is released, the spring of stiffness $k = 5$ N/mm is compressed by 100 mm from its free length. Calculate the maximum height reached by the projectile.

18.10 If the gun in the previous problem is fired horizontally, what is the velocity of the projectile at the point where the spring reaches its free length?
If this gun were fired vertically upwards, would the corresponding velocity be significantly different?

18.11 A spring of stiffness 0.85 N/mm is supported at the top and has a hook at the bottom end. A mass of 2.6 kg is attached to the hook and then released causing the spring to stretch. Calculate the amount by which the spring will be stretched if:
(a) the mass is lowered very gradually until its weight is fully supported by the spring;
(b) the mass is released suddenly, as soon as the hook is engaged.

18.12 A trolley of mass 58 kg, starting from rest, rolls freely 2.5 m down a 19° ramp and then strikes a buffer spring, which has a stiffness modulus of 25 N/mm. Calculate:
(a) the velocity of the trolley just before it makes contact with the spring;
(b) the amount by which the spring will be compressed.

Review questions

1. State the principle of the *conservation of energy.*
2. State the mathematical expression which embodies the principle of the conservation of mechanical energy.
3. State three important limitations of this mathematical expression and discuss their implications.
4. Give some practical examples of how one form of mechanical energy is transformed into another form of mechanical energy.
5. Give some examples of transformation of mechanical energy into non-mechanical form, and vice versa.

PART 7

DYNAMIC SYSTEMS

≡

When we mean to build, we first survey the plot, then draw the model.

William Shakespeare
King Henry IV

CHAPTER 19

The force–acceleration method

This chapter is part of the section of our work that deals with systems of dynamically interacting bodies, which may also be influenced by external forces and various forms of resistance. More particularly, this chapter is devoted to the analysis of such systems based on considerations following directly from Newton's second law, expressed mathematically in the form of two fundamental formulae:

$$F = ma \quad \text{for linear motion}$$
$$T = I\alpha \quad \text{for rotation}$$

We shall call this approach the **force–acceleration method** because it places considerable emphasis on the relation between acceleration and effort (force or torque).

Expected learning outcomes

After carefully studying the material presented in this chapter, working through all numerical examples, and successfully completing all practice problems, students should be able to:

1. identify the relation between accelerations of two components that are constrained to move together;
2. set up simultaneous equations based on the force–acceleration and/or torque–acceleration relations for a system of two interconnected components;
3. solve problems in which accelerations of individual components and/or tension in the rope connecting them must be determined.

19.1 *SYSTEMS OF BODIES IN LINEAR MOTION*

Mechanical devices and simple machines usually consist of two or more components constrained to move together, a situation that can be described as 'a system of bodies in motion'.

Such systems often contain cords or cables and pulleys used to connect the components and to transmit or change the direction of force and motion. It is common practice to neglect the weights and frictional resistances of cords and pulleys, unless specific information is available to describe their effect.

We will limit our discussion to systems involving two separate masses connected in a way which forces them to move simultaneously and imposes a fixed ratio on the distances moved by each mass.

The method of solving problems of this kind consists of first determining the relation between aspects of the motion of the two masses, which depends on how they are interconnected, and then considering each mass as a separate free body. This produces a system of simultaneous equations, which can readily be solved.

Example 19.1

For the system of bodies shown in Figure 19.1(a), determine the acceleration and the force in the cord.

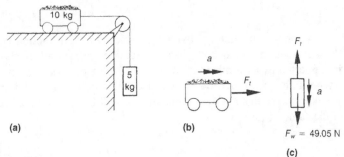

(a)   **(b)**

$F_w = 49.05$ N

Fig. 19.1   **(c)**

Solution

When the 5 kg mass accelerates downwards in this system, the 10 kg mass will accelerate to the right at the same rate because they are simply connected by the cord, running over a pulley. Let us call this acceleration a. Let us also call the tension in the cord F_t. We are now ready to consider each mass as a separate free body.

For the 10 kg mass we can write:

$$F = ma$$
$$\text{or} \quad F_t = 10 \text{ kg} \times a \tag{1}$$

For the 5 kg mass, the net accelerating force is the difference between its weight and the tension in the cord. Therefore, $F = ma$ becomes:

$$F_w - F_t = ma$$
$$\text{or} \quad 5 \times 9.81 - F_t = 5 \times a \tag{2}$$

Solving equations 1 and 2 yields:

$$a = 3.27 \text{ m/s}^2$$
$$\text{and} \quad F_t = 32.7 \text{ N}$$

Example 19.2

Determine the accelerations of bodies A and B, and the force of tension in the cord, for the system in Figure 19.2(a).

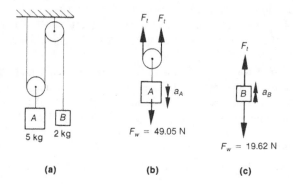

(a)   (b)   (c)

Fig. 19.2

Solution

In this case, it is not immediately obvious in which direction motion will occur. This is decided by considering which body has a greater mass per fall of cord supporting it.* Body A is supported by two falls of cord and has a mass of 5 kg, i.e. 2.5 kg per fall. Body B has 2 kg per fall. Therefore, body A will accelerate downwards and body B upwards.

It is not hard to see that the displacement of A will be one-half that of B. Acceleration will be in the same ratio:

$$a_A = \frac{a_B}{2}$$

---

* This rule is based on the principle that the motion, if any, will take place in the direction of the external pull that produces the greatest static tension in the cord, when all but one body in turn are held immovable while the static tension in the connecting cord is determined.

Consider free-body diagrams:

For body B (Fig.19.2(c)):

$$F = m \times a$$
$$(F_t - F_{wB}) = m_B \times a_B$$
$$F_t - 2 \times 9.81 = 2 \times a_B \tag{1}$$

For body A (Fig. 19.2(b)):

$$F = m \times a$$
$$(F_{wA} - 2F_t) = m_A \times a_A$$
$$5 \times 9.81 - 2F_t = 5 \times a_A$$

Substitute $a_A = \dfrac{a_B}{2}$:

$$5 \times 9.81 - 2F_t = \frac{5 \times a_B}{2} \tag{2}$$

After equations 1 and 2 are simplified, we have:

$$F_t - 19.62 = 2a_B$$
$$49.05 - 2F_t = 2.5a_B$$

Multiplying the first equation by 2, and then adding the left-hand and right-hand sides respectively, eliminates F_t as follows:

$$+ \quad \begin{array}{c} 2F_t - 39.24 = 4a_B \\ 49.05 - 2F_t = 2.5a_B \\ \hline 9.81 = 6.5a_B \end{array}$$

$$\therefore a_B = \frac{9.81}{6.5}$$
$$= 1.51 \text{ m/s}^2$$

It follows that:

$$a_A = \frac{a_B}{2}$$
$$= \frac{1.51}{2}$$
$$= 0.755 \text{ m/s}^2$$

and:

$$F_t = 2a_B + 19.62$$
$$= 2 \times 1.51 + 19.62$$
$$= 22.6 \text{ N}$$

Problems

19.1 Determine the horizontal pull required to accelerate two wagons, each of mass 400 kg, along a level surface at the rate of 0.4 m/s², and the force in the bar connecting the two wagons (Fig. 19.3).

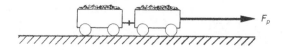

Fig. 19.3

19.2 A mass of 5 kg resting on a smooth horizontal table is connected, by a fine string passing over a smooth pulley on the edge of the table, with a mass of 100 g hanging freely. How far will the mass of 100 g descend in 3 s, starting from rest?

19.3 The trolley *A* shown in Figure 19.4 has a mass of 25 kg and rolls with negligible resistance. Determine the acceleration produced by mass *B*, equal to 1 kg, and the tension in the cord.

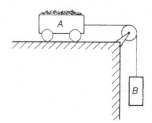

Fig. 19.4

19.4 If the friction in the wheels of the trolley in problem 19.3 increases from lack of lubrication to produce a total tractive resistance of 3 N, what additional mass should be placed on block *B* to maintain the same rate of acceleration as in the previous problem?

19.5 In Figure 19.5, block *A* has a mass of 15 kg and the coefficient of friction between the block and the horizontal surface is 0.3. What is the mass of block *B* that is required to accelerate the system at 2.5 m/s²?

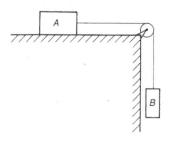

Fig. 19.5

19.6 Two masses of 1.0 kg and 1.1 kg are suspended by a fine string passing over a smooth pulley. With what acceleration will the masses move?

19.7 A 70 kg man hoists himself on a bosun's chair as shown in Figure 19.6. If the pull exerted by the man on the rope is 300 N, what is the acceleration?

Fig. 19.6

19.8 In the arrangement shown in Figure 19.7, the loaded trolley, having a total mass of 100 kg, can remain at rest or move without acceleration. Calculate the mass of the counterweight attached to the rope.

  If the mass of the trolley when empty is 40 kg, how long will it take to reach a velocity of 4 m/s starting from rest? Neglect frictional resistance.

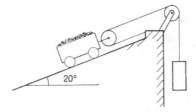

Fig. 19.7

19.2 *SYSTEMS CONTAINING ROTATING COMPONENTS*

Let us now consider systems of bodies in motion which include rotating components. Our discussion will be limited to a pair of components constrained to move simultaneously and connected in a way that imposes a definite relation between their motions.

  The method of solving these problems is similar to the solution of systems of bodies in linear motion. It consists of first determining the relationship between aspects of motion of the two bodies, which depends on how they are interconnected, and then considering each component as a separate free body. This usually results in a system of simultaneous equations which can readily be solved.

Example 19.3

The drum in Figure 19.8(a) has a mass moment of inertia of 25 kg.m². A mass of 2 kg is attached to the cord which is wrapped around the drum. Neglecting frictional resistance, determine the time taken for the mass to drop 2 m after being released from rest, and the tension in the cord.

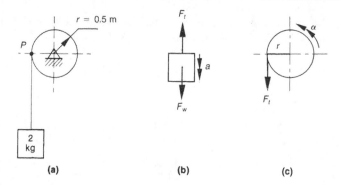

(a)   (b)   (c)

Fig. 19.8

Solution

When the 2 kg mass accelerates downwards, the drum will accelerate in an anticlockwise direction at the rate related to the acceleration of the mass, because they are connected by the cord.

Point P on the cord has an instantaneous linear acceleration a downwards, and an angular acceleration α. These are related by:

$$a = r\alpha$$
$$\text{or} \quad a = 0.5\alpha \qquad (1)$$

For the 2 kg mass as a free body, we can write:

$$F = ma$$
$$F_w - F_t = ma$$

where F_w is the weight, i.e. $2 \text{ kg} \times 9.81 \ \dfrac{\text{N}}{\text{kg}} = 19.62 \text{ N}$

F_t is the tension in the cord

Therefore:

$$19.62 - F_t = 2a \qquad (2)$$

Likewise, for the drum, we write:

$$T = I\alpha$$

where T is the torque, i.e. $F_t \times r = F_t \times 0.5$

$I = 25 \text{ kg.m}^2$

$$\text{or} \quad F_t \times 0.5 = 25\alpha \qquad (3)$$

These three equations can easily be solved. For example, substitute $a = 0.5\alpha$ from equation 1 into equation 2:

$$19.62 - F_t = 2 \times (0.5\alpha)$$

Now divide this new equation by equation 3:

$$\frac{19.62 - F_t}{0.5F_t} = \frac{2 \times 0.5\alpha}{25\alpha}$$

Simplifying and solving for F_t gives the tension:

$$F_t = 19.24 \text{ N}$$

Substituting back into equation 2 gives:

$$19.62 - 19.24 = 2a$$
$$\therefore a = 0.192 \text{ m/s}^2$$

Now $S = v_0 t + \dfrac{at^2}{2}$ can be used to solve for time taken, since the distance dropped is $S = 2$ m and the initial velocity is zero:

$$2 = 0 + \frac{0.192t^2}{2}$$

Hence time taken is 4.56 seconds.

 # Problems

19.9 A mass of 12 kg is attached to a cord wrapped around a horizontal drum of diameter 0.8 m. The drum has a mass moment of inertia of 18 kg.m². Neglecting friction, determine the tension in the cord and the downward acceleration of the mass after its release from rest.

19.10 If, in the previous problem, bearing friction is equivalent to a resistance torque of 9 N.m, what will the acceleration of the mass be?

19.11 A hoist mechanism consists of a 450 mm diameter drum and a 950 mm diameter brake cylinder as shown in Figure 19.9. The mass moment of inertia of the rotating parts is 85 kg.m² and the coefficient of friction between the brake shoe and the cylinder is 0.6. Bearing friction is negligible.

   Determine the normal force F_n which must be applied to each shoe in order to lower a load of 400 kg at constant speed.

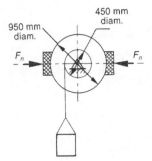

Fig. 19.9

19.12 If the brakes in the previous problem fail when the load is 25 m above the ground and is moving down with a velocity of 2.4 m/s, determine the downward acceleration of the load and the velocity with which it will strike the ground.

19.13 If the hoist in problem 19.11 is used to lift a load of 600 kg, calculate the torque required to give the load an upward acceleration of 1 m/s^2.

19.14 A gear train consists of a pinion with pitch-circle diameter 150 mm and mass moment of inertia 0.1 kg.m^2, and a large gear with diameter 450 mm and mass moment of inertia 1.2 kg.m^2.

Determine the torque which must be applied to the pinion in order to accelerate the large gear from rest to an angular velocity of 955 rpm in 20 s.

Review questions

1. What is meant by a *system of bodies in motion*?
2. Explain how mechanical components can be constrained to move together.
3. Describe the *force–acceleration method* of solving problems involving a system of two bodies in linear motion.
4. How different is the application of this method when rotating components are involved?

CHAPTER 20

The work–energy method

Like the previous chapter, this one is also part of that section of our work which deals with dynamically interacting components. In some problems, the system will comprise two distinctly separate components constrained to move together. In others, the influence of one part of the system on another will be expressed as work input or as work done against some form of resistance.

However, the emphasis in this chapter is on considerations of changes in the total energy possessed by a component, or by a system of two interacting components, when influenced by energy input or output in the form of work. Mathematically, this principle can be generalised as follows:

Initial energy ± work = final energy

This approach to problem solving in mechanics is referred to as the **work-energy method**.

Expected learning outcomes

After carefully studying the material presented in this chapter, working through all numerical examples, and successfully completing all practice problems, students should be able to:
1. identify and calculate all appropriate energy levels for a mechanical component or a system of components, which may include potential energy, kinetic energy and strain energy;
2. identify and calculate appropriate work inputs or outputs;
3. use the work–energy method for calculating an unknown distance, velocity, elevation, or amount of compression in a spring.

20.1 *THE WORK–ENERGY METHOD*

The conservation of energy principle can be extended to situations where the external work done on a body and the work done against friction cannot be neglected and must form a part of the total energy account. If we let:

$$(PE_1 + KE_1) = \text{initial total energy of the body}$$
$$(PE_2 + KE_2) = \text{final total energy of the body}$$
$$\pm W = \text{net work done on the body}$$

then the final energy of the body is equal to the initial energy of the body increased by the amount of net work done on the body:

$$(PE_1 + KE_1) \pm W = (PE_2 + KE_2)$$

The net work is the difference between positive work done on the body by forces acting in the direction of motion, and negative work done by forces, such as friction, which are opposing the motion.

Weight should not be included when determining the external work, since its effect has been allowed for in terms of potential energy.

Example 20.1

A 180 t train climbs an incline of 1.5° for 2 km. Its initial velocity before the climb is 90 km/h. The tractive effort exerted by the engine is 53.2 kN, and tractive resistance is 95 N/t. Determine the final speed after the climb (Fig. 20.1).

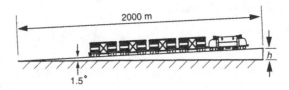

Fig. 20.1

Solution

Calculate all relevant terms for the work–energy equation:

Initial potential energy:
$$PE_1 = 0$$

The final elevation:
$$h = 2000 \text{ m} \times \sin 1.5°$$
$$= 52.35 \text{ m}$$

Final potential energy:
$$PE_2 = mgh$$
$$= 180\,000 \text{ kg} \times 9.81 \frac{N}{kg} \times 52.35 \text{ m}$$
$$= 92.45 \text{ MJ}$$

Postitive work:
$$\text{Effort} \times \text{distance} = 53\,200 \text{ N} \times 2000 \text{ m}$$
$$= 106.4 \text{ MJ}$$

Tractive resistance:
$$95 \frac{N}{t} \times 180 \text{ t} = 17\,100 \text{ N}$$

Negative work:

$$\text{Resistance} \times \text{distance} = 17\,100 \text{ N} \times 2000 \text{ m}$$
$$= 34.2 \text{ MJ}$$

Net work:

$$W = \text{pos.}\,W - \text{neg.}\,W$$
$$= 106.4 - 34.2$$
$$= 72.2 \text{ MJ}$$

Initial velocity:

$$v_1 = 90 \text{ km/h}$$
$$= 25 \text{ m/s}$$

Initial kinetic energy:

$$KE_1 = \frac{mv^2}{2}$$
$$= \frac{180\,000 \text{ kg} \times (25 \text{ m/s})^2}{2}$$
$$= 56.25 \text{ MJ}$$

Substitute into:

$$(PE_1 + KE_1) \pm W = (PE_2 + KE_2)$$
$$0 + 56.25 + 72.20 = 92.45 + KE_2$$

Hence, final kinetic energy is:
$$KE_2 = 36 \text{ MJ}$$

Final velocity can now be calculated:

$$\frac{mv^2}{2} = 36 \text{ MJ}$$
$$= 36 \times 10^6 \text{ J}$$
$$\text{or} \quad v = \sqrt{\frac{36 \times 10^6 \times 2}{180\,000}}$$
$$= 20 \text{ m/s or } 72 \text{ km/h}$$

The work–energy method can be applied to rotational as well as linear motion. Furthermore, it can be used successfully when more than one moving component is involved. The energy of a system of connected bodies, whose motions are related, is equal to the sum of the energies of the separate bodies.

Example 20.2

A block of mass 10 kg is attached to a cord which is wrapped around a drum of diameter 800 mm and mass moment of inertia 15 kg.m². After the block is released and has dropped a distance of 2.5 m, its velocity is 2 m/s. Determine the magnitude of the bearing-friction torque which resists rotation (Fig. 20.2).

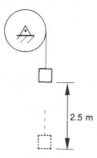

2.5 m

Fig. 20.2

Solution

Using the lowest position of the block as a datum:

$$PE_1 = mgh_1$$
$$= 10 \times 9.81 \times 2.5$$
$$= 245.3 \text{ J}$$
$$PE_2 = 0$$
$$KE_{1(\text{block})} = 0$$
$$KE_{1(\text{drum})} = 0$$

$$KE_{2(\text{block})} = \frac{mv^2}{2}$$
$$= \frac{10 \times 2^2}{2}$$
$$= 20 \text{ J}$$

Final angular velocity of the drum:

$$\omega_2 = \frac{v_2}{r}$$
$$= \frac{2}{0.4}$$
$$= 5 \text{ rad/s}$$
$$KE_{2(\text{drum})} = \frac{I\omega^2}{2}$$
$$= \frac{15 \times 5^2}{2}$$
$$= 187.5 \text{ J}$$

Substitute into:

$$(PE_1 + KE_{1(\text{block})} + KE_{1(\text{drum})}) + W = (PE_2 + KE_{2(\text{block})} + KE_{2(\text{drum})})$$
$$245.3 + 0 + 0 + W = 0 + 20 + 187.5$$
$$\therefore W = -37.8 \text{ J}$$

This is negative work done by resistance torque equal to $W = T \times \theta$, where:

$$\theta = \frac{S}{r}$$

$$= \frac{2.5 \text{ m}}{0.4 \text{ m}}$$

$$= 6.25 \text{ rad}$$

Substitute:

$$37.8 = T \times 6.25$$
$$\therefore T = 6.05 \text{ N.m}$$

Hence bearing friction torque T is 6.05 N.m.

 Problems

20.1 A 25 kg box is pushed 10 m along a horizontal surface by a horizontal force of 80.3 N. If the box acquires a velocity of 1.2 m/s at the end of its travel, determine the coefficient of friction between the box and the surface.

20.2 A car of mass 1.4 t is travelling along a level road when the brakes are suddenly applied, causing the car to skid and come to rest in 51.6 m.
   Assuming the coefficient of friction between the road and the tyres is 0.5, calculate the initial speed of the vehicle.

20.3 A 50 kg block is pushed up an inclined plane, at 25° to the horizontal, by a force of 450 N parallel to the incline. The coefficient of friction is 0.35.
   Determine the distance measured along the plane in which the velocity will change from 1 m/s to 3 m/s.

20.4 Calculate the torque that must be applied to a flywheel of mass moment of inertia 475 kg.m² in order to accelerate it to 650 rpm from rest in 220 turns. The bearing friction is 63 N.m.

20.5 A flywheel of mass moment of inertia 30 kg.m² has constant torque of 170 N.m applied to it in order to accelerate it from 200 rpm to 1000 rpm while it makes a total of 180 revolutions. Calculate the bearing friction torque.

20.6 A mass of 16 kg, resting on a smooth horizontal table, is connected by a fine string passing over a smooth pulley on the edge of a table, with a mass of 3 kg hanging freely.
   With what velocity will the 3 kg block strike the floor if it is released from a height of 1 m above floor level?

20.7 What additional mass must be added to the 3 kg mass in the previous question in order to increase the velocity with which the mass strikes the ground to 2 m/s?

20.8 A system of a 120 kg empty trolley and a 100 kg counterweight is as shown in Figure 20.3.

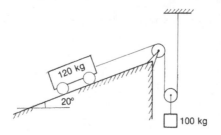

Fig. 20.3

After the counterweight has dropped 5 m from rest, the velocity of the trolley is 3 m/s. If all frictional resistances can be regarded as a single force F acting on the trolley parallel to the plane, calculate the resistance force.

20.9 For the system in problem 20.8, use the work–energy principle to calculate the correct load that can be:
(a) raised in the trolley
(b) lowered in the trolley
at constant velocity.

Assume that the resistance to motion is proportional to the mass of the loaded trolley.

20.10 A load of 150 kg is being lowered with a velocity of 5 m/s, as shown in Figure 20.4. The mass moment of inertia of the rotating assembly is 37.5 kg.m^2.

Determine the force F required to bring the load to rest after travelling an additional 8 m if the coefficient of friction is 0.45.

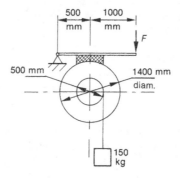

Fig. 20.4

20.2 *THE WORK–ENERGY METHOD INVOLVING SPRINGS*

Although in the previous section no mention was made of elastic strain energy, the energy stored in a compressed spring can easily be included as an additional term in the work–energy equation, as follows:

$$(PE_1 + KE_1 + SE_1) \pm W = (PE_2 + KE_2 + SE_2)$$

Example 20.3

A railway truck of mass 8 t rolls 20.5 m down a 3° incline, starting with an initial velocity of 4.2 m/s, and continues on a level track for 63 m before striking a spring buffer. The total tractive resistance is 1320 N. What is the spring stiffness if the spring is compressed by 400 mm?

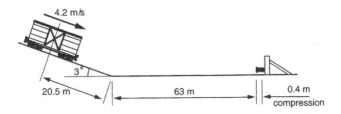

Fig. 20.5

Solution
The energy levels of the truck and spring system in its initial condition are:

$$PE_1 = mgh$$
$$= 8000 \times 9.81 \times 20.5 \sin 3°$$
$$= 84\ 200\ J$$

$$KE_1 = \frac{mv^2}{2}$$
$$= \frac{8000 \times 4.2^2}{2}$$
$$= 70\ 560\ J$$

$$SE_1 = 0 \text{ (the spring is not compressd at this stage)}$$

The final energies, when the truck is brought to rest, are:

$$PE_2 = 0$$
$$KE_2 = 0$$

$$SE_2 = \frac{kx^2}{2}$$
$$= \frac{k \times 0.4^2}{2}$$
$$= 0.08k$$

where k is the unknown spring constant.

The work done against tractive resistance is:

$$W = \text{force} \times \text{distance}$$
$$= 1320 \times (20.5 + 63.0 + 0.4)$$
$$= 110\ 750\ J$$

This work is negative, as it is done against frictional resistance and represents the amount of mechanical energy lost from the system. This energy loss occurs by virtue of conversion of mechanical energy into heat, which is then dissipated into the surroundings, thus diminishing the amount of mechanical energy left in the system.

We can now write the energy balance as follows:

$$PE_1 + KE_1 + SE_1 \pm W = PE_2 + KE_2 + SE_2$$

Substitution yields:

$$84\ 200 + 70\ 560 + 0 - 110\ 750 = 0 + 0 + 0.08k$$
$$\therefore k = 550 \text{ N/mm}$$

Hence the spring stiffness corresponding to maximum compression of 400 mm is $k = 550$ N/mm.

 ## Problems

20.11 When the railway truck of mass 8 t was brought to rest against a spring buffer, as described in Example 20.3, the spring of stiffness 550 N/mm was compresssed by 400 mm. Assuming that total tractive resistance remains unchanged at 1320 N, calculate the horizontal distance the truck will rebound before coming to a final stop.

20.12 A spring of stiffness $k = 20$ N/mm is compressed by 25 mm behind a 2.5 kg block, which is held stationary on a horizontal surface. If the coefficient of kinetic friction between the block and the surface is 0.17, determine the maximum distance the block will slide along the surface when released.

20.13 The 18 kg block shown in Figure 20.6 slides 2.5 m from rest down the 30° incline and then strikes a spring of stiffness 2.7 N/mm. The coefficient of friction between the block and the surface of the inclined plane is 0.2. Calculate the maximum amount by which the spring will be compressed.

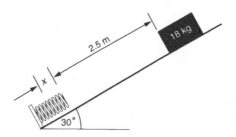

Fig. 20.6

20.14 Calculate the distance that the block in problem 20.13 rebounds up the inclined plane.

20.15 Rotation of a small drum of mass moment of inertia 0.44 kg.m² and with frictional resistance torque 0.3 N.m is initiated by a string wound around its 20 mm diameter shaft and attached to a stretched spring, as shown in Figure 20.7. If the spring, which has a stiffness 2 N/mm, is initially stretched by 120 mm before the drum is released from rest, determine the maximum angular velocity, in revolutions per minute, attained by the drum.

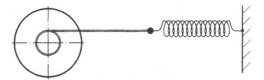

Fig. 20.7

Review questions

1. Briefly outline the *work–energy method* of problem solving.
2. Distinguish between *positive work* and *negative work*.
3. What is the most common cause of negative work?
4. State the general formula for work–energy calculations.
5. What forms of energy does it cover?

CHAPTER 21

The impulse–momentum method

In this chapter we introduce a third method of dealing with systems of dynamically interacting components. This method is based on the concept of linear momentum, and on the related physical quantity called impulse. For this reason, we refer to this way of problem solving as the impulse–momentum method.

Momentum is a concept used in mechanics for the solution of certain types of problems which are difficult to solve by the force–acceleration or by the work–energy methods introduced earlier. These problems usually involve direct relations between force, mass, velocity and time.

Expected learning outcomes

After carefully studying the material presented in this chapter, working through all numerical examples, and successfully completing all practice problems, students should be able to:

1. calculate the momentum of a body of known mass moving with a given velocity, and relate a change in momentum to the impulse of a force;
2. solve problems using the principle of conservation of linear momentum;
3. define *coefficient of restitution*;
4. solve problems involving a direct central impact between two bodies.

21.1 *MOMENTUM*

Momentum, sometimes described as the quantity of motion, is the product of the mass m of a body and its velocity v at any given instant.[*] No special symbol is used here for momentum, hence:

$$\boxed{\text{Momentum} = mv}$$

Unlike energy, momentum is a vector quantity, i.e. it has a direction that corresponds to the direction of the velocity. The unit of momentum is the kilogram metre per second, kg.m/s, when mass and velocity are expressed in SI base units.

---

[*] In this book, only the linear momentum will be considered. A similar concept, called **angular momentum**, may also be used in rotational dynamics where it is expressed in rotational terms.

Example 21.1

A rocket of mass 2.5 t is fired vertically upwards with a velocity of 250 m/s. What is its momentum?

Solution

$$\text{Momentum} = m \times v$$
$$= 2500 \text{ kg} \times 250 \text{ m/s}$$
$$= 625\ 000 \text{ kg.m/s upwards}$$

According to Newton's first law of motion, the velocity of a body does not change unless an external force is applied to change the velocity. This law therefore implies conservation of momentum in the absence of an external force.

This principle enables us to solve certain types of problems directly, where the use of other methods is not convenient.

Example 21.2

A block of wood of mass 2 kg is freely suspended on a string. A bullet of mass 75 g is fired horizontally into the block. If the velocity of the bullet before impact is 415 m/s, calculate the velocity of the block, with the bullet embedded in it, immediately after the impact.

Solution

Momentum before impact:

$$\text{Bullet:} \quad 0.075 \text{ kg} \times 415 \text{ m/s} = 31.13 \text{ kg.m/s}$$
$$\text{Block:} \qquad\qquad\qquad\qquad = 0$$
$$\therefore \text{ Total before impact} = 31.13 \text{ kg.m/s}$$

Momentum after impact:

$$\text{Bullet and block:} \quad (2 + 0.075) \text{ kg} \times v = 2.075v$$

Conservation of momentum requires that the momentum after impact be equal to the momentum before impact:

$$\therefore 2.075v = 31.13$$
$$v = 15 \text{ m/s}$$

Hence velocity immediately after impact is 15 m/s.

21.2 IMPULSE

According to Newton's second law of motion, force and acceleration are related by the formula $F = ma$. If a constant force F is applied during a time interval t, the acceleration produced by the force is:

$$a = \frac{v - v_0}{t}$$

and substitution yields:

$$F = m\left(\frac{v - v_0}{t}\right)$$

This can be rearranged as follows:

$$\boxed{Ft = mv - mv_0}$$

The right-hand side of this equation can be recognised as the change of momentum from mv_0 to mv, where v_0 and v are the initial and final velocities, respectively, of the body on which force F is acting.

The product of the force F and the time t during which it acts is called **impulse**. Because it contains force, impulse is a vector quantity. The unit of impulse is the newton second, N.s.

Example 21.3

When a golf ball of mass 50 g is struck by a club, the ball and club are in contact for 0.001 s. Immediately after impact, the ball travels at 45 m/s. Determine the average force of the collision.

Solution

Momentum before impact:
$$mv_0 = 0$$

Momentum after impact:
$$mv = 0.05 \text{ kg} \times 45 \text{ m/s}$$
$$= 2.25 \text{ kg.m/s}$$

Substitute into $Ft = mv - mv_0$:

$$F \times 0.001 = 2.25 - 0$$
$$\therefore F = 2250 \text{ N}$$
$$= 2.25 \text{ kN}$$

The concept of impulse is also helpful in calculating thrust developed by a continuous flow of fluids, such as exhaust gases from a rocket, steam from a nozzle, or water from a garden hose.

Example 21.4

The exhaust gases from a rocket have a velocity of 1200 m/s and flow at the rate of 5 kg/s. Determine the thrust produced by the gases.

Solution

The exhaust jets accelerate the gases from rest to 1200 m/s.

Initial momentum:

$$mv_0 = 0$$

Final momentum:

$$mv = 5 \text{ kg} \times 1200 \text{ m/s}$$
$$= 6000 \text{ kg.m/s}$$

Substitute into $Ft = mv - mv_0$:

$$F \times 1 \text{ s} = 6000 \text{ kg.m/s} - 0$$
$$\therefore F = 6000 \text{ N}$$
$$= 6 \text{ kN}$$

Hence the thrust is 6 kN.

Problems

21.1 A car of mass 1.4 t is travelling at 60 km/h. Calculate its momentum.

21.2 Determine the change in the momentum of the car in the previous problem if its velocity:
(a) increases by 20 km/h
(b) decreases by 20 km/h

21.3 A vehicle of mass 3.5 t is moving with a velocity of 90 km/h. Determine how long it will take to bring it to rest with a braking effort of 7 kN.

21.4 A gun fires a shell of mass 8 kg in a horizontal direction with a velocity of 375 m/s. The mass of the gun is 2 t and the shell takes 0.012 s to leave the barrel. Calculate the velocity of recoil and the average force on the gun.

21.5 Two men, one of mass 80 kg and the other 60 kg, sit facing each other in two light boats and holding the ends of a rope between them. If they pull with a force of 50 N for 4 s, calculate the velocity acquired by each boat. Neglect water resistance and the mass of the boats.

21.6 An inflated balloon contains 12.8 g of air which is allowed to escape from a nozzle with a velocity of 17.2 m/s. If the balloon deflates at a steady rate in 6.2 s, determine the force exerted on the balloon.

21.7 A rocket of mass 6 t is to be launched vertically. If the flow rate of the gas is 100 kg/s, determine the minimum velocity of the gas to just lift the rocket off the launching pad.
    If the velocity is 900 m/s, what is the net upward accelerating force on the rocket?

21.8 A railway wagon of mass 14 t travelling at 18 km/h collides with a second wagon of mass 12 t which is at rest. If immediately after the collision both wagons travel on coupled together, calculate their common velocity.

21.9 A ballistic pendulum consisting of a large wooden block of mass 3 kg is suspended by cords. When a bullet of mass 7.5 g and unknown velocity v is fired horizontally into it, the block with the bullet embedded in it swings, rising a maximum vertical distance of 200 mm.
    Calculate the increase in the potential energy of the system and the kinetic energy immediately after impact. Hence determine the velocity of the bullet using the conservation of momentum principle.

21.10 A drop hammer of mass 120 kg falls 2.5 m onto a pile of mass 250 kg and drives it 70 mm into the ground. Calculate:
- **(a)** the velocity with which the hammer strikes the pile (use the conservation of energy principle);
- **(b)** the velocity immediately after the impact (use the conservation of momentum principle, assuming the hammer does not rebound on impact);
- **(c)** the average ground resistance (use the work–energy method).

21.3 *IMPACT*

An **impact** is a collision between two bodies that occurs in a very short interval of time and involves relatively large forces which the two bodies exert on each other. We will only consider direct central impact, i.e. the kind of collision in which the two bodies move along the same straight line before and after the collision.

The conservation of momentum principle applies during impact, which enables us to write an equation relating total momentum before impact to total momentum after impact. If we let m_A and m_B be the masses of bodies A and B, and v_{0A} and v_{0B} be their initial velocities, the total momentum before impact is given by:

$$m_A v_{0A} + m_B v_{0B}$$

Similarly, if v_A and v_B are the final velocities, the total momentum after impact is:

$$m_A v_A + m_B v_B$$

For the system, initial momentum equals final momentum:

$$m_A v_{0A} + m_B v_{0B} = m_A v_A + m_B v_B$$

This equation, by itself, is not sufficient for the solution of problems involving the impact between two bodies, as it contains two final velocities which are not usually known.

We have to examine the effect of deformation and subsequent restitution of the colliding bodies during impact. The extent of restoration of the original shape immediately after collision depends on the elastic properties of the material.[*] If the bodies are completely elastic, they will rebound and return to their original shape, like billiard balls. If, on the other hand, the bodies are completely plastic, they will stay permanently deformed and will travel together with the same velocity, like two lumps of putty, after collision.

The measure of the ability of the bodies to regain their original shape is called the **coefficient of restitution**. Mathematically, it is defined in terms of relative velocities before and after impact. The equation defining the coefficient of restitution ε is:[†]

$$\varepsilon(v_{0A} - v_{0B}) = (v_B - v_A)$$

---

[*] See more about elasticity of materials in Chapter 25.
[†] The symbol ε is another Greek letter called 'epsilon'.

The value of the coefficient varies from $\varepsilon = 0$ for completely plastic impact to $\varepsilon = 1$ for completely elastic collisions, e.g. for steel on lead, ε is about 0.12; for lead on lead, 0.2; for glass on glass, 0.93.

Before an illustrative example is attempted, one very important point must be emphasised. We know that velocities are vectors. It is therefore necessary to choose a sign convention, e.g. positive to the right and negative to the left, and to be absolutely consistent when applying the momentum and restitution equations.

Example 21.5

A railway car of mass 18 t, moving at a speed of 10 m/s to the right, collides with another car of mass 12 t which is moving to the left at 3 m/s. The coefficient of restitution is 0.6. Determine the final velocities of the two cars (Fig. 21.1).

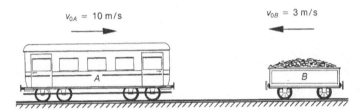

Fig. 21.1

Solution

The following information is given:

$$m_A = 18 \text{ t} \qquad\qquad v_{0A} = 10 \text{ m/s}$$
$$m_B = 12 \text{ t} \qquad\qquad v_{0B} = -3 \text{ m/s}$$
$$\varepsilon = 0.6$$

Substitute:

(a)
$$m_A v_{0A} + m_B v_{0B} = m_A v_A + m_B v_B$$
$$18 \times 10 - 12 \times 3 = 18 v_A + 12 v_B$$
$$144 = 18 v_A + 12 v_B \qquad\qquad (1)$$

(b)
$$\varepsilon(v_{0A} - v_{0B}) = v_B - v_A$$
$$0.6(10 + 3) = v_B - v_A$$
$$7.8 = v_B - v_A \qquad\qquad (2)$$

Solving the two equations yields:

$$v_A = 1.68 \text{ m/s}$$
$$v_B = 9.48 \text{ m/s}$$

Both answers are positive, meaning that the two cars will move to the right after the collision, but with new velocities.

Example 21.6

A tennis ball of mass 150 g is dropped from a height of 1 m and rebounds to a height of 0.8 m (Fig. 21.2). What is the coefficient of restitution between the ball and the ground during the impact?

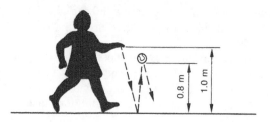

Fig. 21.2

Solution

Kinetic energy before impact:

$$KE_1 = PE_1$$
$$= mgh_1$$
$$= 0.15 \times 9.81 \times 1$$
$$= 1.47 \text{ J}$$

Velocity before impact:

$$v_{0A} = 4.43 \text{ m/s (negative)}$$

Kinetic energy after impact:

$$KE_2 = PE_2$$
$$= mgh_2$$
$$= 0.15 \times 9.81 \times 0.8$$
$$= 1.18 \text{ J}$$

Velocity after impact:

$$v_A = 3.96 \text{ m/s (positive)}$$

Velocity of the ground:

$$v_B = v_{0B} = 0$$

Substitute in $\varepsilon(v_{0A} - v_{0B}) = (v_B - v_A)$:

$$\varepsilon(-4.43 - 0) = (0 - 3.96)$$
$$\therefore \ \varepsilon = 0.89$$

 Problems

21.11 A car of mass 1 t is moving at 15 m/s when a 2 t truck runs into the back of it at 25 m/s. Immediately after the collision, the velocity of the truck is 20 m/s in the same direction. What is the velocity of the car after the collision?

21.12 What is the coefficient of restitution between the two vehicles in the previous problem during the collision?

21.13 Two bodies, *A* of mass 5 kg and *B* of mass 4 kg, collide with initial velocities 6 m/s to the right and 2 m/s to the left respectively. After the collision, the velocity of body *A* is 0.75 m/s to the left. What is the velocity of body B after the collision?

21.14 Calculate the coefficient of restitution for the bodies in the previous problem.

21.15 Two bodies, *A* of mass 8 kg and *B* of mass 14 kg, move in the same direction along a straight line with velocities 10 m/s and 1.2 m/s respectively. If the coefficient of restitution is 0.6, calculate:
(a) the velocities of the bodies after the collision
(b) the amount of kinetic energy lost during the collision

21.16 Repeat problem 21.15 if the collision is fully elastic, i.e. $\varepsilon = 1$.

21.17 Repeat problem 21.15 if the collision is fully plastic, i.e. $\varepsilon = 0$.

21.18 In order to determine the coefficient of restitution of a material, a ball of mass 0.5 kg is dropped onto a hard surface from a height of 2.5 m and is observed to bounce to a height of 1.8 m. What is the value of the coefficient?

Review questions

1. Define *linear momentum.*
2. Define *impulse.*
3. How is impulse related to the change in linear momentum?
4. Explain the principle of the *conservation of linear momentum.*
5. State the relation expressing the conservation of momentum in the case of impact between two bodies.
6. State the mathematical definition of the coefficient of restitution.
7. What is the numerical value of the coefficient of restitution in the case of:
   (a) completely plastic collision?
   (b) completely elastic collision?

PART 8

MECHANICS OF MACHINES

≡

Give me a lever long enough, a fulcrum and a place to stand and I can move the Earth.

Archimedes
on the law of the lever

CHAPTER 22

The law of a machine

This chapter is devoted to discussion of some of the fundamental principles on which the operation of all machines is based.

A **machine** can be defined as a mechanical device, consisting of one or more rigid components, designed and used for transmitting force and motion and for doing work. The general purpose of machines is to augment or replace human effort for the accomplishment of physical tasks. In particular, the fundamental feature of most machines is that a large load is moved, or a large resistance is overcome, by means of a relatively small effort.

In the next chapter these principles will be applied to a range of elementary mechanical devices called simple machines.

Expected learning outcomes

After carefully studying the material presented in this chapter, working through all numerical examples, and successfully completing all practice problems, students should be able to:

1. define and use the terms *velocity ratio*, *mechanical advantage* and *efficiency* of a machine;
2. distinguish between *theoretical effort*, *actual effort* and *frictional effort*;
3. state and use the equation known as the *law of a machine*;
4. define and use the term *limiting efficiency*.

22.1 *MECHANICAL ADVANTAGE AND VELOCITY RATIO*

All machines have an input side and an output side. The force exerted on the machine on the input side is known as the **effort,** F_E. The resistance to be overcome, or the force on the output side of the machine, is called the **load,** F_L.

Machines are usually of such design that by application of a small effort, a large load can be moved. The relationship between the load and the effort, which gives an indication of the advantage that can be obtained by using the machine, is called the **mechanical advantage** of the machine:

$$\text{Mechanical advantage} = \frac{\text{load}}{\text{effort}}$$

$$\boxed{\text{MA} = \frac{F_L}{F_E}}$$

Mechanical advantage is usually greater than one; it depends upon the type of machine which is being used, and it varies with the load.

In order to do work, both the load and the effort must move. In some machines the motion is linear, while in others it is rotational.

The ratio of the distance moved through by the effort on the input side (S_E) to the distance moved through by the load on the output side (S_L) is called the **velocity ratio** of the machine:

$$\text{Velocity ratio} = \frac{\text{distance moved by effort}}{\text{distance moved by load}}$$

$$\boxed{\text{VR} = \frac{S_E}{S_L}}$$

Velocity ratio is usually greater than one and, unlike the mechanical advantage, is constant for a given machine, i.e. it depends only on the arrangement of moving parts and is independent of the load.

Ideally, the mechanical advantage of any given machine should be equal to its velocity ratio. However, for a real machine, the actual mechanical advantage is always less than the ideal, due to the presence of friction between moving parts such as bearing or sliding surfaces.

Example 22.1

A simple machine is represented diagrammatically in Figure 22.1. The load is 450 N and the effort is 50 N. The distances moved by the load and by the effort are 100 mm and 1200 mm respectively. Calculate the mechanical advantage and the velocity ratio.

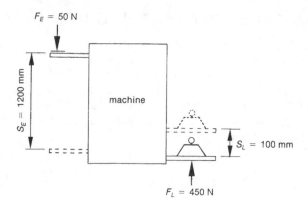

Fig. 22.1

Solution
Mechanical advantage:

$$MA = \frac{F_L}{F_E}$$

$$= \frac{450 \text{ N}}{50 \text{ N}}$$

$$= 9$$

Velocity ratio:

$$VR = \frac{S_E}{S_L}$$

$$= \frac{1200 \text{ mm}}{100 \text{ mm}}$$

$$= 12$$

Note that both mechanical advantage and velocity ratio have no units.

22.2 *WORK AND EFFICIENCY*

Whenever a force moves through a distance, the product of force and distance is the work done.

On the *input* side of a machine, the work done by the effort is equal to:

$$W_E = F_E \times S_E$$

Similarly, on the *output* side, the work done in moving the load is given by:

$$W_L = F_L \times S_L$$

The ratio of the useful work done by the machine in moving the load on the output side to the work put into the machine by the effort on the input side is called the **efficiency** of the machine.[*]

$$\text{Efficiency} = \frac{\text{work done in moving load}}{\text{work done by the effort}}$$

$$\eta = \frac{W_L}{W_E}$$

$$= \frac{F_L \times S_L}{F_E \times S_E}$$

Since MA $= \dfrac{F_L}{F_E}$ and VR $= \dfrac{S_E}{S_L}$, then:

$$\boxed{\eta = \frac{\text{MA}}{\text{VR}}}$$

Usually, efficiency is expressed as a percentage, and in the absence of friction, it should ideally be equal to 100 per cent. In actual machines, efficiency is always less than 100 per cent.

Example 22.2

For the machine in the previous example, calculate the input and output work, and the efficiency.

Solution

Input work:

$$W_E = F_E \times S_E$$
$$= 50 \text{ N} \times 1.2 \text{ m}$$
$$= 60 \text{ J}$$

Output work:

$$W_L = F_L \times S_L$$
$$= 450 \text{ N} \times 0.1 \text{ m}$$
$$= 45 \text{ J}$$

Efficiency:

$$\eta = \frac{W_L}{W_E}$$

$$= \frac{45}{60}$$
$$= 0.75$$
$$= 75\%$$

---

[*] The symbol for efficiency is the Greek letter η, called 'eta'.

Alternatively:

$$\eta = \frac{\text{MA}}{\text{VR}}$$

$$= \frac{9}{12}$$
$$= 0.75$$
$$= 75\%$$

22.3 FRICTION EFFORT

If the machine were perfect, no work would have to be done against friction, and the efficiency would be 100 per cent. This would mean that for a perfect machine:

$$\frac{\text{MA}}{\text{VR}} = 100\%$$

or that the ideal mechanical advantage is equal to the velocity ratio:

$$\text{MA} = \text{VR}$$

It also means that if there were no friction to be overcome, it would take a smaller effort to move the same load. The effort required to move a given load F_L if the machine is 100 per cent efficient is called the **theoretical effort**, F_{Th}. Substituting into $MA = VR$, we have:

$$\frac{F_L}{F_{Th}} = \text{VR} \qquad \text{or} \qquad F_{Th} = \frac{F_L}{\text{VR}}$$

The difference between the actual effort F_E and the theoretical effort F_{Th} is the effort wasted in overcoming friction and is known as the **frictional effort**, F_F.

$$\boxed{F_F = F_E - F_{Th}}$$

Example 22.3
For the machine in the previous examples, calculate the theoretical and frictional efforts.

Solution
Theoretical effort:

$$F_{Th} = \frac{F_L}{\text{VR}}$$

$$= \frac{450 \text{ N}}{12}$$
$$= 37.5 \text{ N}$$

Frictional effort:

$$F_F = F_E - F_{Th}$$
$$= 50 \text{ N} - 37.5 \text{ N}$$
$$= 12.5 \text{ N}$$

22.4 *THE LAW OF A MACHINE*

The **law of a machine** is an equation which expresses the relationship between load F_L and effort F_E. In many cases this relationship, when plotted as a graph of effort against load, is a straight line. Its mathematical equation is of the linear form:

$$F_E = aF_L + b$$

where F_E is the effort
$\quad\quad$ F_L is the load
$\quad\quad$ a is the slope of the graph
$\quad\quad$ b is the value of F_E where the graph cuts the F_E axis

After the constants have been determined for a particular machine, the law of the machine can be used to predict the effort required to move any load by the machine.

Example 22.4

The machine in the previous examples was tested under different loads, and the following efforts were recorded for each of the loading conditions:

Load F_L (N)	0	200	400	600	800	1000
Effort F_E (N)	5	25	45	65	85	105

Plot the load–effort graph and determine the law of the machine. Use the law to estimate the effort required to move a load of 700 N.

Solution

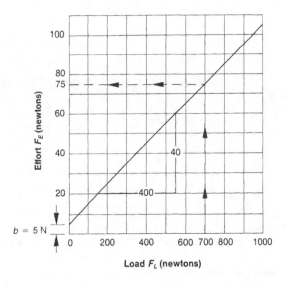

Fig. 22.2

In Figure 22.2, which is the load–effort graph, the line cuts the effort axis at $F_E = 5$. This is the value of b.

The slope is $a = \dfrac{40}{400} = 0.1$.

Therefore the law of the machine is:

$$F_E = 0.1F_L + 5$$

For a load of 700 N, the effort required is:

$$\begin{aligned}
F_E &= 0.1 \times F_L + 5 \\
&= 0.1 \times 700 + 5 \\
&= 75 \text{ N}
\end{aligned}$$

22.5 *LIMITING EFFICIENCY*

If we calculate and plot efficiency for the experimental results under different load conditions, we will find that the efficiency increases with the load. However, the increase is not proportional to the load.

There is a limiting value to the efficiency of a particular machine, which is always less than 100 per cent. The value of the limiting efficiency can be found by combining the law of a machine with the definition of efficiency, as follows:

$$\begin{aligned}
\eta &= \frac{\text{MA}}{\text{VR}} \\
&= \frac{F_L}{F_E \times \text{VR}}
\end{aligned}$$

Also:

$$F_E = aF_L + b$$

Substitute:

$$\begin{aligned}
\eta &= \frac{F_L}{(aF_L + b)\text{VR}} \\
&= \frac{1}{\left(a + \dfrac{b}{F_L}\right)\text{VR}} \\
&= \frac{1}{a\text{VR} + \left(\dfrac{b\text{VR}}{F_L}\right)}
\end{aligned}$$

As the load F_L increases, the term $\dfrac{b\text{VR}}{F_L}$ becomes smaller, tending towards zero at very large loads, when the limiting efficiency becomes:

$$\boxed{\eta = \frac{1}{a\text{VR}}}$$

Example 22.5

For each of the test results in the previous example, calculate efficiency and show that it tends towards a limiting value at large loads.

Solution
Efficiency:

$$\eta_1 = \frac{F_L}{F_E \text{VR}} = \frac{0}{5 \times 12}$$
$$= 0\%$$

$$\eta_2 = \frac{200}{25 \times 12} = 66.7\%$$

$$\eta_3 = \frac{400}{45 \times 12} = 74.1\%$$

$$\eta_4 = \frac{600}{65 \times 12} = 76.9\%$$

$$\eta_5 = \frac{800}{85 \times 12} = 78.4\%$$

$$\eta_6 = \frac{1000}{105 \times 12} = 79.4\%$$

Limiting efficiency:

$$\eta = \frac{1}{a\text{VR}}$$
$$= \frac{1}{0.1 \times 12}$$
$$= 83.3\%$$

This relationship can best be illustrated by a graph as shown in Figure 22.3.

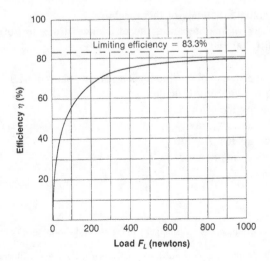

Fig. 22.3

Problems

22.1 The following parts (a) to (e) refer to the same machine:

(a) Given that the effort required to lift a load of 5 t is 343 N, calculate the mechanical advantage.

(b) If the effort moves 200 mm for every millimetre moved by the load, calculate the velocity ratio.

(c) If the load is lifted a total distance of 1.37 m, calculate the output work and the input work.

(d) Calculate the efficiency at this load.

(e) Calculate the frictional effort at this load.

22.2 If the law of the machine in the previous problem is $F_E = \dfrac{F_L}{150} + 16$, in newtons, calculate:

(a) the mechanical advantage and efficiency when the load is (i) 2 t, and (ii) 8 t

(b) the limiting efficiency

22.3 A test on a machine gave the following results:

Load (kN)	0	40	80	120	160	200
Effort (kN)	2	4.5	7.0	9.5	12.0	14.5

Draw a graph and determine the law of the machine, in kilonewtons. If the velocity ratio is 18, calculate the efficiency when the load is 100 kN.

22.4 A load is moved, by a lifting device, a distance of 25 mm for every 0.5 m stroke of the operating lever through which the effort is applied.

(a) What is the velocity ratio of this machine?

(b) Assuming ideal efficiency, find the effort required to lift a 200 kg load.*

(c) If the actual effort required to lift a 200 kg load is 109 N, calculate the mechanical advantage.

(d) What is the actual efficiency of this machine?

(e) If, through lack of maintenance, the efficiency drops to 76%, what effort will be required to lift 200 kg?

22.5 In the operation of a screw jack, which has a velocity ratio of 120, an effort of 20 N raises a load of 208 kg by 200 mm. Calculate:

(a) the work done by the effort

(b) the work done against gravity

(c) the work done against friction

(d) the efficiency of the machine

---

* You will notice here that there is some ambiguity in the use of the term 'load'. Strictly speaking, load is defined as a force F_L and should be expressed in newtons. However, where lifting devices are involved, the term 'load' can refer to the mass being lifted. As such, it is usually given in kilograms or tonnes, necessitating conversion into appropriate force units by using the formula for weight, $F_w = mg$.

22.6 A lifting machine has a velocity ratio of 20. Its performance under different load conditions was observed and recorded as follows:

Load (kg)	60	100	140	180	220	260	300	340
Effort (N)	49	72	95	118	140	163	186	209

(a) Plot the effort against the load (in newtons).
(b) Determine the law of this machine.
(c) What is the efficiency when a load of 200 kg is lifted?
(d) What is the limiting efficiency of the machine?

22.7 For the lifting machine in the previous problem, calculate the efficiency for each of the loading conditions listed and plot the efficiency–load graph. In this case, the load axis may be graduated in kilograms to represent the mass being lifted. Also show the limiting efficiency line. From the graph, determine:
(a) the efficiency when a load of 80 kg is lifted
(b) the efficiency when a load of 320 kg is lifted

Review questions

1. State the definition of a *machine*.
2. Define *mechanical advantage* and *velocity ratio*.
3. How can mechanical advantage and velocity ratio be related to the efficiency of a machine?
4. Explain the terms *theoretical effort* and *frictional effort*, and state the relation between them.
5. Explain what is meant by the *law of a machine*.
6. Explain *limiting efficiency* of a machine, and show how it can be calculated.

CHAPTER 23

Simple machines

In this chapter we are going to examine some examples of simple machines.

From the early beginnings of mechanical engineering science, five fundamental devices have been regarded as the 'mighty five'. They are called **simple machines** and include the lever, inclined plane, wheel-and-axle, pulley and screw. These are the most basic and most important elements of mechanisms used in engineering.

It has been said that each one of the 'mighty five' can rightly be called invaluable, but their combined value is truly infinite. This alludes to the almost infinite variety of applications of simple machines as elements of mechanisms in all kinds of modern machinery.

However, our aim at this stage is more modest. We are going to examine the operating principles behind the five simple machines with respect to their basic attributes, such as velocity ratio, mechanical advantage and efficiency.

Expected learning outcomes

After carefully studying the material presented in this chapter, working through all numerical examples, and successfully completing all practice problems, students should be able to:
1. list and describe the five devices known as *simple machines*;
2. solve problems involving velocity ratio, mechanical advantage and efficiency of simple machines.

23.1 *VELOCITY RATIOS OF SIMPLE MACHINES*

Problems dealing with machines usually involve the calculation of their velocity ratio. The velocity ratio (VR) of a particular machine is independent of load and friction and depends only on the dimensions and arrangement of the moving components. The velocity ratio of a machine can be determined from the dimensions alone, and methods for calculating VR for various simple machines are summarised below.

$$\text{Velocity ratio} = \frac{\text{distance moved by effort}}{\text{distance moved by load}}$$

The lever

The lever is the simplest machine. It consists of a rigid bar pivoted at a point called the **fulcrum**, with a load applied at one point on the bar, and an effort at another point sufficient to move or balance the load.

A lever which is bent is called a **bell-crank lever.** It is used to provide a change in the direction of the applied forces.

The lever has been used in some form since early times for moving or lifting heavy stones and other objects. A beam balance, which originated in Egypt about 5000 BC, is still widely used in its accurate modern forms. Many mechanisms used in modern machinery consist of a combination of straight and bent lever arrangements connected together.

The velocity ratio of a single lever arrangement is equal to the ratio of the perpendicular distances from the fulcrum to the line of action of the effort (d_E) and the load (d_L) (Fig. 23.1).

$$VR = \frac{d_E}{d_L}$$

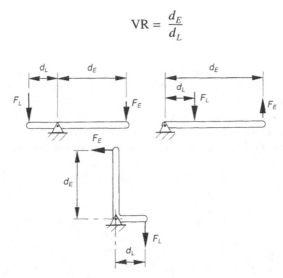

Fig. 23.1   *Single lever arrangements*

The inclined plane

The inclined plane, as a simple machine, is a surface inclined at an angle to the horizontal used to lift a load by an effort acting parallel to the plane (Fig. 23.2).

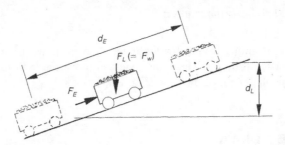

Fig. 23.2   *An inclined plane*

The inclined plane is said to have been used in Egypt for moving heavy stones in the construction of pyramids. Modern machinery often employs the principle of the inclined plane for various sliding surfaces, cams and metal-cutting tools.

The velocity ratio is the ratio of the distance moved by the effort to the distance moved by the load in the same time. When the effort moves along the plane a distance d_E, the load moves through a vertical distance d_L, i.e. in the direction in which the weight of the load acts. Hence:

$$VR = \frac{d_E}{d_L}$$

If the angle of inclination is θ, the ratio d_E/d_L is equal to $1/\sin \theta$. The inclined plane may therefore be regarded as a machine having a velocity ratio of $1/\sin \theta$.

Wheel-and-axle

In a mechanism known as the wheel-and-axle, the effort is applied at the circumference of the wheel while the load is raised by a rope wound around the axle whose diameter is smaller than that of the wheel.

The wheel-and-axle operates basically on a leverage principle. The velocity ratio is equal to the ratio of the radius of the wheel to the radius of the axle. It is usually more convenient to express this ratio as the ratio of the two diameters, D_W and D_A.

$$VR = \frac{D_W}{D_A}$$

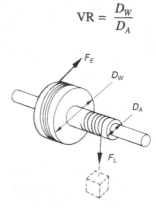

Fig. 23.3   *Wheel-and-axle mechanism*

The situation is similar to that for the lever. A lever, however, can move a load for only short distances, while the wheel-and-axle can move the load for a distance limited only by the available length of the rope (Fig. 23.3).

Pulley and pulley block

The pulley is one of the most useful of the basic simple machines. It consists essentially of a wheel with a grooved rim carrying a rope or chain and supported in either a fixed or a movable bearing block.

A system of fixed and movable pulleys with a continuous rope can provide a useful mechanical advantage.

From the geometry of block-and-tackle arrangements, such as the one shown in Figure 23.4, it can be shown that with pulley blocks using one continuous rope, the velocity ratio is equal to the number of falls of rope supporting the load:

VR = No. of falls of rope supporting the load

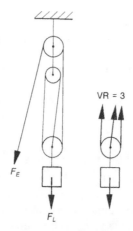

Fig. 23.4 *Pulleys in block-and-tackle arrangement*

Screw jack

A simple screw jack is a portable lifting machine used for raising heavy loads through a short distance. It consists of a screw raised by a nut rotated by the effort applied at the end of a long arm. In some designs, the screw rotates within the nut which is part of the main body of the jack (Fig. 23.5).

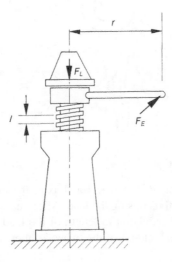

Fig. 23.5 *A screw jack*

In either case, for each revolution of the arm, i.e. the distance moved by the effort, the load is raised by an amount equal to the lead of the screw. Thus when the distance moved by the effort is $d_E = 2\pi r$, where r is the length of the arm, the load moves through a vertical distance l, equal to the lead of the screw.[*] Therefore velocity ratio is given by:

$$\text{VR} = \frac{d_E}{d_L} = \frac{2\pi r}{l}$$

The invention of the screw is attributed to the Pythagorean philosopher Archytas of Tarentum (5th century BC). However, the date of its first use as a mechanical device is not known. The screw press, similar in principle to the operation of the screw jack, was probably invented in Greece in the 2nd century BC. It was used as a wine and olive-oil press in the time of the Roman Empire.

In addition to its use in screw jacks and presses, the screw is used in modern machinery for a variety of purposes, including continuous transmission of motion.

23.2 DEVELOPMENTS FROM SIMPLE MACHINES

It was implied earlier that the real importance of simple machines lies not so much in their direct use in isolated applications, but rather in the multitude of ways they can be incorporated into more complex machinery. We will examine here a familiar lifting device in which a movable pulley block is combined with a wheel-and-axle, just to illustrate how simple machines can be combined to work together.

The device shown in Figure 23.6 is known as the **Weston differential chain block**. It uses a continuous chain which passes around a wheel-and-axle and supports a movable pulley block carrying a load. The wheel-and-axle block consists of two pulley wheels of different diameters, which are cast together to rotate as one. Special pulleys are used with suitable link slots to engage the links in the chain in order to prevent slipping.

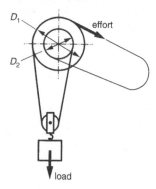

Fig. 23.6   *Weston differential chain block*

As the effort force rotates the wheel-and-axle, the effort loop is lengthened at the expense of the load loop, thus raising the load. After one revolution of the wheel-and-axle in the top block, the total reduction in length of the load loop is equal to the difference between the circumferential lengths of the 'wheel' pulley and the 'axle' pulley:

$$\pi D_1 - \pi D_2$$

---

[*] In the case of a single start thread, which is common in simple screw jacks, the lead is equal to the pitch of the screw, i.e. the distance between corresponding points on the successive convolutions of the thread.

This difference is equally divided between the two falls of chain supporting the load. Therefore, after one revolution of the wheel-and-axle, the load is raised by:

$$\frac{\pi D_1 - \pi D_2}{2}$$

The distance moved by the effort in one revolution is πD_1. Therefore, the velocity ratio of a Weston differential chain block is given by:

$$\text{VR} = \frac{2D_1}{D_1 - D_2}$$

Due to the weight of the chain combined with frictional effects, the efficiency of Weston differential chain blocks is usually low, which has an advantage of preventing the chain from running backwards when the effort is removed. Therefore a heavy load can be supported in an elevated position without any assistance from the effort.

Example 23.1

A Weston differential chain block has the two top pulley diameters equal to 300 mm and 250 mm, and is 45% efficient. Calculate its velocity ratio and the effort required to lift an aluminium casting of mass 60 kg.

Solution

Velocity ratio:

$$\text{VR} = \frac{2D_1}{D_1 - D_2}$$
$$= \frac{2 \times 300}{300 - 250}$$
$$= 12$$

Mechanical advantage:

$$\text{MA} = \eta \text{VR}$$
$$= 0.45 \times 12$$
$$= 5.4$$

The load force is equal to the weight of the load:

$$F_L = 60 \times 9.81$$
$$= 588.6 \text{ N}$$

Hence the effort required is:

$$F_E = \frac{F_L}{\text{MA}}$$
$$= \frac{588.6 \text{ N}}{5.4}$$
$$= 109 \text{ N}$$

Problems

23.1 A mechanism for operating a valve consists of an arrangement of two levers connected together as shown in Figure 23.7.

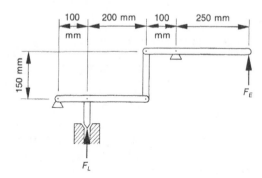

Fig. 23.7

If the load is 30 N and the frictional effort is 20% of the actual effort, determine the velocity ratio, the mechanical advantage, and the actual effort required.

23.2 Determine the effort, parallel to the plane, required to move a body of mass 225 kg up a plane inclined at 20° to the horizontal if the coefficient of friction is 0.25. Hence calculate the efficiency.

23.3 A simple wheel-and-axle mechanism has a wheel of diameter 700 mm and an axle of diameter 100 mm. If an effort of 50 N is required to lift a load of 29 kg, determine the mechanical advantage, efficiency and the frictional effort at this load.

23.4 A lifting tackle consists of two pulley blocks containing two and three pulleys respectively. How must this tackle be used to obtain the greatest velocity ratio? What is the value of this ratio?

23.5 With the best arrangement of pulleys in the previous problem, and an efficiency of 85%, determine the effort required to lift a load of 250 kg.

23.6 A screw jack is used to raise a load of 1.2 t with an effort of 100 N at the end of an arm 250 mm long. The lead of the screw is 10 mm. Calculate the efficiency of the device.

23.7 Determine the load, in tonnes, that can be lifted with a screw jack if an effort of 75 N is applied to the arm of length 300 mm, and the lead of the screw is 5 mm. The efficiency is 65%.

What is the number of revolutions required to lift the load through a vertical distance of 200 mm?

23.8 A Weston differential chain block has the two top pulley diameters equal to 250 mm and 210 mm, and is 48% efficient. Calculate its velocity ratio and the effort required to lift a box of mass 165 kg.

Review questions

1. For each of the following simple machines, draw a diagram and explain how the velocity ratio can be determined:
   (a) lever
   (b) inclined plane
   (c) wheel-and-axle
   (d) pulley block
   (e) screw jack
2. Explain the principle of operation of the Weston differential chain block.
3. State the formula for calculating the velocity ratio of a Weston differential chain block.

CHAPTER 24

Mechanical drives

Mechanical drives are devices, such as gearboxes and belt or chain drives, used for the purpose of mechanical power transmission. Unlike simple machines discussed in the previous chapter, whose action is usually intermittent, mechanical drives are designed for continuous operation involving rotating components.

Transmission of power by mechanical drives is often associated with changes in the speed or direction of rotational motion of mechanical components.

Some of the fundamental principles of simple machines, such as efficiency and velocity ratio, are also applicable to the study of mechanical drives. However, in the case of mechanical drives, instead of work done it is usually more convenient to refer to power transmitted, i.e. to the time rate of doing work, as will be seen from the following discussion.

Expected learning outcomes

After carefully studying the material presented in this chapter, working through all numerical examples, and successfully completing all practice problems, students should be able to:
1. determine velocity ratios of gear trains, and calculate torque, power and efficiency for a gearbox;
2. determine tension in the chain, and torque and power transmitted through a chain drive;
3. determine tension in the belt, and calculate torque and power transmitted through a flat belt or a V-belt drive.

24.1 *MECHANICAL POWER AND DRIVE EFFICIENCY*

All mechanical drives have an input side and an output side, usually involving a rotating shaft, sprocket or pulley on each side.

The torque exerted on the drive on the input, or driver, side is called the **input torque**, T_{in}. The torque transmitted by the drive to its output side is called the **output torque**, T_{out}.

Mechanical power associated with continuous rotation of a component is given by:

$$P = T\omega = \frac{2\pi NT}{60}$$

where P is power in W
$\quad\quad$ T is torque in N.m

ω is rotational speed in rad/s

N is speed in rpm

If the input and output power are calculated using the corresponding values of torque and speed, **drive efficiency** can be defined as the ratio of output power (P_{out}) to input power (P_{in}).

$$\eta = \frac{P_{out}}{P_{in}}$$

Example 24.1

The input shaft of a gearbox rotates at 1450 rpm and transmits a torque of 65.9 N.m. The output shaft rotates at 500 rpm and transmits a torque of 143.3 N.m. Determine the input and output power, and the efficiency of the device.

Solution

Input power:

$$P_{in} = \frac{2\pi NT}{60}$$

$$= \frac{2 \times \pi \times 1450 \times 65.9}{60}$$

$$= 10 \text{ kW}$$

Output power:

$$P_{out} = \frac{2\pi NT}{60}$$

$$= \frac{2 \times \pi \times 500 \times 143.3}{60}$$

$$= 7.5 \text{ kW}$$

Efficiency:

$$\eta = \frac{P_{out}}{P_{in}}$$

$$= \frac{7.5 \text{ kW}}{10 \text{ kW}}$$

$$= 75\%$$

24.2 GEAR DRIVES

Gear wheels operate in pairs to transmit torque and motion from one shaft to another by means of specially shaped projections or teeth. The teeth on one gear mesh with corresponding teeth on the second gear so that motion is transferred from one to the other without any slip taking place.

There are four main types of gears: spur, helical, worm and bevel. Gear types are determined largely by the relative positions of the input and output shafts.

The most common type is the **spur gear**, which has tooth elements that are straight and parallel to its axis. A spur gear pair can be used to connect parallel shafts only.

Parallel shafts, however, can also be connected by gears of another type. **Helical gears**, for example, have a higher load-carrying capacity than spur gears when connecting parallel shafts.

Bevel gears are commonly used for transmitting rotary motion and torque around corners. The connected shafts, whose axes would intersect if extended, are usually, but not necessarily, at right angles to each other.

Worm gear is a gear of high reduction ratio, connecting shafts whose axes are at right angles but do not intersect. It consists of a screw-like component carrying a helical thread of special form, the worm, meshing in sliding contact with a concave face gear wheel. Because of their similarity, the operation and efficiency of a worm and gear depend on the same factors as the operation and efficiency of a screw.

The velocity ratio of a gear drive is equal to the ratio of the revolutions of the driver wheel (the input) to the revolutions of the driven wheel (the output) in the same time. In any interval of time, the same number of teeth from both gears come in contact with each other. Therefore it can be seen that the velocity ratio is the ratio of the number of teeth in the driven wheel to the number of teeth in the driver wheel:

$$\text{VR} = \frac{\text{No. of teeth in driven wheel}}{\text{No. of teeth in driver wheel}}$$

Example 24.2

Figure 24.1 shows a gear drive in a certain machine. If the input shaft rotates at 660 rpm and transmits a torque of 12 N.m, and the efficiency is 80%, determine output speed, torque and power.

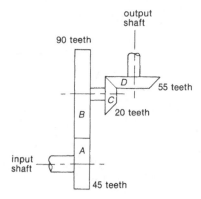

Fig. 24.1

Solution

Gear *A* rotates at 660 rpm.

Gear *B* rotates at 660 rpm × $\dfrac{45}{90}$ = 330 rpm.

Gear *C* rotates at 330 rpm.

Gear *D* rotates at 330 rpm × $\dfrac{20}{55}$ = 120 rpm.

Therefore, the output speed is 120 rpm.

Input power is:

$$P_{in} = \frac{2\pi NT}{60}$$
$$= \frac{2\pi \times 660 \times 12}{60}$$
$$= 829.4 \text{ W}$$

With an efficiency of 80%, output power is:

$$P_{out} = \eta \times P_{in}$$
$$= 0.8 \times 829.4$$
$$= 663.5 \text{ W}$$

Output torque can be calculated from:

$$P_{out} = \frac{2\pi NT_{out}}{60}$$
$$663.5 = \frac{2\pi \times 120 \times T_{out}}{60}$$
$$T_{out} = \frac{663.5 \times 60}{2\pi \times 120}$$
$$= 52.8 \text{ N.m}$$

The following points should be noted very carefully:

1. Velocity ratio is a function of the dimensions and arrangement of the moving parts, e.g. number of teeth in meshing gears. It is independent of torque and efficiency.
2. Efficiency is the ratio of power output to power input; it depends on power losses that occur within the gear drive due to friction.
3. Output torque depends both on velocity ratio and on efficiency. If output torque is not given, it should be calculated from output power and speed.

24.3 CHAIN DRIVES

A chain drive consists of an endless chain of links meshing with the driving and driven sprockets. A very familiar example is a bicycle drive (Fig. 24.2).

The chain fits into specially shaped teeth cut in the sprockets, which prevent the chain from slipping. The tension force F_t in the tight side of the chain is responsible for the transmission of power between the two sprockets.

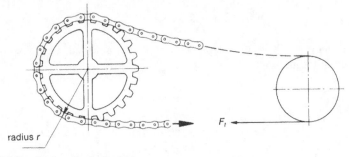

Fig. 24.2   *A bicycle chain drive*

The velocity ratio is inversely proportional to the number of teeth in the sprockets:

$$VR = \frac{\text{No. of teeth in driven sprocket}}{\text{No. of teeth in driver sprocket}}$$

The torque on each sprocket is equal to the force in the chain multiplied by the radius of the sprocket.

In many respects the operation of a chain drive is similar to that of a pair of spur gears, with the exception that the centre distance between two parallel shafts is limited only by the length of the chain.

Efficiency of chain drives is usually high, and for our purposes can be assumed to be 100 per cent.

Example 24.3

A chain drive transmits 20 kW of power from an 80 mm diameter driver sprocket with 18 teeth to a 200 mm diameter driven sprocket with 45 teeth.

If the speed of the driver is 500 rpm, calculate the input and output torque and the force in the chain.

Solution

From $P = \dfrac{2\pi NT}{60}$, input torque is equal to:

$$T_{in} = \frac{60P}{2\pi N}$$
$$= \frac{60 \times 20\,000}{2\pi \times 500}$$
$$= 382 \text{ N.m}$$

The velocity ratio is:

$$VR = \frac{45}{18}$$
$$= 2.5$$

Therefore, the velocity of the driven sprocket is:

$$500 \div 2.5 = 200 \text{ rpm}$$

If the efficiency is 100%, the output power is undiminished, i.e. it is equal to 20 kW. Hence, output torque is:

$$T_{out} = \frac{60P}{2\pi N}$$
$$= \frac{60 \times 20\,000}{2\pi \times 200}$$
$$= 955 \text{ N.m}$$

The force of tension in the chain is found from $T = F_t \times r$, where T is torque and r is the corresponding radius:

$$F_t = \frac{T}{r}$$

$$= \frac{382}{0.04}$$
$$= 9550 \text{ N}$$
$$= 9.55 \text{ kN}$$

24.4 FLAT BELT DRIVES

Power transmission by belts is possible only with sufficient friction between the belt and its pulleys. In order to provide the necessary grip on the pulley, both sides of the belt must be in tension, i.e. there must be a tension force in both sides of the belt (Fig. 24.3).

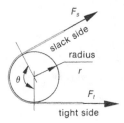

Fig. 24.3   *A flat belt drive*

As a result of applied torque and the friction between the belt and the pulley, there is a difference between the tight-side tension, F_t, and the slack-side tension, F_s. It can be shown that friction limits the ratio between these two belt tensions according to the equation:

$$\frac{F_t}{F_s} = e^{\mu\theta}$$

where e is equal to 2.718*
  μ is the coefficient of friction between the belt and the pulley
  θ is the angle of contact in radians

At the same time, torque is equal to the algebraic sum of the moments of the two forces about the centreline of the pulley:

$$T = F_t r - F_s r = r(F_t - F_s)$$

where r is the radius.

---

* e is a mathematical constant and is usually available as a function key on most calculators.

Example 24.4

Determine the maximum torque that can be transmitted by a flat belt if the maximum tension is 500 N, the coefficient of friction is 0.25, the angle of contact is 150° and the diameter of the pulley is 300 mm.

Solution

Given:

$$\mu = 0.25$$
$$\theta = 150°$$
$$= 150 \times \frac{\pi}{180}$$
$$= 2.618 \text{ rad}$$
$$F_t = 500 \text{ N}$$

we can find the slack-side tension from:

$$\frac{F_t}{F_s} = e^{\mu\theta}$$

$$\frac{500}{F_s} = e^{0.25 \times 2.618}$$

$$\therefore F_s = 260 \text{ N}$$

Therefore, torque is:

$$T = r(F_t - F_s)$$
$$= 0.15(500 - 260)$$
$$= 36 \text{ N.m}$$

24.5 *V-BELT DRIVES*

The V-belt drive benefits from the wedging effect produced by the rubber belt in the specially shaped groove of the pulley (Fig. 24.4). The wedging action of the belt tension increases the normal force on the belt and hence increases friction which provides the grip.

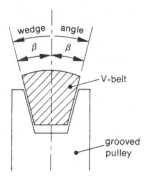

Fig. 24.4   *Grooved pulley for V-belt drive*

The flat belt drive equation given above can be modified by the inclusion of the sine of one-half of the wedge angle as shown.

$$\frac{F_t}{F_s} = e^{(\mu\theta/\sin\beta)}$$

where β is equal to one-half of the wedge angle.[*]

Example 24.5

If in the previous example a V-belt drive with a wedge angle of 40° is used, other conditions being the same, calculate the maximum torque.

Solution

Wedge angle is 40°. Therefore β is 20°.

$$\frac{F_t}{F_s} = e^{\left(\frac{\mu\theta}{\sin\beta}\right)}$$

$$\frac{500}{F_s} = e^{\left(\frac{0.25 \times 2.618}{\sin 20°}\right)}$$

$$\frac{500}{F_s} = e^{1.914}$$

$$\therefore F_s = 73.8 \text{ N}$$

Hence torque is:

$$T = 0.15(500 - 73.8)$$
$$= 63.9 \text{ N.m}$$

Compared with the flat belt drive of the previous example, the torque is almost doubled. Therefore, power transmitted at a given speed would also be nearly doubled.[†]

 Problems

24.1 If the input shaft of a gearbox rotates at 1450 rpm and the input torque is 50 N.m, while the output shaft rotates at 500 rpm transmitting 116 N.m of output torque, calculate the input and output power, and the efficiency.

24.2 A gearbox reduction unit contains four pairs of gears having the following particulars:

Pair 1: 20 teeth, 131 teeth
Pair 2: 20 teeth, 106 teeth
Pair 3: 20 teeth, 72 teeth
Pair 4: 20 teeth, 40 teeth

The input speed is 1000 rpm. Calculate the output speed.

---

[*] β is the letter 'beta' from the Greek alphabet.
[†] The efficiency of a V-belt drive ranges from 90 to 98 per cent, with a generally accepted average of 95 per cent. However, for the sake of simplicity in presentation, we have assumed 100 per cent efficiency for all calculations involving flat belt and V-belt drives in this book.

   Likewise, we have ignored the effect of centrifugal force, which at high velocities tends to lift the belt off the pulley, thus reducing the frictional grip.

24.3 The hoist of a crane consists of a 10 kW electric motor running at 1440 rpm driving a 300 mm diameter drum through a 60:1 gear reduction unit. If the efficiency is 90%, calculate the load, in tonnes that can be lifted at the rated motor capacity, and the lifting speed.

24.4 A bicycle drive has 36 teeth on the crank sprocket and 12 teeth on the wheel sprocket. The cranks are 180 mm long and the road wheel is 600 mm in diameter.

What is the ratio of the road-wheel rim to the pedal in terms of:
(a) rotational speed?
(b) linear speed?

24.5 A small pump is driven through a chain drive by an electric motor running at 950 rpm and developing 1.91 kW of power at this speed. Details of the sprockets are as follows:

Motor: 60 mm diameter, 16 teeth
Pump: 180 mm diameter, 48 teeth

Calculate the input and output torque, and the force in the chain.

24.6 A 400 mm diameter pulley is driven at 750 rpm by a flat belt with a tight-side tension of 300 N and a slack-side tension of 45.3 N. Determine the power transmitted by the belt.

24.7 Determine the number of V-belts required to transmit 45 kW to a 500 mm diameter pulley, at 850 rpm, if the tight-side tension in each belt is not to exceed 775 N, and the tight-side to slack-side tension ratio is estimated to be 8:1.

24.8 A flat belt makes contact with a 350 mm diameter pulley over an angle of 160°. The coefficient of friction is 0.3 and the speed of the pulley is 1200 rpm. If the maximum allowable tension in the belt is 550 N, calculate the maximum torque, and the maximum power that can be transmitted by the belt.

24.9 If instead of a flat belt in the previous problem a V-belt with a wedge angle of 28° is used, other conditions being equal, what will the maximum torque and power be?

24.10 A V-belt drive consists of a driven pulley of diameter 250 mm and a driver pulley of diameter 100 mm, at a centre-to-centre distance of 270 mm. The driver pulley rotates at 1440 rpm, transmitting 30 kW through 4 belts with a wedge angle of 40° and $\mu = 0.35$.

Determine the total torque on each pulley and the maximum tension in each belt.

Review questions

1. State the formula used for calculating mechanical power associated with continuous rotation of a component.
2. Define *drive efficiency*.
3. Briefly describe different types of gear drives.
4. Explain how velocity ratio of a gear train is related to the number of teeth.
5. What is meant by *gearbox efficiency*?
6. Show how tension in a bicycle chain is related to the power transmitted to the wheel.
7. What is the formula relating belt tensions in:
   (a) a flat belt drive?
   (b) a V-belt drive?
8. What are the advantages of V-belt drives over flat belt drives?

STRENGTH OF MATERIALS

PART 9

STRESS AND ELASTICITY

A name known to all engineers is that of Robert Hooke, because the proportionality of the pull to the stretch in the tensile loading of elastic materials is the very foundation of elastic theory on which so much mechanical design depends.

Aubrey F. Burstall
A History of Mechanical Engineering

CHAPTER 25

Tensile stress

One of the major concerns of the engineer lies in estimating the ability of engineering materials to withstand conditions such as tension, compression, torsion and bending without failure or excessive deformation. The branch of engineering science which deals with those properties of materials that relate directly to problems of strength and stability of structures and of mechanical components is known as 'strength of materials'.[*]

This is the opening chapter in a series of topics on strength of materials that are covered in this book. It introduces some fundamental concepts, such as stress, strain and factor of safety, in the context of elastic behaviour of engineering materials in tension.

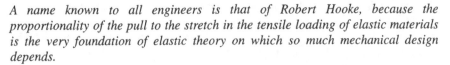

Expected learning outcomes

After carefully studying the material presented in this chapter, working through all numerical examples, and successfully completing all practice problems, students should be able to:

1. define and use the terms *ultimate tensile strength*, *tensile stress*, *strain* and *Young's modulus*;
2. use the concept of factor of safety to establish allowable tensile stress;
3. solve problems involving engineering components that are subjected to direct axial tensile loads.

---

[*] Strength of materials should not be confused with other branches of applied science, such as chemistry and metallurgy, which are concerned with the manufacture, properties and uses of engineering materials, from their own different perspectives.

25.1 *TENSILE STRENGTH*

Strength is the ability of the material to withstand applied force without failure. As such, strength is one of the most important mechanical properties of engineering materials. Strength varies, depending on the nature of the load applied to the material, and we often have to distinguish between tensile, compressive and shear strengths of a particular material.

The tensile strength of a material is determined by the tension test which consists of the gradual application of an axial tensile force, i.e. the pulling apart of the specimen, until fracture occurs.

The equipment for tensile testing of materials usually consists of a test specimen, generally cylindrical, with a middle section of smaller diameter than the ends (Fig. 25.1), an appropriate set of grips to grasp the test piece, and a machine that applies, measures and records the load.

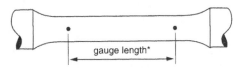

Fig. 25.1 *A specimen for testing tensile strength*

It is obvious that for the result to be meaningful, it should depend on the strength of the material and be independent of the actual size of the test piece. It is therefore expressed not as a force but as force per unit of cross-sectional area of the test specimen.

Much engineering design is based on this measure of tensile strength, known as **ultimate tensile strength** (UTS),[†] which is defined as the ratio of the maximum tensile force applied before fracture occurs to the initial cross-sectional area of the test specimen.

$$UTS = \frac{\text{maximum tensile force}}{\text{initial cross-sectional area}}$$

Since diameters of test pieces are usually measured in millimetres, it is often convenient to calculate cross-sectional areas in square millimetres and therefore to express ultimate tensile strength in units of force (newtons) per square millimetre of area. Typical average values of ultimate tensile strengths of some materials are given in Table 25.1.

---

* Gauge length is the exact length between two marks on the specimen measured along its axis before loading is begun.
† Also known as ultimate tensile **stress**.

Table 25.1   *Ultimate tensile strength*

Material	UTS (N/mm^2)
tool steel	1000
high-tensile steel	590
mild steel	470
copper wire	415
copper sheet	210
brass	190
cast iron	180
aluminium	150
timber (pine)	105[a]
nylon	70

[a] Wood strength varies considerably with the direction of application of the load. The strength given here is for parallel (i.e. along the grain) direction of the force. The transverse (i.e. perpendicular to the grain) strength for pine timber is only 3 N/mm^2.

Example 25.1

A steel test specimen, 10 mm in diameter, ruptures under a tensile load of 37 kN. What was the ultimate tensile strength of the steel?

Solution

$$\text{Cross-sectional area} = \frac{\pi \times 10^2}{4}$$
$$= 78.54 \text{ mm}^2$$
$$\text{Tensile force} = 37\,000 \text{ N}$$
$$\therefore \text{UTS} = \frac{37\,000 \text{ N}}{78.54 \text{ mm}^2}$$
$$= 471 \text{ N/mm}^2$$

25.2 *DIRECT AXIAL STRESS IN TENSION*

So far we have discussed ultimate strength of materials, i.e. strength up to the instant of failure. We must now consider the behaviour of materials under loads which do not cause rupture, crushing or any form of permanent damage or deformation. We must also learn how to calculate the allowable load on a structure or a component which will ensure a sufficient degree of safety.

Consider a bar of solid material, forming a component of a machine or structure, subjected to an axial pull, i.e. tension. The force acting on the bar is called the **direct axial load**, *F*.

If the bar is of uniform cross-sectional area, *A*, the force *F* may be assumed to be distributed uniformly over the cross-section, requiring a certain degree of adhesion between the particles of the material in order to keep the material intact.

The intensity of force distribution over the cross-sectional area of the material subjected to direct load is called **direct stress**. Thus direct stress can be defined as the share of the total axial load carried by each unit of cross-sectional area:

$$f = \frac{F}{A}$$

where f is the symbol for direct stress, and can be identified as tensile stress by means of an appropriate subscript, as in f_t, if necessary.

The SI unit of stress is the **pascal**, with the symbol Pa. The definition of the pascal follows from the definition of stress, i.e. force per unit area, or newton per square metre. The pascal is a very small unit, not very suitable for measuring the stresses normally encountered in engineering applications. For the majority of such applications, a larger prefixed unit called the **megapascal** (MPa) is used (1 MPa = 1 000 000 Pa).

It is also very useful to remember that if force is expressed in newtons and area in square millimetres, the stress will automatically be found in megapascals. This is convenient, first because linear dimensions of mechanical components are normally measured in millimetres, and so area can easily be expressed in square millimetres, and second, the order of magnitude of stress will conveniently be between zero and 1000 MPa. This helps to avoid the use of small decimal fractions or very large numbers.[*]

Example 25.2

If a bar of mild steel, 20 mm × 10 mm in cross-section, is subjected to a tensile force of 18.8 kN, determine the stress in the material.

Solution

$$\text{Stress } f_t = \frac{F}{A}$$
$$= \frac{18\,800 \text{ N}}{20 \text{ mm} \times 10 \text{ mm}}$$
$$= 94 \text{ MPa}$$

25.3 *FACTOR OF SAFETY*

It is interesting to compare the result of the above example with the ultimate tensile strength of the material (UTS for mild steel is 470 N/mm², i.e. 470 MPa). Obviously, the stress is not sufficient to rupture the material of the bar. The bar is safe under the applied load. But how safe? If we compare the UTS and the actual stress in the material of the bar as a ratio, it is:

$$\frac{470 \text{ MPa}}{94 \text{ MPa}} = 5$$

---

[*] For all calculations that involve stress in this and the following chapters, we shall use newtons (N) for force, millimetres (mm) for linear dimensions, square millimetres for area and megapascals (MPa) for stress. The student may prefer to work in base units; the results should be the same.

This ratio is an indication of the degree of safety built into the situation by virtue of the fact that the actual working stress in the material is considerably below the ultimate tensile strength. Thus we say that the factor of safety in the previous example is equal to 5.

The **factor of safety** (FS) can be defined as the ratio of the ultimate tensile strength to the actual working stress in the material.[*]

$$FS = \frac{\text{ultimate strength}}{\text{working stress}}$$

The factor of safety is a dimensionless number always greater than one. It depends on a number of considerations: possible defects in materials or workmanship, the exactness with which probable loads are known, possibility of shock or impact loads, safety to human life or property; and an allowance for decay, wear, corrosion etc.

Typical factors of safety used in design practice are given in Table 25.2.

Table 25.2   *Factors of safety*

Material	Static load	Cyclic load
steel, ductile materials	3–4	8
cast iron, brittle materials	5–6	10–12
timber	7	15

In design problems the aim may be to determine a suitable cross-section of a material for a given load using a recommended safety factor. The working stress under these conditions is usually referred to as **design stress** or **allowable stress**.

Example 25.3

Determine the minimum required diameter of a high-tensile steel rod to carry a tensile load of 26 kN with a safety factor of 3.5.

Solution

From $FS = \dfrac{\text{UTS}}{f}$, allowable stress is:

$$f = \frac{\text{UTS}}{\text{FS}}$$
$$= \frac{590 \text{ N/mm}^2}{3.5}$$
$$= 168.6 \text{ MPa}$$

---

[*] This definition of the factor of safety is equally valid for tension, compression or shear, provided that the appropriate ultimate strength is used. For example, the factor of safety in tension is $FS = \dfrac{\text{UTS}}{f_t}$.

Now, from $f = \dfrac{F}{A}$, the area required is:

$$A = \frac{F}{f}$$
$$= \frac{26\,000\ \text{N}}{168.6\ \text{MPa}}$$
$$= 154.2\ \text{mm}^2$$

But area $A = \dfrac{\pi D^2}{4} = 154.2\ \text{mm}^2$, hence:

$$D = \sqrt{\frac{154.2 \times 4}{\pi}}$$
$$= 14.0\ \text{mm}$$

 # Problems

Refer to Tables 25.1 and 25.2 if required.

25.1 What is the approximate ratio of the ultimate tensile strength of mild steel to that of cast iron? Which is a stronger material in tension?

25.2 A test specimen of aluminium alloy, diameter 12 mm, ruptures under a tensile load of 14.7 kN. What is the ultimate tensile strength of this alloy?

25.3 In order to determine its ultimate tensile strength, a strip of Teflon, 20 mm × 5 mm in cross-section, is subjected to an axial force until rupture occurs at 1.7 kN. What is the ultimate tensile strength of Teflon?

25.4 What would be the diameter of a copper wire if it ruptured under a tensile load of 264 N?

25.5 A tensile test on a 10 mm diameter specimen registered failure under a load of 32.6 kN. What force would be sufficient to break a 1.5 mm diameter wire made from this material?

25.6 A tensile-testing machine is to test specimens of diameter 15 mm which have ultimate tensile strengths of up to 820 N/mm². What is the maximum tensile force the machine must be capable of?

25.7 A tensile-testing machine is capable of a maximum pull of 125 kN. What is the maximum diameter that test specimens can have to ensure that materials of ultimate tensile strengths up to 1100 N/mm² can be tested to the point of failure?

25.8 Which is stronger in tension: a 20 mm diameter aluminium rod or a 15 mm × 15 mm brass bar?

25.9 A mild steel bar, 30 mm × 20 mm, is subjected to a tensile load of 141 kN. Calculate the stress in the rod and the factor of safety.

25.10 What would be a typical safety factor used in the design of a timber structure for supporting a large water tank?

25.11 What is the maximum allowable stress in a component made from high-tensile steel, designed for a cyclic load?

25.12 Determine the required cross-sectional dimensions of a square-section bar of high-tensile steel if it must carry a tensile load of 2950 N with a safety factor of 5.

25.4 *AXIAL STRAIN IN TENSION*

When a material is subjected to direct axial load, its size is changed in the direction of the applied force. A member subjected to a tensile force tends to stretch. It is found that for a given magnitude of tensile load, the elongation produced is proportional to the length of the member.

The elongation per unit of original length, called the **axial strain,** is a convenient relative measure of the change in the longitudinal dimension. If the total elongation produced in a member is designated by x, and the original length by l, then the axial strain, e, is:

$$e = \frac{x}{l}$$

It can easily be shown that since strain is the ratio between two lengths, it is dimensionless, provided that both x and l are expressed in the same units, usually millimetres.

Example 25.4

If the mild steel bar in Example 25.2 is 2.7 m long and extends 1.27 mm under the load, what is the axial strain?

Solution

$$\text{Strain } e = \frac{x}{l}$$

$$= \frac{1.27 \text{ mm}}{2700 \text{ mm}}$$

$$= 0.47 \times 10^{-3} \text{ (or 0.47 per 1000)}$$

25.5 *HOOKE'S LAW*

All solid materials exhibit some degree of stiffness, or the ability to resist deformation under load. Load–extension diagrams recorded during tensile testing of metals and other materials all indicate that in order to produce elongation, a force must be applied. Furthermore, the greater the force, the greater the amount of elongation produced by that force.

The property of stiffness is closely related to the elastic and plastic behaviour of materials. It is found that up to a certain point, known as the **elastic limit,** the material can be stretched without taking up any permanent deformation. It can be said that within the elastic limit, the material possesses the property of elasticity, or the ability to return to its original size after the force has been removed.

If, after the material is stretched to its elastic limit, the force continues to increase, the material ceases to be elastic and displays plasticity, i.e. it will suffer permanent deformation until eventually rupture occurs at what we already know to be the ultimate stress.

A further observation reveals that there is another limit, known as the **limit of proportionality**, within which the elongation produced by a tensile force is directly proportional to the force.[*]

The concept of proportionality between load and elongation of elastic materials was originally expounded in 1676 by the famous English physicist Robert Hooke (1635–1703). Hooke, who was the curator of experiments to the Royal Society of London, and a distinguished member of the Society, conducted scientific research in a remarkable variety of fields, ranging from astronomy and mathematics, of which he was a professor, to the study of microscopic fossils, which made him one of the first proponents of a theory of evolution. Hooke's experimental study of elastic materials led to his discovery of the law,[†] bearing his name, which laid the foundation for our understanding of the relationship between stress and strain.

In its original form, Hooke's law simply stated that for a bar of elastic material of uniform cross-section, subjected to a progressively increasing tensile load, the elongation is directly proportional to the deforming force. This formulation is limited to a particular bar of known dimensions. Furthermore, it is only true provided that the elongation produced does not exceed the limit of proportionality. A more general statement of **Hooke's law**, expressed in terms of stress ($f = F/A$) and strain ($e = x/l$), is that within the limit of proportionality, the strain is directly proportional to the stress producing it. The mathematical equation expressing Hooke's law is:

$$\frac{\text{stress}}{\text{strain}} = E$$

$$\text{or} \quad \frac{f}{e} = E$$

The constant E, relating stress and strain, is a measure of a material's ability to resist stretching, within the limit of proportionality, and is sometimes appropriately described as the **modulus of stiffness**. Historically, however, this modulus was introduced as a result of experimental study of elastic materials by another English scientist, Thomas Young (1773–1829), and was named in his honour as **Young's modulus of elasticity**.

From definitions of stress and strain, it follows that:

$$\boxed{E = \frac{f}{e} = \frac{F/A}{x/l} = \frac{Fl}{Ax}}$$

where E is Young's modulus of elasticity in MPa
  F is axial force in N
  A is cross-sectional area in mm^2
  x is elongation in mm
  l is original length in mm
  f is stress in MPa
  e is strain, dimensionless (mm/mm)

---

[*] For most materials, particularly metals, the elastic limit and the limit of proportionality almost coincide and are regarded as one for most practical purposes; often known as the 'proportional elastic limit'.

[†] In 1676, not yet sure of all the facts but wishing to establish priority of dates, Hooke tucked the anagram 'CEIIINOSSTTUV' into a scientific publication on an entirely different subject. Unscrambled, this reads *Ut tensio, sic vis*, which translated from Latin means 'As the stretch, so is the force', implying proportionality between elongation and force.

The modulus of elasticity can be determined experimentally. Table 25.3 gives typical values for common engineering materials.

Table 25.3 *Young's modulus of elasticity*

Material	E (MPa)
aluminium	70 000
brass	90 000
bronze	105 000
cast iron	120 000
copper	112 000
steel	200 000
timber	12 000

Example 25.5

The stress in the mild steel bar of Example 25.2 is 94 MPa, and the corresponding strain is 0.47×10^{-3}. What is the value of Young's modulus?

Solution

$$E = \frac{f}{e}$$
$$= \frac{94 \text{ MPa}}{0.47 \times 10^{-3}}$$
$$= 200\,000 \text{ MPa}$$

Example 25.6

A 40 mm diameter aluminium tie-rod, 500 mm long, is turned down to a diameter of 30 mm over 200 mm of its length. The rod is then subjected to a 20 kN axial pull. Determine the total amount of elongation and the safety factor.

Solution

From Tables 25.1 and 25.3, the ultimate tensile strength of aluminium is 150 N/mm^2 and Young's modulus is 70 000 MPa.

For the 40 mm diameter section, which is 300 mm long:

Stress:

$$f_1 = \frac{F}{A_1}$$
$$= \frac{20\,000 \text{ N}}{\dfrac{\pi \times 40^2}{4} \text{ mm}^2}$$
$$= 15.92 \text{ MPa}$$

Strain:

$$e_1 = \frac{f_1}{E}$$

$$= \frac{15.92}{70\,000}$$

$$= 0.227 \times 10^{-3}$$

Elongation:

$$x_1 = e_1 l_1$$

$$= 0.227 \times 10^{-3} \times 300 \text{ mm}$$

$$= 0.0682 \text{ mm}$$

Similarly, for the 30 mm diameter section of length 200 mm:

Stress:

$$f_2 = \frac{F}{A_2}$$

$$= \frac{20\,000 \text{ N}}{\dfrac{\pi \times 30^2}{4} \text{ mm}^2}$$

$$= 28.29 \text{ MPa}$$

Strain:

$$e_2 = \frac{f_2}{E}$$

$$= \frac{28.29}{70\,000}$$

$$= 0.404 \times 10^{-3}$$

Elongation:

$$x_2 = e_2 l_2$$

$$= 0.404 \times 10^{-3} \times 200 \text{ mm}$$

$$= 0.0808 \text{ mm}$$

The total elongation (x) of the rod is made up of the extensions of the two parts.

Total elongation:

$$x = x_1 + x_2$$

$$= 0.0682 \text{ mm} + 0.0808 \text{ mm}$$

$$= 0.149 \text{ mm}$$

The safety factor must be based on the maximum stress in the material, which occurs in the reduced section and is equal to 28.29 MPa.

Safety factor:

$$FS = \frac{UTS}{f}$$

$$= \frac{150}{28.29}$$

$$= 5.3$$

Problems

Refer to Tables 25.1 and 25.3 if required.

25.13 What is the strain in a structural member, 3.5 m long, if its elongation is 1.05 mm?

25.14 What is Young's modulus of a material in which a stress of 20 MPa produces a strain of 0.167 per 1000?

25.15 A steel measuring tape, 5 mm wide $\times$ 0.3 mm thick, is stretched with a pull of 50 N when used to measure a length of 40 m. Determine the stress in the material and the elongation produced.

25.16 What tensile load would produce an elongation of 0.18 mm in an aluminium rod 12 mm in diameter and 1.5 m long?

25.17 What is the diameter of a copper wire if a tensile force of 475 N produces an elongation of 6 mm in 10 m of its length?

25.18 If a light fitting of mass 59.8 kg is suspended from a 2.7 m length of wire of the type described in the previous problem, calculate the stress in the wire, the factor of safety, and the elongation of the wire.

25.19 A brass bar, with cross-section 25 mm $\times$ 25 mm and length 350 mm, is to carry an axial tensile load with a safety factor of 4. Calculate the maximum load it may carry and the elongation produced under this load.

25.20 During a test on a new synthetic material, a tensile force of 4170 N acting on a 10 mm diameter specimen produced an elongation of 0.17 mm on a gauge length of 100 mm.

What should be the square section of a component made from this material if it is to carry a tensile load of 12.5 kN with a strain not exceeding 0.001?

25.21 A brass rod is 500 mm long and 20 mm in diameter. If it is turned down to a diameter of 15 mm for a length of 100 mm at each end, calculate the total elongation when subjected to an axial pull of 7 kN. Calculate also the factor of safety at this load.

25.6 *A NOTE ON USEFUL TERMINOLOGY*

No material is perfectly rigid. When subjected to a force, no matter how small, every material will in fact suffer some degree of deformation. However, different materials respond differently to applied forces, or to the manner in which forces are applied.

Apart from ultimate strength, engineering materials have many other measurable properties related to their behaviour under load. The science of engineering materials has its own vocabulary of precisely defined terms used to describe the ability of different materials to undergo, or to resist, deformation.

While this is not a book on properties and testing of engineering materials, it is useful to be aware of some of these terms. Here is a brief summary.

> **elasticity**   the ability of a material to return to its original dimensions after having been deformed, upon removal of the deforming force
> **plasticity**   the ability of a material to undergo permanent deformation without failure
> **ductility**   the ability of a material to be permanently deformed by predominantly tensile forces

malleability   the ability of a material to be permanently deformed by predominantly compressive forces

stiffness   the ability of a material to resist deformation under load

toughness   the ability of a material to absorb energy when being deformed and therefore resist deformation and failure

hardness   the ability of a material to resist surface scratching, abrasion or indentation

impact strength   the ability of a material to withstand sudden application of a load

fatigue   the tendency of a material to fail under repeated application of a relatively small cyclic load

Review questions

1. Explain what is meant by the *ultimate tensile strength* of a material.
2. Define *tensile stress*.
3. What is the *factor of safety* in design?
4. List some criteria which influence selection of appropriate factors of safety.
5. Define *tensile strain*.
6. State *Hooke's law*.
7. Define *Young's modulus*.

CHAPTER 26

Compressive and thermal stresses

After learning about tensile stress in the preceding chapter, we direct our attention now to two other forms of direct axial stress, namely compressive stress and thermal stress.

Compression may be regarded as a phenomenon which is opposite to tension, in so far as an axial force is applied in the opposite direction, tending to compress, rather than stretch, the material.

Thermal stresses can be tensile or compressive. They are caused by the tendency of most materials to expand when heated, or to contract when cooled. If such thermal expansion or contraction is fully or partially prevented, stresses are set up within the material.

Expected learning outcomes

After carefully studying the material presented in this chapter, working through all numerical examples, and successfully completing all practice problems, students should be able to:
1. calculate the compressive stress in a component subjected to a direct axial compressive load;
2. calculate the compressive stress due to the weight of a column;
3. define *Poisson's ratio* and use it to estimate the amount of lateral deformation in a material subjected to an axial load;
4. calculate thermal stresses in components constrained between two supports.

26.1 COMPRESSIVE STRENGTH

Most engineering materials can be divided broadly into three groups: ductile, brittle and composite.

Ductile materials, such as copper, aluminium and steel, tend to have approximately the same strength in compression as they have in tension. They are seldom used in applications where compression is the predominant type of loading condition, their main usefulness being in their ability to withstand considerable tensile stress.

Brittle materials, which include cast iron, bricks, masonry and concrete, have relatively low tensile strength and are not used in structures and components subjected to large tensile forces. However, these materials can usually withstand high compressive loads and are employed in heavy masses for building foundations, piers and dams (brick, masonry and low-grade concrete), and for machine-tool framework (cast iron). For these applications, the relatively high **ultimate compressive strength** (UCS), which these materials possess, is far more important than their tensile strength. UCS is defined as follows:

$$\text{UCS} = \frac{\text{force causing crushing failure}}{\text{cross-sectional area}}$$

As the name suggests, **composite materials** represent some combination of individual materials from the previous categories. The best known examples are fibreglass and reinforced concrete. Reinforced concrete, which is used extensively in bridge and building construction, combines high compressive strength of concrete with superior tensile properties of steel reinforcement.

It is clear that knowledge of compressive strength of some materials, as distinct from their tensile strength, is useful from the point of view of the design engineer.

Compressive strength of cast iron is generally about four times its tensile strength, depending on the composition of alloy elements. Its average compressive strength is 700 N/mm^2.

Compressive strength of concrete, and especially that of bricks, varies so greatly that no single figure can be given. Typical values lie somewhere between 10 N/mm^2 and 80 N/mm^2, with the average somewhere around 20 N/mm^2.

Compressive strength of timber, loaded parallel to the grain, is in the order of 40 N/mm^2 to 55 N/mm^2; when loaded perpendicular to the grain, it can be as low as 5 N/mm^2.

Example 26.1

A portable testing machine, used for quality control on a large construction site, carries out crushing tests on concrete by applying an axial force of 433 kN. This causes compression failure in a concrete specimen, 150 mm in diameter and 300 mm high (Fig. 26.1). What is the ultimate compressive strength of concrete?

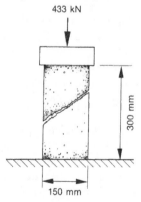

433 kN

300 mm

150 mm

Fig. 26.1

Solution

$$\text{UCS} = \frac{\text{force causing crushing failure}}{\text{cross-sectional area}}$$

$$= \frac{433\,000 \text{ N}}{\left(\dfrac{\pi \times 150^2}{4}\right) \text{ mm}^2}$$

$$= 24.5 \text{ N/mm}^2$$

26.2 *DENSITY, WEIGHT AND STRESS*

In static structures, direct axial compressive stress is often caused by the weight of the structure itself, or by the weights of various loads supported by the structure. Therefore, it is often necessary to calculate the weights of structural members, such as beams and columns, or to estimate the weight of a load, such as a tank full of water, before compressive stresses can be evaluated.

Let us refresh our memory on how weight can be related to the density of the material from which a component is made. Density, ρ,* of a material is defined as its mass per unit volume. The SI unit of density, derived from the units of mass (kg) and volume (m^3), is kilogram per cubic metre (kg/m^3).

$$\rho = \frac{m}{V}$$

Alternatively, tonnes per cubic metre may be used as a convenient unit of density.

Typical values of density for some solid materials are given in Table 26.1. It should be understood that these are approximate or typical values only; the actual density of each material will depend on its exact composition.

Table 26.1   *Densities of various solids*

Material	Density (kg/m^3)
aluminium	2780
balsa wood	160
brass	8250
brick	2080
bronze	8670
cast iron	7200
concrete	2240
copper	8870
ice	920
oregon pine timber	530
rubber	920
sand	1470
steel	7800
zinc	7020

Example 26.2

Determine the weight of a tubular steel column, 3 m high, with outside diameter 120 mm and inside diameter 100 mm.

---

* ρ is the Greek letter 'rho'.

Solution
Cross-sectional area:

$$A = \frac{\pi \times 0.12^2}{4} - \frac{\pi \times 0.1^2}{4}$$
$$= 3.456 \times 10^{-3} \text{ m}^2$$

Volume:

$$A \times L = 3.456 \times 10^{-3} \text{ m}^2 \times 3 \text{ m}$$
$$= 0.0104 \text{ m}^3$$

Mass:

$$V \times \rho = 0.0104 \text{ m}^3 \times 7800 \ \frac{\text{kg}}{\text{m}^3}$$
$$= 80.86 \text{ kg}$$

Weight:

$$mg = 80.86 \text{ kg} \times 9.81 \ \frac{\text{N}}{\text{kg}}$$
$$= 793.3 \text{ N}$$

Example 26.3
What is the stress at the foot of the column in the previous example if there is no other load resting on the column?

Solution
The cross-sectional area of the column, in square millimetres, is:

$$A = 3.456 \times 10^{-3} \text{ m}^2$$
$$= 3456 \text{ mm}^2$$

Its own weight resting on this area is 793.3 N. Hence compressive stress at the foot of this column, due to its own weight, is:

$$f_c = \frac{F_w}{A}$$
$$= \frac{793.3 \text{ N}}{3456 \text{ mm}^2}$$
$$= 0.23 \text{ MPa}$$

It is quite clear that the magnitude of stress in this case is quite insignificant. But let us consider another example.

Example 26.4

Four tubular columns, as described in Example 26.3, form four legs of a platform, supporting a 4.2 m diameter cylindrical water tank, filled to a depth of 3.5 m (Fig. 26.2). If the total mass of the structure and the tank without water is 2300 kg, and water density is 1000 kg/m^3, calculate the stress in the tubular legs.

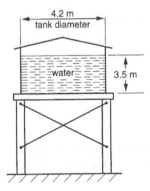

Fig. 26.2

Solution

$$\text{Mass of water in tank} = \frac{\pi \times 4.2^2}{4} \times 3.5 \times 1000$$
$$= 48\,490 \text{ kg}$$

$$\text{Total mass} = 48\,490 + 2300$$
$$= 50\,790 \text{ kg}$$

$$\text{Weight of this mass} = 50\,790 \times 9.81$$
$$= 498\,250 \text{ N}$$

The total cross-sectional area of the four columns is:

$$4 \times 3456 \text{ mm}^2 = 13\,824 \text{ mm}^2$$

Therefore, the stress in the tubular legs is:

$$f_c = \frac{F_w}{A}$$
$$= \frac{498\,250 \text{ N}}{13\,824 \text{ mm}^2}$$
$$= 36 \text{ MPa}$$

26.3 *AXIAL STRAIN IN COMPRESSION*

The mechanism of failure of materials under compression is complex and different from that of tensile rupture. Furthermore, it varies quite significantly between ductile and brittle materials.

Brittle materials, such as concrete or glass, tend to suffer relatively sudden crushing failure, with little observable axial deformation before the failure occurs.

On the other hand, ductile materials under compression exhibit similar behaviour to that seen under tension. In particular, provided the applied stress is below the elastic limit in compression, gradual contraction in size in proportion to the applied compressive force is observed. Therefore, in the case of ductile materials in compression, the formula and numerical values of Young's modulus can be applied, with suitable changes in the directions of force and deformation.

$$E = \frac{f}{e} = \frac{F/A}{x/l} = \frac{Fl}{Ax}$$

Example 26.5

An axial compressive load of 190 kN is applied to a cylindrical component, 80 mm in diameter and 300 mm long, made of brass. Calculate the compressive stress, the compressive strain and the amount of axial deformation under load.

Solution
Cross-sectional area:

$$A = \frac{\pi \times 80^2}{4}$$
$$= 5027 \text{ mm}^2$$

Compressive stress:

$$f = \frac{F}{A}$$
$$= \frac{190\,000 \text{ N}}{5027 \text{ mm}^2}$$
$$= 37.8 \text{ MPa}$$

Young's modulus for brass is 90 000 MPa (see Table 25.3). Hence axial strain is:

$$e = \frac{f}{E}$$
$$= \frac{37.8}{90\,000}$$
$$= 0.000\,42$$

Therefore, the amount of axial deformation under load is:

$$x = el$$
$$= 0.000\,42 \times 300 \text{ mm}$$
$$= 0.126 \text{ mm}$$

26.4 *POISSON'S RATIO*

When a piece of material is compressed in the axial direction, the axial contraction is always accompanied by some lateral expansion at right angles to the applied force, i.e. the material expands sideways. These lateral deformations can be expressed in relative terms, in millimetres per millimetre, and are called **lateral strains**.

It has been shown by experimental evidence that for any given material within its elastic limit, lateral strains have a fixed relation to the axial strains caused by the force. This constant is known as **Poisson's ratio**, named after the French scientist who formulated this concept. It is a definite property of a material, and its value applies equally to materials in tension or in compression. Poisson's ratio is given the symbol v.[*]

$$\text{Poisson's ratio } v = \frac{\text{lateral strain}}{\text{axial strain}}$$

Example 26.6

It has been observed that when an 80 mm × 40 mm × 40 mm piece of rubber is compressed along the 80 mm axis by 16 mm, its dimensions increase by 3.6 mm in each of the other two directions (Fig. 26.3). What is the value of Poisson's ratio of this material?

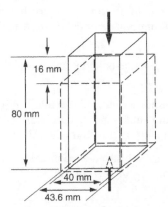

Fig. 26.3

Solution

Axial strain:

$$e_A = \frac{16\,\text{mm}}{80\,\text{mm}}$$
$$= 0.2$$

Lateral strain:

$$e_L = \frac{3.6\,\text{mm}}{40\,\text{mm}}$$
$$= 0.09$$

---

[*] v is the Greek letter 'nu'.

Poisson's ratio:

$$v = \frac{e_L}{e_A}$$
$$= \frac{0.09}{0.2}$$
$$= 0.45$$

Rubber is known to have an exceptionally high value of Poisson's ratio (0.5 is the highest value theoretically possible). For some grades of concrete, it can be as low as 0.1, and for cork it is practically zero. However, for the majority of engineering materials, the value of Poisson's ratio lies within the narrow range of 0.25 to 0.35, mostly around 0.3.

 # Problems

26.1 An aluminium column has an outside diameter of 160 mm, an inside diameter of 140 mm, and is 2.5 m high. Determine its mass and weight.

26.2 What is the ultimate compressive strength of brick if a test cube, 50 mm × 50 mm × 50 mm, is crushed by a force of 77.5 kN during a compression test?

26.3 What compressive force would be required to crush a cylindrical test specimen of length 100 mm and diameter 70 mm, made of concrete, if the ultimate compressive strength of concrete is 65 N/mm^2?

26.4 Determine the stress in a concrete column, 150 mm × 150 mm, if it supports an axial load of 112.5 kN.

26.5 A short cast-iron column has an outside diameter of 100 mm and an inside diameter of 75 mm. Calculate the maximum load it may carry if the factor of safety is 6 and the ultimate compressive strength of cast iron is 700 N/mm^2.

26.6 A vertical timber column, with cross-section 200 mm × 200 mm and height 3 m, is subjected to an axial compression load of 120 kN. Calculate the stress in the timber and the amount of axial deformation.

26.7 An axial compressive load of 85 kN is applied to a cylindrical component, 50 mm in diameter and 250 mm long, made of aluminium. Calculate the compressive stress, the compressive strain and the amount of axial deformation under load.

26.8 If the aluminium cylinder in problem 26.7 is machined down to a diameter of 40 mm over a length of 150 mm, the rest remaining at the 50 mm diameter, and is subjected to the same compressive force of 85 kN, what will be the total amount of axial deformation under load?

26.9 What is the value of Poisson's ratio for an aluminium alloy if its lateral strain under axial load is found to be 0.40 in 1000, while its axial strain is 1.25 in 1000?

26.10 A cylindrical piece of rubber, initially 136 mm long and 75 mm in diameter, is subjected to axial compression so that its final length under stress is reduced to 117 mm. Assuming a Poisson's ratio of 0.43, calculate the final diameter.

26.11 For the rubber cylinder in the previous problem, calculate the percentage increase in cross-sectional area and the percentage decrease in volume.

26.5 *THERMAL EXPANSION OF SOLIDS*

In general, when solid materials are heated, they expand, and when cooled, they contract. This behaviour of materials can lead to some very useful applications, such as shrink fitting gear wheels on shafts, or they can lead to damaging effects of restricted thermal expansion, such as on steampipes or bridge spans.

Experimental data have been collected that describe the change in length per unit of the original length of a bar of material for each degree of temperature rise. This measure is known as the **coefficient of thermal expansion**, α, of the material. Typical values of the coefficient are shown in Table 26.2.

Table 26.2   *Coefficients of thermal expansion*

Material	α (mm/mm.K)
aluminium	0.000 024
brass	0.000 021
bronze	0.000 020
copper	0.000 017
steel	0.000 012
cast iron	0.000 010
concrete	0.000 010
glass	0.000 008

We should note that the unit of the coefficient of thermal expansion, when expressed fully, represents the amount of elongation as a fraction of a millimetre for every millimetre of original length, which is due to a temperature rise of one kelvin (mm/mm.K). Since a temperature difference of one kelvin is exactly equal to a temperature difference of one degree Celsius, this unit is often stated as mm/mm.°C, without affecting the numerical value of the constant. Furthermore, since the millimetres can be cancelled out, the unit is sometimes referred to as 'per degree'.

It follows from the definition of the coefficient of linear expansion that the increase in length (ΔL) is proportional to the coefficient (α), to the original length (L_0) and to the temperature rise (Δt):

$$\Delta L = \alpha L_0 \, \Delta t$$

Example 26.7

If an aluminium rod is 1.2 m long at 15°C, what will be its free length (Fig. 26.4(a)) at 40°C?

Solution

The amount of thermal expansion is:

$$\Delta L = \alpha L_0 \, \Delta t$$
$$= 0.000\,024 \times 1200 \times (40 - 15)$$
$$= 0.72 \text{ mm}$$

Therefore, the final length of the rod will be:

$$L = L_0 + \Delta L$$
$$= 1200 \text{ mm} + 0.72 \text{ mm}$$
$$= 1200.72 \text{ mm}$$

On the face of it, this does not seem to be a very significant increase in the length of the bar. However, there are two possible circumstances that can make thermal expansion very important. The first is when accurate or delicate instruments are involved—thermal expansion can introduce significant errors into their operation. The second is when the tendency for expansion is restricted by rigid supports—thermal stresses are then set up within the material (Fig. 26.4(b)).

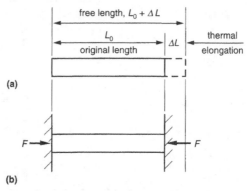

Fig. 26.4   **(a)** *Free thermal expansion*   **(b)** *Thermal expansion contained between unyielding supports*

26.6 *THERMAL STRESS*

In situations where thermal expansion is fully prevented by unyielding supports, we can regard ΔL as the amount by which the material has to be compressed in order to fit it back into its original size.

Example 26.8

If thermal expansion of the rod in Example 26.7 is fully prevented by being constrained between two unyielding supports, what will be the compressive stress induced in the material?

Solution

$$\text{Compressive strain} = \frac{\text{amount of prevented deformation}}{\text{original length}^*}$$

$$\therefore e = \frac{0.72}{1200}$$
$$= 0.0006$$

---

* It is common practice in this calculation to use the original length before heating ($L_0 = 1200$ mm) rather than the theoretically more correct value of 1200.72 mm. However, the minuscule improvement in accuracy, which can be achieved by using the more precise value, does not usually justify the complexities which this refinement often brings.

Young's modulus for aluminium is 70 000 MPa. Therefore, stress is:

$$f = Ee$$
$$= 70\,000 \times 0.0006$$
$$= 42 \text{ MPa}$$

This is quite a significant compressive stress for a modest temperature rise of only 25 degrees. In a long slender member, it can easily lead to distortion by buckling.

In practice, we seldom find absolutely rigid unyielding supports. There is usually some amount (ΔY) by which the supports will yield against the pressure exerted by the member constrained between them.

Example 26.9

If in the previous example the supports yield by a total amount of 0.48 mm, what will be the stress in the aluminium rod?

Solution

$$\text{Compressive strain} = \frac{\text{amount of prevented deformation}}{\text{original length}}$$

$$e = \frac{\Delta L - \Delta Y}{L_0}$$

$$= \frac{0.72 - 0.48}{1200}$$

$$= 0.0002$$

Therefore, stress is:

$$f = Ee$$
$$= 70\,000 \times 0.0002$$
$$= 14 \text{ MPa}$$

All this can be summarised as follows:

$$\text{Stress } f = Ee = E\frac{\Delta L - \Delta Y}{L_0}$$

where ΔY is the total amount by which the supports yield.

In the case of completely prevented expansion, $\Delta Y = 0$. Then:

$$\text{Stress } f = E\frac{\Delta L - \Delta Y}{L_0}$$

$$= E\frac{\Delta L}{L_0}$$

$$= E\frac{\alpha L_0 \Delta t}{L_0}$$

Hence the formula for compressive stress due to *completely prevented* thermal expansion is:

$$f = E\alpha\,\Delta t$$

 Problems

26.12 Prove that, for the same temperature rise, an aluminium rod will expand twice as much as a steel rod, and three times as much as a glass rod, if all three rods have the same original length.

26.13 Calculate the amount of free thermal expansion of a steel beam, which is precisely 15 m long at 0°C, when the temperature rises to 45°C.

26.14 A surveyor's steel tape measures exactly 40 m, between end markings, at 15°C. What error will be made in measuring a distance of 400 m when the temperature of the tape is 40°C?

26.15 A continuous concrete road pavement was laid in 15 m sections with expansion joints, at the end of each section, measuring 10 mm at 10°C. What will be the size of the joint when the temperature of the concrete reaches 50°C?

26.16 At what temperature will the gap between the concrete slabs in problem 26.15 close up completely?

26.17 A brass cylinder, 35 mm in diameter and 0.5 m long, is held endwise between two unyielding supports while its temperature is raised by 50 degrees. Determine the stress in the bar and the force it exerts on each of the supports.

26.18 If the supports in problem 26.17 yield by 0.32 mm, determine the new stress in the brass cylinder and the force it continues to exert on each of the supports.

26.19 The steel rails of a continuous, straight railway track consist of sections, each of which is 12 m long, laid with 9 mm spaces between their ends at 0°C. At what temperature will the rails touch end to end?

26.20 What compressive stress will be produced in the rails in problem 26.19 if the temperature of the rails rises to 70°C?

26.21 Calculate the maximum temperature rise which may be allowed for a metal bar constrained between two unyielding supports if the stress in the material of the bar is not to exceed 60 MPa. Take the metal bar to be made of:
(a) steel
(b) copper
(c) cast iron

Review questions

1. Name some materials which are stronger in compression than in tension.
2. What is the most common cause of compressive stress in static structures?
3. Explain the meaning of *Poisson's ratio*.
4. Define *coefficient of thermal expansion*, and state its unit.
5. State the formula for calculating the amount of free thermal expansion for a solid material.
6. State the formula for calculating compressive stress in materials constrained between two unyielding end supports.

CHAPTER 27

Shear and torsional stresses

It is now time for us to look at a very different type of stress which is *not* due to an axial load. This type of stress is called **shear stress** and manifests itself primarily as direct shear or as torsion. Later you will learn that shear stress also plays a significant part in other more complex types of loading, such as bending of beams, and in various forms of combined stress situations.

However, in this chapter, we start at the very beginning by first considering direct shear stress, and then discussing pure torsion.

Expected learning outcomes

After carefully studying the material presented in this chapter, working through all numerical examples, and successfully completing all practice problems, students should be able to:

1. define *shear stress*, and distinguish between *direct shear* and *torsional shear*;
2. calculate direct shear stress and solve related problems involving shearing of plate and punching of holes;
3. define *modulus of rigidity*, and explain its relationship to Young's modulus and Poisson's ratio;
4. calculate torsional shear stress and solve related problems involving components subjected to pure torsion.

27.1 *SHEAR STRENGTH*

When a steel plate is cut by a guillotine, or a hole is punched in it, the failure of the material occurs not in a plane normal to the force, as is the case in tension or compression, but as a sliding failure parallel to the load applied. The material is said to shear under the action of a shear force.

Apart from direct applications of shear forces for cutting or punching materials, consideration of **shear strength** has relevance to the design of bolted and welded connections, bending of beams, and torsion in shafts.

Typical values of **ultimate shear strength** (USS) of some materials are given in Table 27.1:

Table 27.1 *Ultimate shear strength of some materials*

Material	USS (N/mm^2)
aluminium	125
brass	150
mild steel	360

Example 27.1

Determine the force required to punch a 12 mm diameter hole in a mild steel plate 6 mm thick.

Solution

The area resisting shear is measured by the product of the circumference (πD) and the plate thickness (t). (See Fig. 27.1.)

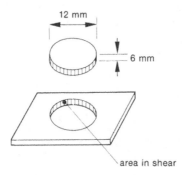

Fig. 27.1

$$A = \pi \times D \times t$$
$$= \pi \times 12 \text{ mm} \times 6 \text{ mm}$$
$$= 226.2 \text{ mm}^2$$

The ultimate strength of mild steel is 360 N/mm², i.e. a force of 360 N is required to shear through every square millimetre of area subjected to shear. Therefore, the total force required to punch the hole is:

$$F = 360 \text{ N/mm}^2 \times 226.2 \text{ mm}^2$$
$$= 81\,432 \text{ N}$$
$$= 81.4 \text{ kN}$$

Example 27.2

Determine the greatest thickness of brass sheet, 850 mm wide, that can be sheared by a straight-cutting* guillotine capable of applying a force of 510 kN.

Solution

The ultimate shear strength of brass is 150 N/mm².
The area in shear is given by the product of the width and thickness of the sheet:

$$A = 850 \text{ mm} \times t$$
$$= 850t \text{ mm}^2$$

The force available is 510 000 N. Therefore:

$$510\,000 \text{ N} = 150 \text{ N/mm}^2 \times 850t \text{ mm}^2$$
$$\therefore t = 4 \text{ mm}$$

---

* If the guillotine knife edge cuts obliquely, the cutting action progressing gradually along the line of the cut requires a somewhat different force.

27.2 SHEAR STRESS

In the previous section, some examples of shear force were considered, including punching of holes and cutting of metal plate by a guillotine. There are also numerous applications where shear force is not sufficient to overcome the ultimate shear strength of a material and consequently no failure occurs. However, shear stress may exist. **Shear stress** is produced by equal and opposite forces whose lines of action do not coincide.

Figure 27.2(a) shows a simple pin coupling in which the pin is placed in double shear by the force F_s along the two shear planes as indicated. Shear stress acts along the planes, parallel to the lines of action of the applied forces. The tendency is to shear through the pin as shown in Figure 27.2(b). Naturally, if the size and strength of the pin are sufficient, no such failure would occur.

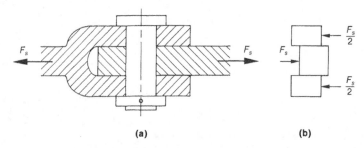

(a)   (b)

Fig. 27.2   **(a)** *A simple pin coupling in double shear*   **(b)** *Tendency to shear through the pin*

Shear stress (f_s) is assumed to be distributed uniformly over the total area subjected to shear force, and may be defined as shear force per unit of the total area:

$$f_s = \frac{F_s}{A_s}$$

Care should be exercised to distinguish between areas in single shear, such as those of a bolt or rivet holding two plates in a simple lap joint, and areas in double shear, as illustrated by the following examples.

Example 27.3

If the diameter of the mild steel pin in Figure 27.2 is 10 mm, and the maximum force applied to the coupling is 11.3 kN, what is the shear stress in the material of the pin and the factor of safety?

Solution

This is a case of double shear. Therefore, the total area is:

$$A_s = 2 \times \frac{\pi \times 10^2}{4}$$
$$= 157 \text{ mm}^2$$

Shear stress:

$$f_s = \frac{F_s}{A_s}$$

$$= \frac{11\,300 \text{ N}}{157 \text{ mm}^2}$$

$$= 72 \text{ MPa}$$

Since the ultimate shear strength of mild steel is 360 N/mm², the factor of safety is:

$$FS = \frac{USS}{f_s}$$

$$= \frac{360}{72}$$

$$= 5$$

Example 27.4

Determine the required diameter of a single mild steel bolt, holding two overlapping strips of metal, against a shear force of 4.5 kN, if the allowable stress in shear is 90 MPa.

Solution

From $f_s = F_s/A_s$, the required area is:

$$A_s = \frac{F_s}{f_s}$$

$$= \frac{4500 \text{ N}}{90 \text{ MPa}}$$

$$= 50 \text{ mm}^2$$

This is a case of single shear; only one cross-sectional area need be considered. Hence:

$$D = \sqrt{\frac{4A_s}{\pi}}$$

$$= \sqrt{\frac{4 \times 50}{\pi}}$$

$$= 7.98 \text{ mm}$$

Therefore an 8 mm diameter bolt will be satisfactory.

 Problems

Where appropriate, refer to Table 27.1 for ultimate shear strength values.

27.1 Determine the force required to punch a 15 mm diameter hole in a 6 mm thick aluminium plate.

27.2 Determine the force required to punch 20 mm × 15 mm rectangular holes in a mild steel plate 8 mm thick.

27.3 A press designed to punch 25 mm diameter holes in 10 mm thick brass plate is to be used to punch 15 mm × 12 mm rectangular holes in mild steel using the same force. What is the maximum thickness of mild steel plate that can be used?

27.4 Determine the force required to shear an aluminium strip, 250 mm wide by 3 mm thick, using a straight-cutting guillotine knife.

27.5 A mild steel bolt, 14 mm in diameter, is subjected to double shear by a force of 27.7 kN. Determine the shear stress and the factor of safety.

27.6 Determine the required number of 10 mm diameter mild steel bolts to hold two overlapping strips of metal against a total shear force of 75.4 kN if the allowable stress in shear is 120 MPa.

27.7 Calculate the required diameter of a brass pin which is to be subjected to double shear under a load of 1.7 kN. The factor of safety is to be 5.

27.8 A 10 mm diameter specimen of hardened tool steel was tested in double shear and failed under a load of 106.8 kN. If a machine-tool component of square cross-section 12 mm × 12 mm is made from the same material, determine the maximum shear load in single shear allowed if the factor of safety is to be 4.

27.9 A flanged coupling is used to transmit a torque of 1500 N.m between two shafts. The coupling has four 8 mm diameter bolts equally spaced on a pitch-circle diameter of 175 mm. Determine the shear stress in the bolts.

27.10 Determine the maximum torque that can be transmitted by a flanged coupling using six 10 mm diameter mild steel bolts equally spaced on a pitch-circle diameter of 250 mm, if the allowable stress is 90 MPa.

27.11 The following specification is used for the design of a flanged coupling between two coaxial shafts:

Speed:	650 rpm
Power transmitted:	550 kW
Bolt diameter:	12 mm
Pitch-circle diameter:	200 mm
Material:	mild steel
Factor of safety:	4

Determine the number of bolts required, assuming the bolts are equally loaded.

27.12 A rectangular hole, 15 mm × 10 mm, is to be punched in a 4 mm thick brass sheet. Calculate the force required and the compressive stress in the punch.

If the ultimate compressive strength of the material of the punch is 650 MPa, what is the factor of safety during punching?

27.13 Determine the maximum thickness of an aluminium sheet in which a 50 mm diameter hole is to be punched, if the allowable compressive stress in the punch is not to exceed 80 MPa.

27.3 *SHEAR MODULUS OF RIGIDITY*

Elasticity of solid materials is not limited to tension and compression. When an elastic material is subjected to moderate shear force, some degree of deformation occurs. However, unlike tension or compression, shear stress is accompanied by deformation of shape, rather than size.

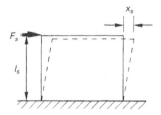

Fig. 27.3   *Shape deformation from shear force*

Let us consider, for example, a solid block, shown in Figure 27.3, subjected to shear force F_s. The distortion of its shape, indicated by dotted lines, can be described in terms of deformation x_s relative to a dimension l_s at right angles to x_s. The ratio of x_s to l_s is called **shear strain** e_s.

$$e_s = \frac{x_s}{l_s}$$

It should be clearly understood that although the form of the definition and the equation arising from it are almost identical, shear strain is distinctly different from axial strain in that deformation is that of shape and not size.

Having defined shear stress (f_s) and shear strain (e_s), we can relate the two as a ratio called the **modulus of rigidity,** usually denoted by G.

$$G = \frac{f_s}{e_s}$$

The modulus of rigidity for any given material is found to be constant, within its limit of proportionality. Furthermore, it is found that the numerical value of the modulus of rigidity of a particular material is equal to a fixed percentage of its Young's modulus (E). For isotropic materials, i.e. materials which have the same elastic properties in all directions,[*] this percentage, determined theoretically by using the molecular theory of structure of the material, is 40 per cent. Therefore:

$$G = 0.4E$$

Example 27.5
What is the value of the modulus of rigidity of steel?

Solution
Young's modulus for steel is $E = 200\,000$ MPa. Therefore:

$$
\begin{aligned}
G &= 40\% \text{ of } E \\
&= 0.4 \times 200\,000 \text{ MPa} \\
&= 80\,000 \text{ MPa}
\end{aligned}
$$

---

[*] This limitation obviously excludes timber.

A more precise relationship between shear modulus of rigidity (G), Young's modulus of elasticity (E) and Poisson's ratio (v) is given by the following formula:

$$G = \frac{E}{2(1 + v)}$$

It can be shown, by simple substitution, that since the value of Poisson's ratio for most materials seldom varies beyond its narrow range of 0.25 to 0.35, the value of G will usually be within 37 to 40 per cent of the Young's modulus. This variation is not very significant, and 40 per cent may be used as sufficiently accurate for our purposes.

Example 27.6

If Poisson's ratio for steel is 0.29, what is a more accurate estimate of its shear modulus of rigidity?

Solution
Modulus of rigidity:

$$G = \frac{E}{2(1 + v)}$$

$$= \frac{200\,000 \text{ MPa}}{2(1 + 0.29)}$$

$$= 77\,500 \text{ MPa}$$

Example 27.7

An antivibration mounting for a machine is in the form of rubber pads, 200 mm × 200 mm × 10 mm thick, each subjected to a periodic horizontal force, reaching a maximum of 600 N, distributed over its area.

Determine the amount of shear deformation, x_s, corresponding to the maximum force, if the modulus of rigidity of rubber is 1.5 MPa.

Solution
Shear stress in the material of the pads is:

$$f_s = \frac{F_s}{A_s}$$

$$= \frac{600 \text{ N}}{200 \text{ mm} \times 200 \text{ mm}}$$

$$= 0.015 \text{ MPa}$$

From $G = f_s/e_s$, shear strain is:

$$e_s = \frac{f_s}{G}$$

$$= \frac{0.015}{1.5}$$

$$= 0.01$$

Therefore, from $e_s = x_s/l_s$, shear deformation, x_s, is:

$$x_s = e_s l_s$$
$$= 0.01 \times 10 \text{ mm}$$
$$= 0.1 \text{ mm}$$

27.4 *TORSIONAL STRESS*

Torsion is a type of loading condition which results from the twisting action of two equal and opposite torques acting on a component, such as on a power transmission shaft, or on a torsion bar used in some vehicles for wheel suspension.

Our discussion is confined to the treatment of components with cylindrical or tubular shape which are subjected to pure twisting action about their longitudinal axis.

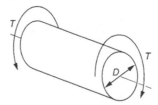

Fig. 27.4 *Torsion on a cylindrical bar*

Let us consider the cylindrical bar shown in Figure 27.4. When action and reaction torques are applied, a certain amount of twisting occurs so that each circular cross-section of the bar rotates slightly relative to the next cross-section immediately adjacent to it. This induces internal shear stresses, within the material of the bar, in the plane of the cross-section. However, unlike direct shear, stresses due to torsion are not distributed uniformly over the cross-sectional area of the bar. In fact there is no stress at all at the centre of the cross-sectional area. On the other hand, torsional shear stress reaches its maximum value at the maximum distance from the centre, i.e. at the circumference of the cross-section.

In this book we are concerned only with the maximum shear stress produced by the twisting action of the applied torque, and with the formula that will enable us to calculate its numerical value. It has been shown that **maximum torsional shear stress** in a solid or hollow cylinder is given by the following relation:

$$\boxed{f_{ts} = \frac{Tr}{J}}$$

where T is torque
 r is the radius of the cylinder
 J is a geometrical property of the cross-section, known as its **polar moment of inertia**

For a *solid* cylinder:

$$r = \frac{D}{2}$$

and:

$$J = \frac{\pi D^4}{32}$$

Example 27.8

Calculate the torsional shear stress in a 50 mm diameter shaft if it is subjected to a torque of 1400 N.m.

Solution

It is important to ensure that all parameters are expressed in a consistent set of units, e.g. newtons–millimetres–megapascals.

Torque:

$$T = 1400 \text{ N.m}$$
$$= 1400 \times 10^3 \text{ N.mm}$$

Radius:

$$r = \frac{D}{2}$$
$$= 25 \text{ mm}$$

Polar moment of inertia:

$$J = \frac{\pi D^4}{32}$$
$$= \frac{\pi \times 50^4}{32}$$
$$= 613.6 \times 10^3 \text{ mm}^4$$

Hence the maximum value of torsional shear stress is:

$$f_{ts} = \frac{Tr}{J}$$
$$= \frac{1400 \times 10^3 \text{ N.mm} \times 25 \text{ mm}}{613.6 \times 10^3 \text{ mm}^4}$$
$$= 57 \text{ MPa}$$

For a *hollow* cylinder, with outside diameter D_o and inside diameter D_i:

$$r = \frac{D_o}{2}$$

and:

$$J = \frac{\pi(D_o^4 - D_i^4)}{32}$$

Example 27.9

Calculate the torsional shear stress in a hollow shaft, which has an outside diameter $D_o = 55$ mm and an inside diameter $D_i = 38$ mm, if it is subjected to a torque of 1400 N.m.

Solution

Torque:

$$T = 1400 \text{ N.m}$$
$$= 1400 \times 10^3 \text{ N.mm}$$

Extreme radius:

$$r_o = \frac{D_o}{2}$$
$$= 27.5 \text{ mm}$$

Polar moment of inertia:

$$J = \frac{\pi(D_o{}^4 - D_i{}^4)}{32}$$
$$= \frac{\pi(55^4 - 38^4)}{32}$$
$$= 693.7 \times 10^3 \text{ mm}^4$$

Hence the maximum value of the torsional shear stress is:

$$f_{ts} = \frac{Tr}{J}$$
$$= \frac{1400 \times 10^3 \text{ N.mm} \times 27.5 \text{ mm}}{693.7 \times 10^3 \text{ mm}^4}$$
$$= 55.5 \text{ MPa}$$

27.5 *TORSIONAL DEFORMATION—ANGLE OF TWIST*

Let us now examine torsional deformation, known as the **angle of twist**. If torque is applied at one end of a cylindrical bar while it is rigidly held at the other end, the bar will be twisted so that the two cross-sections at the ends will rotate relative to each other through an angle θ, as shown in Figure 27.5.

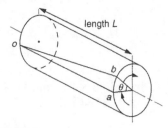

Fig. 27.5   *Angle of twist, θ*

Torsional shear strain, e_{ts}, is defined as the ratio of the circumferential distance *ab* at radius *r* to the corresponding length *L*. Distance $ab = r\theta$, if the angle is expressed in radians. Hence:

$$e_{ts} = \frac{r\theta}{L}$$

We have previously determined that torsional shear stress is given by:

$$f_{ts} = \frac{Tr}{J}$$

It follows from the definition of shear modulus of rigidity that:

$$G = \frac{\text{stress}}{\text{strain}} = \frac{f_{ts}}{e_{ts}} = \frac{Tr}{J} \times \frac{L}{r\theta} = \frac{TL}{J\theta}$$

If this expression is transposed to make the angle of twist as its subject, we get:

$$\theta = \frac{TL}{JG}$$

Therefore, the amount of angular deformation produced by the twisting action of the torque, i.e. the angle of twist θ, is directly proportional to the applied torque T and to the length of the bar L, and is inversely proportional to the shear modulus of rigidity G of the material from which the bar is made and to the polar moment of inertia J of its cross-section.

It should always be remembered that the angle of twist calculated by this formula is in *radians*.

Example 27.10

Calculate the angle of twist of the shaft in Example 27.8 if the material is steel and the shaft is 1530 mm long.

Solution

The modulus of rigidity of steel is $G = 80\,000$ MPa (see Example 27.5).
Substitute:

$$\theta = \frac{TL}{JG}$$

$$= \frac{1400 \times 10^3 \text{ N.mm} \times 1530 \text{ mm}}{613.6 \times 10^3 \text{ mm}^4 \times 80\,000 \text{ MPa}}$$

$$= 0.043\,64 \text{ rad}$$

Converting radians to degrees gives $\theta = 2.5°$.

Problems

27.14 Using the 40 per cent rule, estimate the approximate values of the moduli of rigidity for:
(a) aluminium
(b) brass
(c) cast iron

27.15 Mechanical tests have been conducted on an engineering polymer called Nylon-66 which established the following results:

$$\text{Young's modulus:} \quad E = 2750 \text{ MPa}$$
$$\text{Modulus of rigidity:} \quad G = 975 \text{ MPa}$$

The theoretical relation was then used to calculate Poisson's ratio for this material. What was the answer?

27.16 A square block of rubber, 350 mm × 350 mm × 25 mm thick, is fastened by adhesive between two parallel horizontal plates 25 mm apart. If a horizontal force of 29.4 kN is applied to one of the plates while the other is held firmly in its original position, what will be the magnitude of relative motion between the plates?

Under these conditions, state the values of shear stress and shear strain in the rubber. (Assume $G = 1.5$ MPa.)

27.17 Calculate the polar moment of inertia for a 75 mm diameter solid shaft.

27.18 Calculate the polar moment of inertia for a tubular section with outside diameter 20 mm and inside diameter 12 mm.

27.19 Calculate the stress developed in the extreme fibres of a 35 mm diameter solid bar due to an applied torque of 320 N.m.

27.20 If the tube in Problem 27.18 was subjected to a torque of 123 N.m, what would be the maximum stress in the material?

27.21 A torque of 30.7 N.m is applied to a copper rod, 250 mm long and 10 mm in diameter. Assuming a modulus of rigidity of 44 800 MPa, calculate the angle of twist.

27.22 What should be the diameter of the rod in Problem 27.21 for the angle of twist to be limited to 3.5°?

27.23 Out of the three materials listed in Problem 27.14, which is the most suitable for the following specification of a solid cylindrical rod?

Diameter:	8 mm
Length:	750 mm
Applied torque:	5 N.m
Angle of twist:	15°

Review questions

1. Define *shear stress*, and distinguish between *direct shear* and *torsional shear*.
2. Explain what area is subjected to direct shear when a round hole is punched in a metal plate.
3. Define *shear modulus of rigidity*.
4. State the approximate relation between the modulus of rigidity and Young's modulus for a typical engineering material.
5. What is the relationship between modulus of rigidity and Poisson's ratio?
6. State the formula for calculating torsional shear stress.
7. What is the *angle of twist* and what does it depend on?

PART 10

BENDING OF BEAMS

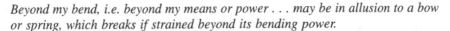

Beyond my bend, i.e. beyond my means or power . . . may be in allusion to a bow or spring, which breaks if strained beyond its bending power.

Brewer's Dictionary of Phrase and Fable

CHAPTER 28

Shear force and bending moment

In Chapter 7 we introduced beams as structural members, or machine components, used to support transverse loads due to weight or other causes. In particular, we learned how to calculate reactions at beam supports if all loads are known. We are about to discuss the effects of applied loads on a beam and the relation between the conditions of loading, the cross-sectional dimensions of the beam, and the internal resistance in the material of the beam.

In order to understand the behaviour of beams under load, we must introduce the concepts of **shear force** and **bending moment** produced in a beam by the applied load, and we must learn to draw shear force and bending moment diagrams for beams subjected to various loading conditions.

Expected learning outcomes

After carefully studying the material presented in this chapter, working through all numerical examples, and successfully completing all practice problems, students should be able to:
1. calculate transverse shear forces for different cross-sections of a beam, and represent results in conventional form as a shear force diagram;
2. calculate bending moments for different cross-sections of a beam, and represent results in conventional form as a bending moment diagram.

28.1 *SHEAR FORCE*

We are now turning our attention to internal forces that exist within the material of a beam subjected to loads.

Consider two portions of a beam cut by an imaginary section transverse to the beam as shown in Figure 28.1(a). The effect of the unbalanced forces on each part of the beam is to move the left-hand portion upwards relative to the right-hand portion, as shown in Figure 28.1(b). The reason that we never actually observe such movement between parts of a solid beam is because internal forces exist within the material of the beam in the plane of the imaginary cross-section which resist the tendency for such movement. The magnitude of the internal force at any given cross-section depends on the sum of the external forces acting on each portion of the beam to one side of the cross-section. This internal resistance force is called **shear force**.

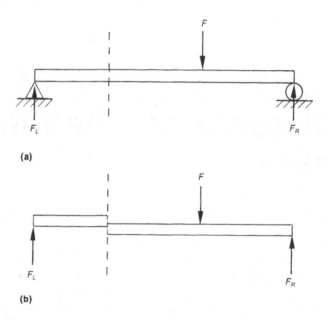

Fig. 28.1   **(a)** *Beam cut by imaginary transverse section*   **(b)** *Left-hand portion tends to move up relative to right-hand portion*

The magnitude of the shear force at any cross-section of a beam is equal to the algebraic sum of all external forces, i.e. loads and reactions, acting on either portion of the beam to one side of the section only.

The sign convention commonly adopted is that shear force at a cross-section is positive if it tends to push the left-hand portion of the beam upwards in relation to the right-hand portion, as shown in Figure 28.2(a). The opposite effect would then be regarded as negative, as in Figure 28.2(b).

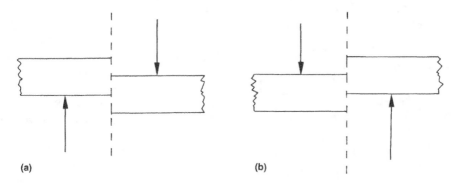

Fig. 28.2   **(a)** *Positive shear*   **(b)** *Negative shear*

Example 28.1

Determine the shear forces at the three cross-sections for the beam and loading shown in Figure 28.3.

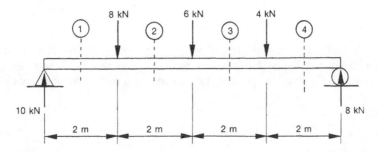

Fig. 28.3

Solution

It is usually convenient to start from the left-hand side and consider all forces to the left of the respective cross-sections.

Shear force at cross-section 1:

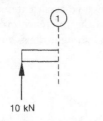

$$SF_1 = \Sigma F$$
$$= 10 \text{ kN (positive shear force)}$$

Shear force at cross-section 2:

$$SF_2 = \Sigma F$$
$$= 10 \text{ kN} - 8 \text{ kN}$$
$$= 2 \text{ kN (positive shear force)}$$

8 kN

10 kN

Shear force at cross-section 3:

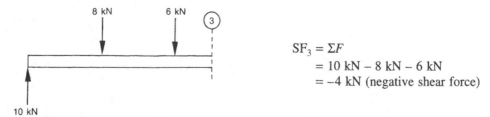

$$SF_3 = \Sigma F$$
$$= 10 \text{ kN} - 8 \text{ kN} - 6 \text{ kN}$$
$$= -4 \text{ kN (negative shear force)}$$

Shear force at cross-section 4:

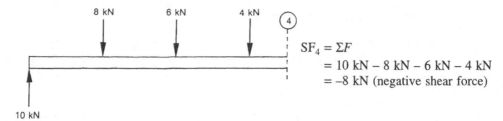

$$SF_4 = \Sigma F$$
$$= 10 \text{ kN} - 8 \text{ kN} - 6 \text{ kN} - 4 \text{ kN}$$
$$= -8 \text{ kN (negative shear force)}$$

It is useful to recognise that if we always work consistently from left to right, i.e. consider forces on the left-hand portion of the beam only, and use the usual sign convention of forces, i.e. positive up and negative down, the answers obtained give the correct sign for the shear force automatically.

28.2 *SHEAR FORCE DIAGRAMS*

To avoid long and unwieldy statements of results obtained in problems similar to that in Example 28.1, particularly if a large number of forces is involved, a graphical method of representing results, called a **shear force diagram**, has been developed. It consists simply of plotting the values obtained by calculation against the distance measured along the beam.

It should be noted that in the space between the two adjacent external forces, shear force is constant irrespective of the actual position of the reference cross-section. The shear force diagram will therefore consist of horizontal straight lines, with step changes under each load as illustrated by the following example.

Example 28.2
Draw the shear force diagram for the beam in Example 28.1.

Solution
The shear force diagram should be plotted directly below the space diagram of the beam, the horizontal distance representing the length of the beam. Vertical lines are drawn to some convenient scale, up from the zero line to represent positive shear force, and down for negative shear force. Horizontal lines represent constant shear force between loads. A step change in magnitude equal to the load occurs under each load. Vertical crosshatching is common practice, but not absolutely necessary. Magnitudes can be shown along the lines for clarity. (See Fig. 28.4.)

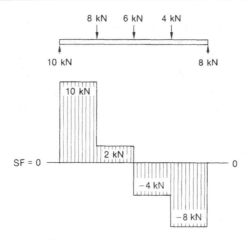

Fig. 28.4   *Shear force diagram*

28.3 *SHEAR FORCES DUE TO DISTRIBUTED LOADS*

When all or part of a beam is subjected to a uniformly distributed load of known intensity, the shear force under the distributed load varies uniformly along the beam and can be illustrated on the shear force diagram by an inclined straight line.

Example 28.3

Calculate the shear forces and draw the shear force diagram for the beam shown in Figure 28.5.

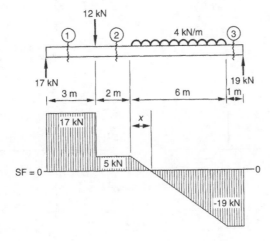

Fig. 28.5

Solution

Shear forces are calculated for cross-sections 1, 2 and 3 as follows:

$$SF_1 = 17 \text{ kN}$$
$$SF_2 = 17 \text{ kN} - 12 \text{ kN}$$
$$= 5 \text{ kN}$$
$$SF_3 = 17 \text{ kN} - 12 \text{ kN} - (4 \text{ kN/m} \times 6 \text{ m})$$
$$= -19 \text{ kN}$$

When constructing the shear force diagram, draw an inclined straight line under the distributed load, connecting the ordinates just to the left (SF$_2$ = 5 kN) and just to the right (SF$_3$ = –19 kN) of the load distribution.

Notice carefully where the shear force line intersects with the horizontal zero base line. This point has a special significance which will be discussed later in this chapter. Its precise location x with respect to point 2 can be calculated simply by dividing shear force at point 2 by the intensity of the distributed load:

$$x = \frac{5 \text{ kN}}{4 \text{ kN/m}}$$
$$= 1.25 \text{ m}$$

Therefore, shear force is equal to zero at the point 6.25 m from the left-hand support.

 Problems

28.1 In each of the diagrams (a) to (j) in Figure 28.6, beams are shown with all external forces, i.e. loads and reactions. Do all the necessary calculations and draw the shear force diagram for each beam shown.

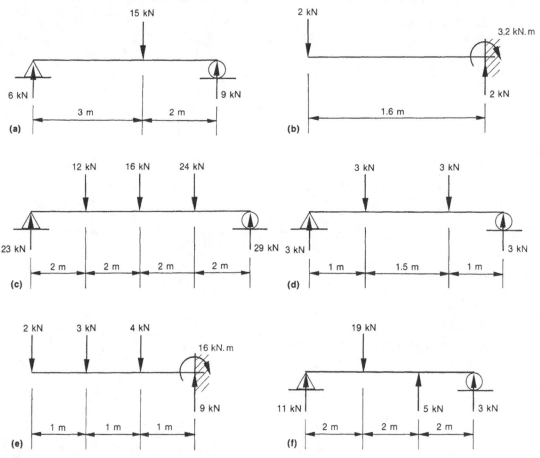

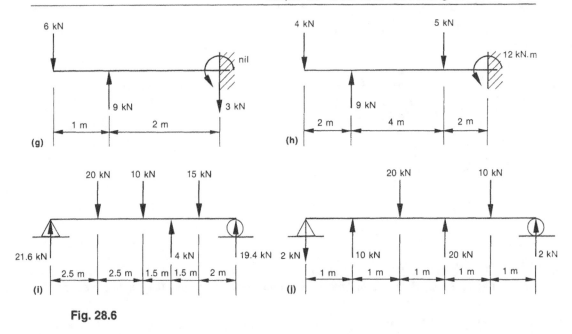

Fig. 28.6

28.2 Calculate shear forces for the beams shown in Figure 28.7, draw the shear force diagrams, and locate the position of zero shear force accurately.

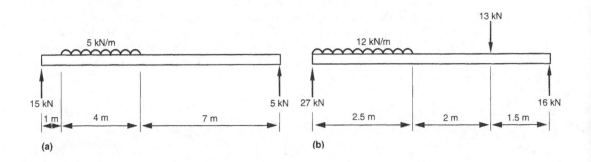

Fig. 28.7

28.4 BENDING MOMENT

In addition to internal shear forces, every cross-section in a beam may also experience an internal moment called a bending moment. Consider the same two portions of a beam, as before, cut by an imaginary reference cross-section. (See Fig. 28.8(a)).

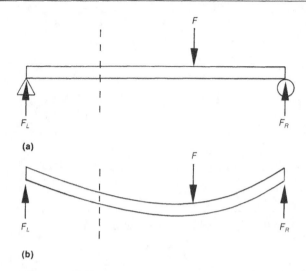

Fig. 28.8 **(a)** *Beam cut by imaginary reference cross-section* **(b)** *The beam bends*

Another effect of the unbalanced forces on each part of the beam, not considered previously, is that the beam bends, as shown in Figure 28.8(b). The amount of bending tendency at the reference cross-section is measured by the summation of the moments about the cross-section of all external forces on a portion of the beam to one side of the cross-section. The sum of the moments is called the **bending moment**.

The magnitude of the bending moment at any cross-section of a beam is equal to the algebraic sum of the moments, about the cross-section, of all external forces, i.e. loads and reactions, acting on either portion of the beam to one side of the section only.

The usual sign convention is that bending moment is positive if it produces bending in a beam which is convex downwards (Fig. 28.9(a)). The opposite effect is negative (Fig. 28.9(b)).

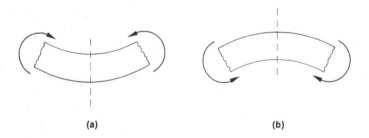

Fig. 28.9 **(a)** *Positive bending* **(b)** *Negative bending*

Example 28.4

For the beam and loading shown in Figure 28.10, determine the bending moments at the three points under the applied forces.

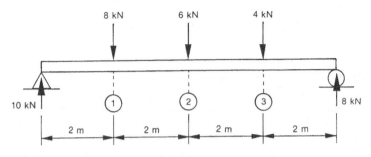

Fig. 28.10

Solution

In the case of bending moment calculations, it is convenient to select the reference cross-sections at the points of application of external loads. Again, we will start from the left-hand side and consider all forces to the left of the respective cross-sections.

Bending moment at cross-section 1:

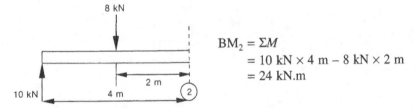

$$BM_1 = \Sigma M$$
$$= 10 \text{ kN} \times 2 \text{ m}$$
$$= 20 \text{ kN.m}$$

Bending moment at cross-section 2:

$$BM_2 = \Sigma M$$
$$= 10 \text{ kN} \times 4 \text{ m} - 8 \text{ kN} \times 2 \text{ m}$$
$$= 24 \text{ kN.m}$$

Bending moment at cross-section 3:

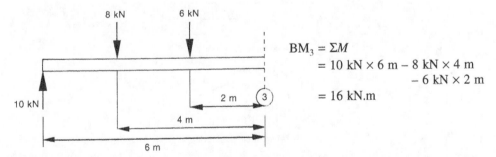

$$BM_3 = \Sigma M$$
$$= 10 \text{ kN} \times 6 \text{ m} - 8 \text{ kN} \times 4 \text{ m}$$
$$- 6 \text{ kN} \times 2 \text{ m}$$
$$= 16 \text{ kN.m}$$

Summation of moments about each of the two supports will show that the bending moment at the supports of a simply supported beam is always zero. It is left to the student to verify this statement by actual calculations.

Here again, it is useful to recognise that if we always work consistently from left to right, i.e. consider moments due to forces acting on the left-hand portion of the beam only, and use the usual sign convention for moments, i.e. positive—clockwise, negative—anticlockwise, the answers obtained give the correct sign for the bending moments automatically.

28.5 *BENDING MOMENT DIAGRAMS*

Information about bending moments can be conveniently represented by means of a **bending moment diagram**, which is a plot of the values obtained by calculation, against the distance measured along the beam.

It can be shown mathematically that in the spaces between two adjacent external forces, bending moment varies directly with the distance measured along the beam. The bending moment diagram will therefore consist of straight lines joining the points representing the computed magnitude of the bending moment at each load-bearing point.

Example 28.5

Draw the bending moment diagram for the beam in Example 28.4.

Solution

The bending moment diagram is usually plotted directly below the shear force diagram using the same horizontal scale along the length of the beam. Vertical lines are drawn to some convenient scale, up from the zero line for positive bending moments and down for negative bending moments, under each of the external forces acting on the beam. The points obtained in this manner are joined by straight lines. Magnitudes of bending moments can be indicated on the diagram and vertical crosshatching can be used for clarity. (See Fig. 28.11.)

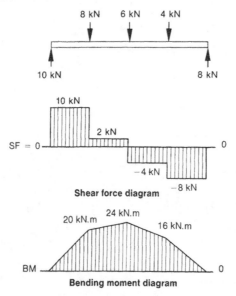

Fig. 28.11   *Bending moment diagram*

28.6 *POSITION OF MAXIMUM BENDING MOMENT*

The problem of beam design consists essentially of selecting the cross-section which will offer the most effective resistance to shear forces and bending moments produced by the loads, without excessive stresses or bending of the beam.

Shear forces, by themselves, in a long slender beam are seldom a major criterion in beam design. Bending moments, on the other hand, can be responsible for very high stresses in the material of the beam and contribute to the curvature of the beam distorted by bending, which can lead to unacceptably high deflections.

In many engineering applications, only the maximum value of the bending moment needs to be known corresponding to the point where the beam is most likely to fail under bending. However, if a beam carries more than two or three loads, all of different magnitude, it is not always possible to tell at a glance where the maximum bending moment occurs.

It has already been observed that when a beam is subjected to several concentrated loads, the shear force is of constant value between the loads, and changes abruptly at each load by an amount equal to the load. This property greatly facilitates the construction of shear force diagrams. We have also learned that when a beam carries a uniformly distributed load, the shear force varies uniformly under the load, and the portion of the shear force diagram which lies directly under the distributed load is drawn as an inclined straight line.

A further observation shows that the bending moment is at a maximum at points where the shear force diagram changes from positive to negative, i.e. passes through the zero axis.* The beam in Figure 28.11, for example, has a maximum bending moment of 24 kN.m at the point under the 6 kN load, where the shear force changes its value from +2 kN to −4 kN.

This relation between the shear force diagram and the position of maximum bending moment is particularly handy when we need to draw a bending moment diagram for a beam with a distributed load. It often enables us to locate and determine the maximum bending moment without laborious calculation and plotting of intermediate points.

28.7 *BENDING MOMENTS DUE TO DISTRIBUTED LOADS*

When part of a beam is subjected to a uniformly distributed load of known intensity, the bending moment under the distributed load varies as the square of the distance from the reference cross-section. This variation is illustrated on bending moment diagrams by a second order curve, a parabola, as will be seen from the following example. For most practical purposes, it is not necessary to actually plot an accurate parabola, provided the position of its maximum point can be established and the corresponding value of the bending moment indicated on the bending moment diagram.

Let us return now to one of the previous examples, in which part of the total load was uniformly distributed.

Example 28.6

Refer to the beam in Example 28.3. Calculate the bending moments and draw the bending moment diagram.

---

* Those familiar with differential calculus will be interested to know that shear force is equal to the first derivative of bending moment with respect to distance, x, measured along the beam:

$$SF = \frac{d(BM)}{dx}$$

Hence, where $SF = \dfrac{d(BM)}{dx} = 0$, BM is at a maximum.

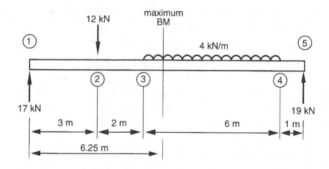

Fig. 28.12

Solution

Bending moments for points 1, 2 and 3 are obtained by simple calculations:

$$BM_1 = 0$$
$$BM_2 = 17 \text{ kN} \times 3 \text{ m}$$
$$= 51 \text{ kN.m}$$
$$BM_3 = 17 \text{ kN} \times 5 \text{ m} - 12 \text{ kN} \times 2 \text{ m}$$
$$= 61 \text{ kN.m}$$

When calculating the bending moment at point 4, we take into account the entire distributed load, which is equivalent to:

$$4 \text{ kN/m} \times 6 \text{ m} = 24 \text{ kN}$$

with the centre of its distribution 3 m to the left of point 4. Hence:

$$BM_4 = 17 \text{ kN} \times 11 \text{ m} - 12 \text{ kN} \times 8 \text{ m} - 24 \text{ kN} \times 3 \text{ m}$$
$$= 19 \text{ kN.m}$$
$$BM_5 = 0$$

We know from previous discussion that maximum bending moment must occur where SF = 0, i.e. where the shear force diagram crosses the zero base line. Furthermore, we were able to establish that this point happens to be 6.25 m from the left-hand support. All that remains now is to calculate the bending moment at this point, and it will have to be the maximum value of the bending moment in the beam.

Now, let us take moments of all forces which lie to the left of the 6.25 m point. This calculation must include the moment of that portion of the distributed load which is also to the left of the point.

$$BM_{max} = 17 \text{ kN} \times 6.25 \text{ m} - 12 \text{ kN} \times 3.25 \text{ m} - (4 \text{ kN/m} \times 1.25 \text{ m}) \times \frac{1.25 \text{ m}}{2}$$
$$= 64.1 \text{ kN.m}$$

This value should be shown on the bending moment diagram along with all other principal values of bending moment (Fig. 28.13). The bending moment diagram is then completed by drawing a smooth curve to connect the maximum bending moment with the bending moments on either side of it, i.e. at points 3 and 4.

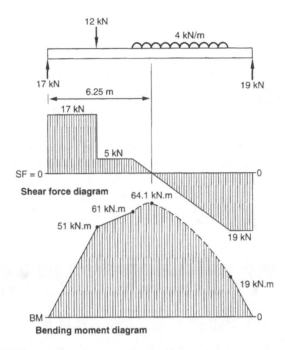

Fig. 28.13

 Problems

28.3 For the beams and loadings shown in Figure 28.6(a) to (j) (Problem 28.1), do all the necessary calculations and draw the bending moment diagrams. In each case, note the position of the maximum bending moment.

Compare these answers with your answers to Problem 28.1 to verify that bending moment is at a maximum when shear force is zero. (The best way to do this is by drawing the BM diagram directly below the SF diagram for each beam.)

28.4 For the two beams with distributed loads as shown in Figure 28.7(a) and (b) (Problem 28.2), do all the necessary calculations and draw the bending moment diagrams. Clearly show the position and magnitude of the maximum bending moment in each case.

 Review questions

1. Explain the concept of *shear force* in a loaded beam.
2. State the sign convention for shear forces.
3. Explain the concept of *bending moment* in a loaded beam.
4. State the sign convention for bending moments.
5. How can the shear force diagram be used to locate the maximum bending moment?

CHAPTER 29

Bending stress

This chapter deals with the relation between the stresses induced in a beam by bending moments and certain geometrical properties of the cross-section of the beam.

In the first part of the chapter, we will discuss the geometrical property of a plane area called the **centroid,** and learn how to locate it. Next we will discuss the **moment of inertia** of a plane area, and learn how to calculate it for a given cross-section of a beam.

In the second part of the chapter, we will learn how to evaluate **bending stresses** in beams.

Expected learning outcomes

After carefully studying the material presented in this chapter, working through all numerical examples, and successfully completing all practice problems, students should be able to:
1. locate the position of the centroid for a plane area of specified shape and size;
2. calculate the moment of inertia for a plane area, of specified shape and size, about its centroidal axis;
3. calculate bending stresses in beams.

29.1 *CENTROIDS OF PLANE AREAS*

The **centroid** of a plane area is the unique point which is the geometrical centre of the area distribution. The concept of the centroid is often explained in terms of the centre of gravity of a thin homogeneous plate of uniform thickness. Although this analogy is helpful, one should keep in mind that the centroid is a geometrical concept and not a mass-related concept.

Centroids of most of the common shapes have been determined by integration and are available in tables and handbooks. The positions of centroids of symmetrical shapes such as circles, squares and rectangles are easily identifiable. In a right-angled triangle, the centroid is located one-third of the side length from the 90° angle in each direction. This is also the point of intersection of the medians (see Table 29.1).

Centroids of composite shapes can be calculated by dividing the area into its simple component parts and then applying the principle of **first moments of area**, which states that the moment of an area about any reference line equals the algebraic sum of the moments of its component areas about the same line. Mathematically, this principle can be expressed as:

$$\bar{x} = \frac{\Sigma(Ax)}{\Sigma(A)}$$

where $\Sigma(A)$ is the sum of all the component areas: $A_1 + A_2 + A_3 + \dots$
$\Sigma(Ax)$ is the sum of all the area moments: $A_1x_1 + A_2x_2 + A_3x_3 + \dots$

Similarly in the y-direction:

$$\bar{y} = \frac{\Sigma(Ay)}{\Sigma(A)}$$

$\bar{x}$ and $\bar{y}$ are the locating coordinates relative to the chosen axes.

Example 29.1

Locate the centroid of the composite area shown in Figure 29.1(a) with respect to the x–x and y–y axes.

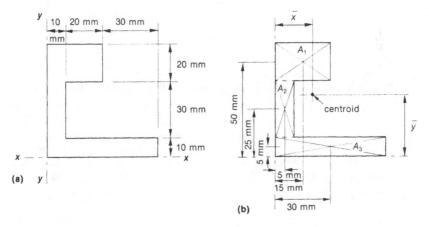

Fig. 29.1

Solution

Divide the area into three rectangular elements, A_1, A_2 and A_3, and locate the centroid of each element as in Figure 29.1(b).

Calculate each elementary area:

$$A_1 = 30 \times 20 = 600 \text{ mm}^2$$
$$A_2 = 30 \times 10 = 300 \text{ mm}^2$$
$$A_3 = 60 \times 10 = 600 \text{ mm}^2$$
$$\therefore \text{ Total area } \Sigma(A) = 1500 \text{ mm}^2$$

Calculate the area moments in the x–x direction:

$$A_1x_1 = 600 \times 15 = 9000 \text{ mm}^3$$
$$A_2x_2 = 300 \times 5 = 1500 \text{ mm}^3$$
$$A_3x_3 = 600 \times 30 = 18\,000 \text{ mm}^3$$
$$\therefore \text{ Total area moment } \Sigma(Ax) = 28\,500 \text{ mm}^3$$

The position of the centroid with respect to the y–y axis, i.e. located in the x–x direction, is:

$$\bar{x} = \frac{\Sigma(Ax)}{\Sigma(A)}$$

$$= \frac{28\,500}{1500}$$

$$= 19 \text{ mm}$$

Similarly in the y–y direction:

$$\bar{y} = \frac{\Sigma(Ay)}{\Sigma(A)}$$

$$= \frac{600 \times 50 + 300 \times 25 + 600 \times 5}{600 + 300 + 600}$$

$$= 27 \text{ mm}$$

It is obvious from this example that the centroid of a composite area can lie outside the outline of the area itself.

It is often convenient to enter all intermediate values in a simple table as illustrated below:

Element	Area A	Distance		Area moment	
		x	y	Ax	Ay
1	600	15	50	9 000	30 000
2	300	5	25	1 500	7 500
3	600	30	5	18 000	3 000
$\Sigma =$	1500	—	—	28 500	40 500

Hence:

$$\bar{x} = \frac{28\,500}{1500}$$

$$= 19 \text{ mm}$$

and:

$$\bar{y} = \frac{40\,500}{1500}$$

$$= 27 \text{ mm}$$

Example 29.2

Locate the centroid of the shape shown in Figure 29.2.

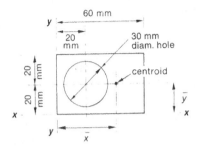

Fig. 29.2

Solution

Areas removed, e.g. holes, can be regarded as negative areas. In all other respects, the method of solution remains as explained before:

$$\bar{x} = \frac{\Sigma(Ax)}{\Sigma(A)}$$

$$= \frac{(60 \times 40) \times 30 + \left(-\dfrac{\pi \times 30^2}{4}\right) \times 20}{(60 \times 40) + \left(-\dfrac{\pi \times 30^2}{4}\right)}$$

$$= 34.2 \text{ mm}$$

From symmetry, $\bar{y} = 20$ mm.

29.2 *MOMENTS OF INERTIA OF PLANE AREAS*

The **moment of inertia** (I) of a plane area, more appropriately called the **second moment of area**, is the sum of all elementary products of area elements and the square of each respective distance from the centroidal axis. As such, it is a mathematical concept which is best illustrated by examples, such as Examples 29.3 to 29.5 below.

For our purposes, the moment of inertia of a cross-section of a beam, combined with another dimension of that cross-section, is a measure of the resistance offered by the cross-section to the bending moment and can be related to the bending stresses induced in the beam by bending.

Moments of inertia of most common shapes have been calculated by means of integral calculus and can be found in tables and handbooks. The moments of inertia of simple geometrical shapes are given in Table 29.1.

Example 29.3

Determine the moment of inertia of a rectangular area, with base 30 mm and height 20 mm, about its horizontal centroidal axis.

Table 29.1   *Centroids and moments of inertia of elementary plane areas*

Shape		Area A	Position of centroid	Centroidal moment of inertia I_c
circle		$\dfrac{\pi D^2}{4}$	at centre	$\dfrac{\pi D^4}{64}$
square		a^2	at intersection of diagonals	$\dfrac{a^4}{12}$
rectangle		bh	at intersection of diagonals	$\dfrac{bh^3}{12}$
right-angled triangle		$\dfrac{bh}{2}$	at intersection of medians ($\frac{1}{3}$ of altitude)	$\dfrac{bh^3}{36}$

Solution

From Table 29.1, the moment of inertia of a rectangle is given by:

$$I_c = \frac{bh^3}{12}$$

Substitute and solve:

$$I_c = \frac{30 \times 20^3}{12}$$
$$= 20\,000 \text{ mm}^4$$

For a given shape, the moment of inertia is not unique but depends on the particular reference axis. Table 29.1 gives the moments of inertia for each shape about its horizontal centroidal axis.

The moment of inertia about any other axis parallel to the centroidal axis is given by the relation known as the **parallel axis theorem**:

$$\boxed{I = I_c + Ad^2}$$

where I is the moment of inertia about any axis
I_c is the moment of inertia about the centroidal axis
A is the area
d is the distance between the axes

Example 29.4

Determine the moment of inertia of the rectangular area shown in Figure 29.3 about a parallel axis 23 mm away from its centroidal axis.

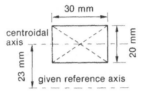

Fig. 29.3

Solution

The moment of inertia about the centroidal axis of the rectangular area is:

$$I_c = \frac{bh^3}{12}$$

$$= \frac{30 \times 20^3}{12}$$

$$= 20\,000 \text{ mm}^4$$

The transfer term is:

$$Ad^2 = (30 \times 20) \times 23^2$$

$$= 317\,400 \text{ mm}^4$$

Therefore, the required moment of inertia about the given reference axis is:

$$I = I_c + Ad^2$$

$$= 337\,400 \text{ mm}^4$$

Example 29.5

Determine the moment of inertia of the composite area given in Example 29.1 about its horizontal centroidal axis.

Solution

The moment of inertia of a composite area about its centroidal axis is equal to the sum of the individual moments of inertia of its component areas relative to the common centroidal axis:

$$I = I_1 + I_2 + I_3$$

A tabulated solution is again helpful:

Element	Area A	Distance from centroid d	Centroidal moment of inertia I_c	Transfer term Ad^2	Transferred moment of inertia $I = I_c + Ad^2$
1	30×20 $= 600$	23	$\dfrac{30 \times 20^3}{12} = 20\,000$	600×23^2 $= 317\,400$	$337\,400$
2	30×10 $= 300$	2	$\dfrac{10 \times 30^3}{12} = 22\,500$	300×2^2 $= 1200$	$23\,700$
3	60×10 $= 600$	22	$\dfrac{60 \times 10^3}{12} = 5000$	600×22^2 $= 290\,400$	$295\,400$

Total moment of inertia of the area about its centroidal axis $\Sigma(I) = 656\,500$ mm^4

 Problems

29.1 For each of the areas shown in Figure 29.4, locate the centroid relative to the x–x and y–y axes.

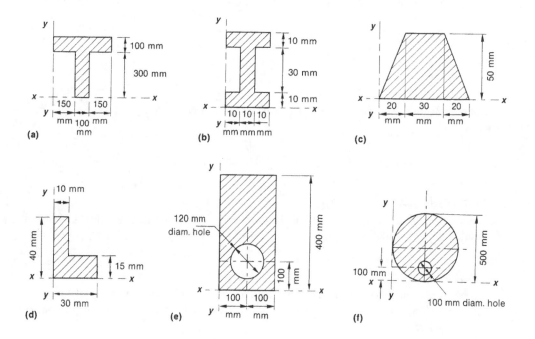

Fig. 29.4

29.2 Determine the moment of inertia of each area (Fig. 29.5) about its horizontal centroidal axis:

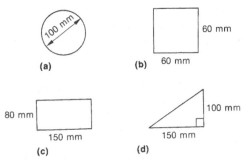

(a)   (b) 60 mm

(c)   (d)

Fig. 29.5

29.3 For each of the areas in problem 29.1, determine the moment of inertia about the horizontal centroidal axis.

29.3 *BENDING STRESS*

This part of the chapter deals with the relation between bending moment in a beam, the stress produced in the material of the beam, and certain geometrical properties of the beam's cross-section.

Let us consider a particular cross-section in a beam at which a bending moment (M) is applied. It can be shown that the magnitude of **bending stress** (f_b) produced in the material of the beam at the cross-section is *directly* proportional to the bending moment.[*] On the other hand, the size and shape of the cross-section of the beam give rise to a resistance which balances the applied bending moment. Bending stress is *inversely* proportional to the moment of inertia of the cross-section.

One important difference between bending stress and some other types of stress considered previously is that, unlike direct tension or compression, bending stress is not distributed uniformly over the cross-sectional area of the beam. Every beam can be considered as being made up of a large number of horizontal layers or fibres held together by the internal adhesion between them. When a beam is subjected to bending, the fibres on the convex side are extended, while those on the concave side are compressed, as shown in Figure 29.6(a). Somewhere between the stretched and the compressed fibres there is a longitudinal plane along which there is no deformation of length. This plane is known as the **neutral plane** (NP).

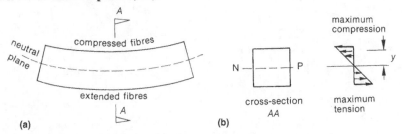

Fig. 29.6   (a) *Beam in bending*   **(b)** *Stress distribution*

---

[*] The full mathematical proof of this and following propositions, which is omitted here, can be found in more advanced texts on Strength of Materials.

It can be shown mathematically that the neutral plane passes through the centroid of the cross-section of the beam.

The further a particular fibre is from the neutral plane, the greater is the amount of elongation or compression experienced by the fibre. We also know that according to Hooke's law, tensile and compressive stress in a material is proportional to the amount of change in length. Therefore, the stress in any one fibre is proportional to its distance (y) from the neutral plane (Fig. 29.6(b)).

The important conclusions that follow from the distribution of stress in the cross-section of a beam can be summarised as follows:

1. There is no stress at the neutral plane.
2. The maximum tensile stress occurs in the extreme fibre on the convex side of the beam.
3. The maximum compressive stress occurs in the extreme fibre on the concave side of the beam.

The formula which embodies the relations discussed above is:

$$f_b = \frac{My}{I}$$

where f_b is the bending stress in MPa

M is the bending moment at a given cross-section, in N.mm

y is the distance from the neutral plane to a particular fibre, in mm

I is the moment of inertia of the cross-section, in mm^4

This formula is applicable to any cross-section in a beam and any fibre within the beam. However, in most cases, designers are only interested in the maximum values of stress. Therefore, they would use the maximum value of M and the distance to the extreme fibres, i.e. the maximum value of y, for their calculations.

If the cross-section is not symmetrical about its centroidal (neutral) axis, a distinction has to be made between tension and compression sides by using different values of y.

Example 29.6

A beam of rectangular cross-section, 300 mm deep by 100 mm wide, is subjected to a positive bending moment of 67.5 kN.m. Determine the maximum value of bending stress.

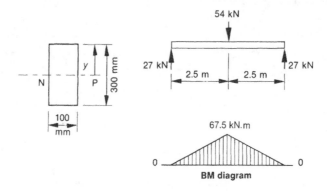

Fig. 29.7

Solution

The cross-section of the beam is as shown in Figure 29.7.

Moment of inertia:

$$I = \frac{bh^3}{12}$$
$$= \frac{100 \times 300^3}{12}$$
$$= 225 \times 10^6 \text{ mm}^4$$

Distance to the extreme fibre:

$$y = 150 \text{ mm}$$

Bending moment:

$$M = 67.5 \text{ kN.m}$$
$$= 67.5 \times 10^6 \text{ N.mm}$$

Bending stress:

$$f_b = \frac{My}{I}$$
$$= \frac{67.5 \times 10^6 \text{ N.mm} \times 150 \text{ mm}}{225 \times 10^6 \text{ mm}^4}$$
$$= 45 \text{ MPa}$$

In this case, because of the symmetrical cross-section, the extreme fibres on the tension and compression sides are at the same distance from the neutral plane. Therefore, the answer represents the maximum compressive stress in the top fibre as well as the maximum tensile stress in the bottom fibre.

Example 29.7

For the cantilever beam shown in Figure 29.8, determine the maximum value of stress.

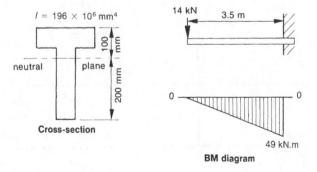

Fig. 29.8

Solution

Maximum bending moment occurs at the fixed end of the beam.

$$M = 14 \text{ kN} \times 3.5 \text{ m}$$
$$= 49 \text{ kN.m}$$
$$= 49 \times 10^6 \text{ N.mm}$$

This is negative bending moment. Therefore, the convex or tension fibre is on the top and the concave or compression fibre is on the bottom. Hence, for maximum tension, $y_t = 100$ mm, and for maximum compression, $y_c = 200$ mm.

Maximum stresses can now be calculated.

Tensile stress in the top fibre:

$$f_t = \frac{My_t}{I}$$
$$= \frac{49 \times 10^6 \text{ N.mm} \times 100 \text{ mm}}{196 \times 10^6 \text{ mm}^4}$$
$$= 25 \text{ MPa}$$

Compressive stress in the bottom fibre:

$$f_c = \frac{My_c}{I}$$
$$= \frac{49 \times 10^6 \text{ N.mm} \times 200 \text{ mm}}{196 \times 10^6 \text{ mm}^4}$$
$$= 50 \text{ MPa}$$

Such non-symmetrical cross-sections are sometimes used for beams made from materials which are stronger in compression than they are in tension, such as concrete.

 # Problems

29.4 Determine the maximum stress for the beam and loading shown in Figure 29.9 if the beam is rectangular, 250 mm deep by 75 mm wide.

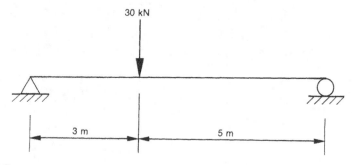

Fig. 29.9

29.5 Determine the maximum load, in kilonewtons, that can be applied at the free end of a 3 m cantilever beam of universal rolled-steel beam cross-section, 356 mm deep, with a moment of inertia of 142×10^6 mm^4, if the allowable stress is 76 MPa.

29.6 How wide should a rectangular cross-section of a beam be if its depth is 150 mm and it is to carry a load of 15 kN in the middle of an 8 m span with an allowable stress of 80 MPa?

29.7 Determine the stress at midspan of the beam in problem 29.6, if the load is reduced to 12 kN.

29.8 A beam of T-section has a moment of inertia 350×10^6 mm^4. The distance from the neutral plane to the extreme fibre on the compression side is 150 mm, and to the extreme fibre on the tension side is 200 mm.

　　Determine the stresses at midspan if the beam is 14 m long and carries a load of 40 kN in the middle.

29.9 Determine the maximum allowable bending moment for a timber beam, 200 mm deep by 100 mm wide, if the allowable stress is 20 MPa.

29.10 Repeat problem 29.9, but this time the beam is placed so that it is 100 mm deep by 200 mm wide. Compare the answers. Which is the better way to place the beam in order to get the maximum load-carrying capacity?

29.11 For the loading as in problem 28.1(c) (Fig. 28.6(c)), select a suitable size of rectangular timber beam if the allowable stress is 25 MPa and the cross-section is to be twice as deep as it is wide.

Review questions

1. Define the term *centroid*.
2. Outline the steps for calculating moments of inertia of plane areas.
3. State the formula for calculating bending stress.
4. Explain the significance of the neutral plane.
5. Describe the pattern of stress distribution on a beam cross-section.

CHAPTER 30

Deflection of beams

In the preceding chapter, our attention was directed towards the analysis of bending stresses because of their obvious importance in the design of beams and beam-like machine parts. However, there is another factor that is of almost equal importance for the proper functioning of a beam as a structural or mechanical component. This factor is **rigidity under load**. For example, a plastered ceiling may suffer cracking if the joists supporting it are too flexible. Likewise, damaging vibrations may develop in rotating machinery if a shaft exhibits excessive flexibility under transverse loads.

There are two measures that can be used to describe the extent of deformation suffered by a beam subjected to bending. The first is the radius of curvature of the beam's neutral axis, originally straight, but distorted into a curve by the applied loads. The second, and perhaps more useful, is the amount of deflection of the neutral axis from its original position in the unloaded beam.

In this chapter, after a brief look at the concept of the radius of curvature, we are going to examine some beam-deflection formulae, commonly used in practical design work.

Expected learning outcomes

After carefully studying the material presented in this chapter, working through all numerical examples, and successfully completing all practice problems, students should be able to:

1. calculate the radius of curvature of a beam under pure bending;
2. calculate the deflection at the free end of a cantilever beam;
3. calculate the deflection at midspan of a symmetrically loaded simply supported beam;
4. calculate the maximum deflection for a simply supported beam due to a single non-symmetrical concentrated load.

30.1 *RADIUS OF CURVATURE*

When a beam is subjected to bending, its shape is distorted into a curve. The **radius of curvature**, R, is not constant along the beam but depends on the magnitude of the bending moment that exists at each cross-section. The radius also depends on the moment of inertia (I) of the cross-section and on Young's modulus of elasticity (E) of the material of the beam. The expression relating these variables is:

$$R = \frac{EI}{M}$$

where R is the radius of curvature in mm
M is the bending moment in N.mm
E is Young's modulus in MPa
I is the moment of inertia in mm^4

It should be understood that the radius of curvature is the inverse measure of distortion in bending, i.e. the greater the curvature, the smaller the radius. For the undistorted (straight) beam, the radius is infinite.

Example 30.1

Determine the radius of curvature at the point of maximum bending moment:
(a) for the beam in Example 29.6, if the material is steel, $E = 200\,000$ MPa (N/mm^2);
(b) for the beam in Example 29.7, if the material is concrete, $E = 23\,000$ MPa (N/mm^2).

Solution

(a)
$$R = \frac{EI}{M}$$
$$= \frac{200\,000 \text{ MPa} \times 225 \times 10^6 \text{ mm}^4}{67.5 \times 10^6 \text{ N.mm}}$$
$$= 666\,700 \text{ mm}$$
$$\therefore R = 666.7 \text{ m}$$

(b)
$$R = \frac{EI}{M}$$
$$= \frac{23\,000 \text{ MPa} \times 196 \times 10^6 \text{ mm}^4}{49 \times 10^6 \text{ N.mm}}$$
$$= 92\,000 \text{ mm}$$
$$\therefore R = 92 \text{ m}$$

As expected, the radii are large, because under moderate conditions of loading, the curvature produced in a solid beam is relatively small.

30.2 *SUMMARY OF BEAM-DEFLECTION FORMULAE*

Various mathematical methods have been used to develop a range of special-case practical formulae for calculating maximum deflection in beams subjected to bending.

In all cases, deflection is found to be directly proportional to the applied load, expressed as a force (F), and inversely proportional to the moment of inertia of the beam cross-section (I) and to the Young's modulus of elasticity of its material (E). In addition, deflection is always dependent on the beam's length (L), sometimes in combination with some other dimension (a) which locates the applied force on the beam.

In this chapter, we are going to examine deflection of beams in the following three groups:

1. cantilever beams carrying concentrated loads or full-length uniformly distributed loads;
2. simply supported beams with symmetrically located, concentrated loads or with full-length uniformly distributed loads;
3. simply supported beams with one non-symmetrically located concentrated load.

Table 30.1 *Deflection formulae for variously loaded beams*

Case	Beam and load	Maximum deflection	Occurs at:
1		$y = \dfrac{FL^3}{3EI}$	free end
2		$y = \dfrac{Fa^2(3L - a)}{6EI}$	free end
3	$F = wL$	$y = \dfrac{FL^3}{8EI}$	free end
4		$y = \dfrac{FL^3}{48EI}$	midspan
5		$y = \dfrac{Fa(3L^2 - 4a^2)}{24EI}$	midspan
6	$F = wL$	$y = \dfrac{5FL^3}{384EI}$	midspan
7		$y = \dfrac{Fab(a + 2b)\sqrt{3a(a + 2b)}}{27EIL}$ (where $a > b$)	$x = \sqrt{\dfrac{a(a + 2b)}{3}}$

Note: To obtain deflection in mm, F must be in N, E must be in MPa, I must be in mm⁴, and L, a and b must be in mm.

Table 30.1 is a summary of the seven most common beam-deflection formulae. In practice, a surprisingly large number of beam-deflection problems encountered by engineers can be solved by the use of these seven formulae. Additional formulae, which cover more complex cases of irregular loading or a different arrangement of beam supports, can be found in various engineering handbooks.

30.3 *DEFLECTION OF CANTILEVER BEAMS*

Cantilever beams are represented in Table 30.1 by the loading diagrams and formulae in cases 1, 2 and 3. With vertical downward loads, deflection will always be downwards and maximum at the free end of a cantilever beam. Study the following examples carefully.

Example 30.2

A steel rod ($E = 200\,000$ MPa), 1.5 m long and 50 mm in diameter, is loaded with a single concentrated load of 270 N at its free end, as shown in Figure 30.1. Calculate the maximum deflection due to the applied load. The weight of the rod itself may be neglected.

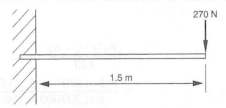

Fig. 30.1

Solution

Here we have:

$$F = 270 \text{ N}$$
$$L = 1500 \text{ mm}$$
$$E = 200\,000 \text{ MPa}$$

and:

$$I = \frac{\pi D^4}{64} \text{ (see Table 29.1)}$$
$$= \frac{\pi \times 50^4}{64}$$
$$= 306.8 \times 10^3 \text{ mm}^4$$

Substitute into the formula for the maximum deflection of a cantilever beam with a concentrated load at the free end (case 1):

$$y = \frac{FL^3}{3EI}$$
$$= \frac{270 \times 1500^3}{3 \times 200\,000 \times 306.8 \times 10^3}$$
$$= 4.95 \text{ mm}$$

This is the amount of downward deflection of the free end of the rod in question.

Example 30.3

If the same steel rod ($E = 200\,000$ MPa, $L = 1500$ mm and $I = 306.8 \times 10^3$ mm^4) carries a single load of 740 N located 0.85 m from the support, as shown in Figure 30.2, what is the maximum deflection? Ignore the weight of the rod.

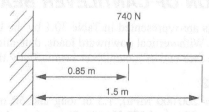

Fig. 30.2

Solution

Substitute into the formula for the maximum deflection of a cantilever beam with a concentrated load located at a specified distance ($a = 850$ mm) from the fixed support (case 2).

$$
\begin{aligned}
y &= \frac{Fa^2(3L - a)}{6EI} \\
&= \frac{740 \times 850^2 \times (3 \times 1500 - 850)}{6 \times 200\,000 \times 306.8 \times 10^3} \\
&= 5.30 \text{ mm}
\end{aligned}
$$

This is the amount of maximum deflection of the rod which occurs at its free end.

Example 30.4

What is the deflection of the same steel rod ($E = 200\,000$ MPa, $L = 1500$ mm and $I = 306.8 \times 10^3$ mm^4) under its own weight, but without any other loads? Assume density of steel is 7800 kg/m^3.

Solution

Calculate the weight of the rod:

$$
\begin{aligned}
F_w &= mg \\
&= \frac{\pi \times 0.05^2}{4} \times 1.5 \times 7800 \times 9.81 \\
&= 225.4 \text{ N}
\end{aligned}
$$

For a rod of constant diameter, the weight is distributed uniformly along its length L. (The equivalent intensity of load distribution in this case is $w = 225.4$ N/1.5 m = 150 N/m) (see Fig. 30.3). Therefore, we have an example of a cantilever beam with a uniformly distributed load over its entire length (case 3).

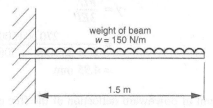

Fig. 30.3

Choose the appropriate formula from the Table and substitute:

$$y = \frac{FL^3}{8EI}$$

$$= \frac{225.4 \times 1500^3}{8 \times 200\,000 \times 306.8 \times 10^3}$$

$$= 1.55 \text{ mm}$$

This is the amount of deflection at the free end of the rod caused by its own weight.

In the case of combined loading, it is possible to find the resultant effect of several loads acting on a beam simultaneously as the sum of the contributions from each of the loads applied individually. This is known as the **principle of superposition**. It states that deflections at the same point on the beam may be computed separately, and then added, provided that a linear relation exists between the deflections and the applied loads, and that the material of the beam is not stressed beyond its proportional elastic limit.

Example 30.5

What is the maximum deflection of the same steel rod if both concentrated loads, located as described in previous examples, act on the rod (Fig. 30.4), and the effect of its own weight is also taken into account?

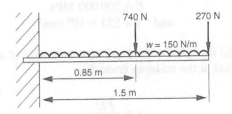

Fig. 30.4

Solution

The results for the separate loads are available from our previous calculations:

Deflection due to 270 N load: $y_1 = 4.95$ mm
Deflection due to 740 N load: $y_2 = 5.30$ mm
Deflection due to own weight: $y_3 = 1.55$ mm

Therefore, the deflection of the free end of the rod under combined loading conditions, including its own weight, is found by summation of the individual results:

$$y = 4.95 + 5.30 + 1.55$$
$$= 11.8 \text{ mm}$$

30.4 *DEFLECTION OF SYMMETRICALLY LOADED SIMPLE BEAMS*

Simply supported beams with symmetrical loading are represented in Table 30.1 by loading diagrams and formulae of cases 4, 5 and 6. With vertical downward loads, deflection is always downwards and maximum at the midspan. Here are some more examples for you to study.

Example 30.6

A steel structural beam ($E = 200\,000$ MPa and $I = 554 \times 10^6$ mm⁴) is 12 m long between simple supports at each end, and carries a concentrated load of 40 kN at its midspan point, as shown in Figure 30.5. Calculate the deflection due to this load.

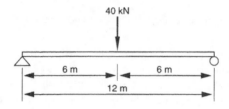

Fig. 30.5

Solution

Here we have:

$$F = 40\,000 \text{ N}$$
$$L = 12\,000 \text{ mm}$$
$$E = 200\,000 \text{ MPa}$$
$$\text{and} \quad I = 554 \times 10^6 \text{ mm}^4$$

Substitute into the formula for the maximum deflection of a simply supported beam with a concentrated load at the midspan (case 4):

$$y = \frac{FL^3}{48EI}$$
$$= \frac{40\,000 \times 12\,000^3}{48 \times 200\,000 \times 554 \times 10^6}$$
$$= 13.0 \text{ mm}$$

This is the maximum deflection at the midpoint.

Example 30.7

If the same steel beam ($E = 200\,000$ MPa, $L = 12\,000$ mm and $I = 554 \times 10^6$ mm⁴) carries two symmetrically located concentrated loads of 30 kN each, located 2.5 m from the supports as shown in Figure 30.6, what is the maximum deflection due to this loading?

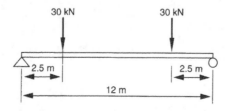

Fig. 30.6

Solution

Substitute into the equation for case 5 (two symmetrical concentrated loads):

$$y = \frac{Fa(3L^2 - 4a^2)}{24EI}$$

$$= \frac{30\,000 \times 2500 \times (3 \times 12\,000^2 - 4 \times 2500^2)}{24 \times 200\,000 \times 554 \times 10^6}$$

$$= 11.5 \text{ mm}$$

This is the maximum deflection, which occurs at midspan.

Example 30.8

If the same simply supported steel beam ($E = 200\,000$ MPa, $L = 12\,000$ mm and $I = 554 \times 10^6$ mm^4) was used to carry a uniformly distributed load of 8 kN/m over its entire length (Fig. 30.7), what would be the midspan deflection?

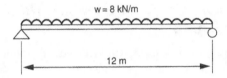

w = 8 kN/m

12 m

Fig. 30.7

Solution

If the load intensity is $w = 8$ kN/m, then the total load on the beam is:

$$F = 8 \text{ kN/m} \times 12 \text{ m}$$
$$= 96 \text{ kN}$$
$$= 96\,000 \text{ N}$$

Substitute into the equation for case 6:

$$y = \frac{5FL^3}{384EI}$$

$$= \frac{5 \times 96\,000 \times 12\,000^3}{384 \times 200\,000 \times 554 \times 10^6}$$

$$= 19.5 \text{ mm}$$

Here again, this is the maximum deflection, which occurs at midspan.

Example 30.9

Determine the maximum deflection if the same simply supported steel beam ($E = 200\,000$ MPa, $L = 12\,000$ mm and $I = 554 \times 10^6$ mm^4) has to carry three symmetrically located loads as shown in Figure 30.8.

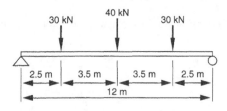

Fig. 30.8

Solution

If we recognise components of the load as those already discussed in Example 30.6 ($y = 13.0$ mm) and Example 30.7 ($y = 11.5$ mm), the midspan deflection in this case is easily found by superposition:

$$\text{Deflection } y = 13.0 \text{ mm} + 11.5 \text{ mm}$$
$$= 24.5 \text{ mm}$$

30.5 *DEFLECTION OF NON-SYMMETRICALLY LOADED SIMPLE BEAMS*

When the loading on a beam is non-symmetrical, solutions are complicated by the fact that the position of the maximum deflection is not immediately predictable. It does not correspond to the midspan or to the position of the load, but lies somewhere between these two points and has to be located by a separate calculation.

Example 30.10

A simply supported timber beam, 5 m long and 100 mm deep × 50 mm wide, carries a single concentrated load of 550 N located 1.5 m from one of the supports (Fig. 30.9). Assuming $E = 12\,000$ MPa, calculate and locate the maximum deflection of the beam.

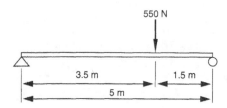

Fig. 30.9

Solution

Examine the formulae for case 7:

$$y = \frac{Fab(a + 2b)\sqrt{3a(a + 2b)}}{27EIL} \qquad x = \sqrt{\frac{a(a + 2b)}{3}}$$

There is a condition: $a > b$

Therefore, we must choose $a = 3.5$ m $= 3500$ mm

and $b = 1.5$ m $= 1500$ mm

Calculate the moment of inertia of the rectangular cross-section:

$$I = \frac{bh^3}{12}$$

$$= \frac{50 \times 100^3}{12}$$

$$= 4.167 \times 10^6 \text{ mm}^4$$

Now substitute and evaluate:

$$y = \frac{Fab(a + 2b)\sqrt{3a(a + 2b)}}{27EIL}$$

$$= \frac{550 \times 3500 \times 1500(3500 + 2 \times 1500)\sqrt{3 \times 3500(3500 + 2 \times 1500)}}{27 \times 12\,000 \times 4.167 \times 10^6 \times 5000}$$

∴ Maximum deflection $y = 23$ mm.

Now we can locate the position along the beam where the maximum deflection would occur:

$$x = \sqrt{\frac{a(a + 2b)}{3}}$$

$$= \sqrt{\frac{3500(3500 + 2 \times 1500)}{3}}$$

$$= 2753 \text{ mm}$$

The distance from the left-hand support (where dimension a is also measured from) is:

$$x = 2.75 \text{ m}$$

 # Problems

Except where otherwise indicated, neglect the weight of the beam.

30.1 Calculate the radius of curvature for a timber beam 150 mm deep and 75 mm wide, subjected to a bending moment of 560 N.m, if the modulus of elasticity for timber is 12 000 MPa.

30.2 Determine the radius of curvature of a 10 mm × 10 mm steel bar, with $E = 200\,000$ MPa, if during an experiment it is subjected to a maximum bending stress of 100 MPa.

30.3 A cantilever beam made of structural steel ($E = 200\,000$ MPa and $I = 23.6 \times 10^6$ mm⁴) is 5 m long and carries a 1.25 kN load located at the free end. What is the maximum deflection of the beam?

30.4 If the load on the cantilever beam in the previous problem is relocated to a point 4 m from the support, what is the maximum deflection?

30.5 What could be the maximum intensity, in newtons per metre, of a uniformly distributed load on a 2 m long cantilever beam, made from timber ($E = 12\,000$ MPa) with a square cross-section 75 mm × 75 mm, if the maximum deflection under the load is not to exceed 12 mm?

30.6 What is the longest span over which a simply supported structural steel beam ($E = 200\,000$ MPa and $I = 215 \times 10^6$ mm^4) can carry a single concentrated load of 14 kN, located at midspan, without its maximum deflection exceeding 15 mm?

30.7 A 3 m length of copper pipe (outside diameter 20 mm and inside diameter 15 mm) is simply supported at each end. Calculate the maximum deflection under its own weight. The density of copper is 8870 kg/m^3 and the modulus of elasticity is 112 000 MPa.

30.8 A simply supported timber beam ($E = 12\,000$ MPa), 8 m long, 200 mm deep and 100 mm wide, carries three concentrated loads of 0.6 kN each, located 2 m, 4 m and 6 m from the left-hand support. Determine the maximum deflection of the beam.

30.9 In a materials-testing laboratory, a 10 mm diameter aluminium rod is set up as a simply supported beam, 900 mm long, and is subjected to a single concentrated load of 5 N located precisely at one-third of its length from one of the supports. Given that the modulus of elasticity of aluminium is 70 000 MPa, calculate and locate the maximum deflection of the rod.

Review questions

1. Name two measures that can be used to describe the extent of deformation of a beam.
2. Explain the term *radius of curvature*.
3. State how the radius of curvature of a beam can be calculated.
4. Explain the term *deflection*.
5. Where does maximum deflection occur in a cantilever beam?
6. Where does maximum deflection occur in a symmetrically loaded simply supported beam?
7. Explain the *principle of superposition*.

PART 11

TOWARDS ENGINEERING DESIGN

The subject of mechanics of materials cuts broadly across all branches of the engineering profession with remarkably many applications.

Egor P. Popov
Mechanics of Materials

CHAPTER 31

Bolted and welded joints

In this chapter we apply the concept of stress to the analysis of connections used to join together components of building frames, such as beams and columns, and used in the construction of boilers and pressure vessels.

The methods of joining two structural members, or two plates of metal, together can be divided into two groups:

1. bolted connections
2. welded connections

Bolted connections have the advantage of being semipermanent, i.e. they are capable of being disassembled. Bolts are the basic screw fasteners generally used with through holes and the appropriate nuts.

When a permanent structure or a tight boiler joint is required, construction methods presently used in industry tend to favour welding in preference to bolted connections. Welding is often the least expensive process and has the advantage of not requiring holes to be drilled in the parts being joined, and therefore not reducing their original strength.

However, the choice of the most suitable construction method depends on many design variables, which the engineer must understand. This chapter provides an introduction to the analysis of bolted and welded connections.

31.1 *BOLTED JOINTS*

We shall consider two types of joints here. A **lap joint** is formed when two plates are lapped over one another and the bolt goes through both plates as in Figure 31.1(a). In a **butt joint**, the plates are placed edge to edge, and the joint is made with the use of straps (Fig. 31.1(b)).

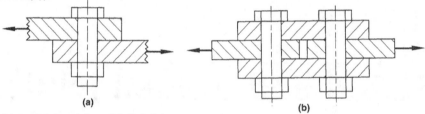

Fig. 31.1   **(a)** *Lap joint*   **(b)** *Butt joint*

There are single, double and multiple joints, according to whether there are one or more rows of bolts. Our discussion will be limited to single-row joints only.

When a bolted joint is subjected to a force which is in excess of its strength, the joint will usually fail in one of three ways (Fig. 31.2):

1. the shearing of the bolts in single or double shear;
2. the tearing apart of the plate weakened by the presence of holes;
3. the compression or crushing failure between the bolts and the plate.

Other types of failure, e.g. bending of the bolts, are possible but less common and are not discussed in this book.

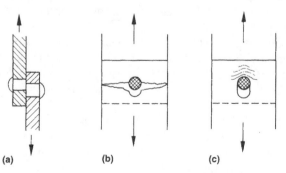

Fig. 31.2   **(a)** *Shearing failure*   **(b)** *Tearing failure*   **(c)** *Crushing failure*

The analysis of a bolted joint involves the calculation of each type of stress, i.e. in shear, tension (tearing) and compression (crushing), and then comparing these with allowable stresses for the materials used. For the purposes of illustration in this chapter, we shall adopt the following typical values of allowable stress:[*]

Allowable stress in shear: 90 MPa
Allowable stress in tension: 110 MPa
Allowable stress in compression: 220 MPa

In calculations, it is convenient to consider the allowable load per bolt calculated as the product of the area under stress and the corresponding value of the allowable stress. Each joint must be checked separately with respect to shear, tearing and crushing, and the least allowable load resulting from these calculations can then be taken to be the allowable strength of the joint.

Example 31.1

A simple lap joint is composed of two steel straps, 40 mm wide × 8 mm thick, held together by a single bolt 16 mm in diameter. Calculate the allowable strength of the joint.

Solution

(a) *Strength of bolt in shear*
    The area in single shear:

$$A_s = \frac{\pi D^2}{4}$$

$$= \frac{\pi \times 16^2}{4}$$

$$= 201.1 \text{ mm}^2$$

Allowable shear stress:

$$f_s = 90 \text{ MPa}$$

Allowable load in shear:

$$A_s \times f_s = 201.1 \times 90$$
$$= 18\ 100 \text{ N}$$
$$= 18.1 \text{ kN}$$

(b) *Strength of straps in tension (tearing)*
    Net area of the plate subjected to tearing:

$$A_t = \text{net width} \times \text{thickness}$$
$$= (40 - 16) \text{ mm} \times 8 \text{ mm}$$
$$= 192 \text{ mm}^2$$

---

[*] These are only average values based on permissible stress recommended for structural joints in certain types of steel construction. For any other material, or for detailed design purposes, appropriate codes and handbooks should be consulted.

Allowable stress:

$$f_t = 110 \text{ MPa}$$

Allowable load in tension:

$$A_t \times f_t = 192 \times 110$$
$$= 21\ 120 \text{ N}$$
$$= 21.1 \text{ kN}$$

(c) *Bearing strength*
The projected area of the bolt subjected to crushing:

$$A_c = \text{plate thickness} \times \text{bolt diameter}$$
$$= 8 \text{ mm} \times 16 \text{ mm}$$
$$= 128 \text{ mm}^2$$

Allowable compressive stress:

$$f_c = 220 \text{ MPa}$$

Allowable bearing load:

$$A_c \times f_c = 128 \times 220$$
$$= 28\ 160 \text{ N}$$
$$= 28.2 \text{ kN}$$

The joint can only be as strong as its weakest element, in this case the bolt in shear. Therefore the allowable load on the bolt represents the strength of the joint.

$$\therefore \text{ Strength of joint} = 18.1 \text{ kN}$$

31.2 *EFFICIENCY OF BOLTED JOINTS*

The strength of a bolted joint is always somewhat less than that of the unpunched parent metal.[*] If the strength of the joint is compared with the original strength of the unpunched plate, the efficiency of the joint can be calculated from the following definition:

$$\text{Joint efficiency} = \frac{\text{strength of joint}}{\text{strength of unpunched plate}}$$

Example 31.2
Determine the efficiency of the joint in Example 31.1.

---

[*] Depending on the plate thickness and other design and manufacturing criteria, the holes may be punched or drilled. For our purposes here, this difference is not significant.

Solution

Cross-sectional area of the unpunched plate in tension is:

$$40 \text{ mm} \times 8 \text{ mm} = 320 \text{ mm}^2$$

Therefore, the strength of unpunched plate, using the same allowable stress in tension as before, is:

$$f \times A = 110 \text{ MPa} \times 320 \text{ mm}^2$$
$$= 35\ 200 \text{ N}$$
$$= 35.2 \text{ kN}$$

The strength of the joint, as calculated previously, is 18.1 kN. Therefore, the efficiency of the joint is:

$$\text{Efficiency} = \frac{18.1 \text{ kN}}{35.2 \text{ kN}}$$
$$= 0.514$$
$$= 51.4\%$$

It should be understood that in a continuous joint, bolts are usually equally spaced, or form a pattern which repeats itself along the length of the joint. In such a case, a section equal in length to a repeating section is used as a unit for calculations.

Problems

Use allowable stress in shear, tension and compression of 90 MPa, 110 MPa and 220 MPa respectively.

31.1 Determine the shear stress in the bolts of a lap joint held by eight 10 mm diameter bolts if the total force on the joint is 31.4 kN.

31.2 Determine the tearing stress in a 200 mm width of a 10 mm thick plate with a single row of four 14 mm diameter bolt holes drilled in it, if the total force is 115.2 kN.

31.3 Determine the crushing stress in the four bolts in the previous problem. Does it exceed the allowable compressive stress?

31.4 Determine the strength per section, and the efficiency, of a lap joint in which each repeating section 100 mm wide is held by a 30 mm diameter bolt holding two plates, each 10 mm thick.

31.5 A butt joint between two 8 mm thick plates is made by using two 6 mm thick straps with a single row of 10 mm diameter bolts spaced at a repeating interval of 50 mm along the joint.

If a force on the joint is 10 kN for every repeating section of 50 mm, determine the magnitudes of the existing stresses in bolt shear, plate tearing and crushing of the bolts.

31.6 Determine the strength and efficiency of the joint described in the previous problem.

31.7 Design a connection composed of two steel straps held together by a single bolt. The straps are to be of the same width and thickness and should transmit a force of 2500 N. Use the following sequence in your calculations:

(a) Determine the required bolt diameter for the allowable shear stress of 90 MPa. Round up to the next millimetre.

(b) Using the result of step (a), calculate the required plate thickness for the allowable compression (crushing) stress of 220 MPa. Round up to the next millimetre.

(c) Using diameter and thickness selected in steps (a) and (b), calculate the minimum width of the straps required for plate tension (tearing) stress of 110 MPa. Round up to the next millimetre.

31.8 Calculate the efficiency of the joint designed in problem 31.7.

31.9 A single bolted butt joint is to be designed for a force of 81 kN per 100 mm pitch. If the plates which are placed edge to edge are to be twice as thick as each of the two straps, determine:

(a) the minimum diameter, to the nearest millimetre, of the bolts required for allowable shear;

(b) the minimum thickness, to the nearest millimetre, of the plate and straps (check both crushing and tearing).

31.10 Determine the efficiency of the joint constructed in accordance with the results of problem 31.9.

31.3 *WELDED CONNECTIONS*

Welding is a manufacturing process for the permanent joining of metal parts by fusion. The most common method of joining is achieved by striking an electric arc betwen a rod of similar metal and the pieces to be joined, metal being melted from the electrode into the joint. As a method of construction, welding is widely used in structural work and for repair and fabrication of boilers, pressure vessels and heavy machinery.

The two types of welds most frequently used are **fillet welds** and **butt welds**, as illustrated in Figure 31.3.

(a)   (b)

Fig. 31.3   **(a)** *Double-fillet lap joint*   **(b)** *Butt weld joint*

Butt welds

The plates for butt welds may be unbevelled for thin plates or bevelled on one or both sides. They are most frequently used for manufacturing boiler shells, air receivers etc., and are usually subjected to tension and compression, not to shear. The thickness of the weld is at least equal to the thickness of the plates joined, and its strength is thought of in relation to the strength of the plate.

Tests show that good butt welds have about the same strength as the plates being joined. In practice, it is safe to assume an efficiency of the joint of approximately 90 per cent. However, for our purposes, it is possible to regard the strength of a butt weld as equal to that of the plates joined.

Fillet welds

A standard full fillet weld has a section of an isosceles right triangle as shown in Figure 31.4, with the legs of the triangle equal to the thickness of the plate. The size of the weld is its leg length, and the throat is 0.707 times that length.[*] In a fillet weld, the throat is the critical dimension.

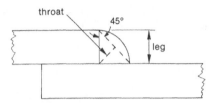

Fig. 31.4   *A standard full fillet weld*

It is common practice to take one-third of the nominal tensile strength of electrode used as the permissible working stress, i.e. allowable stress, in fillet welds, and to refer it to the area based on the length of weld and thickness of the throat. The strength of the fillet weld is therefore given by:

$$F = flt = 0.707fls$$

where F is the maximum allowable load on the weld
f is the allowable stress
l is the length of the weld
s is the nominal size of the weld
t is the throat thickness

It should be understood that stress distribution within the weld may be a complex combination of shear, tension, and sometimes bending. However, for many practical purposes, it can be assumed that stress in the weld is uniformly distributed shearing stress.

Example 31.3

For a fillet weld of 8 mm nominal size and electrode strength of 410 MPa, determine:
(a) the throat thickness
(b) the allowable stress
(c) the allowable load per millimetre of length
(d) the length of weld required to carry a load of 52.6 kN

Solution

(a)   Throat thickness $= 0.707s$
$= 0.707 \times 8$ mm
$= 5.66$ mm

(b)   Allowable stress $f = \dfrac{410}{3}$
$= 136.7$ MPa

---

[*] $\sin 45° = 0.707$

(c) Allowable load per millimetre $= \dfrac{F}{l}$

$$= ft$$
$$= 136.7 \text{ MPa} \times 5.66$$
$$= 773 \text{ N/mm}$$

(Note that here, as before, if stress is expressed in megapascals and linear dimensions in millimetres, force will be in newtons.)

(d)   Length of weld $l = \dfrac{F}{ft}$

$$= \frac{52\,600}{136.7 \times 5.66}$$
$$= 68 \text{ mm}$$

It is customary to add an allowance for end-craters, i.e. for starting and stopping, equal to about twice the nominal weld size. Therefore, the required length of weld specified in the previous example would be:

$$68 \text{ mm} + 2 \times 8 \text{ mm} = 84 \text{ mm}$$

In this book, we will omit this allowance and will simply regard the computed answers, e.g. 68 mm in the above example, as representing the effective length of weld required in each particular case.

Eccentricity of welded joints is a common problem in structural design, e.g. when non-symmetrical members, such as structural angles, are welded to gusset plates. The total load is presumed to act along the centroidal axis, while the required resistances offered by the welds are inversely proportional to the distances d_1 and d_2 from the axis, determined by summation of moments as illustrated by the following example.

Example 31.4

A structural steel angle is to be welded to a gusset plate as shown in Figure 31.5. Determine the length of 8 mm weld required to withstand a load of 100 kN, if the allowable stress is 136.7 MPa.

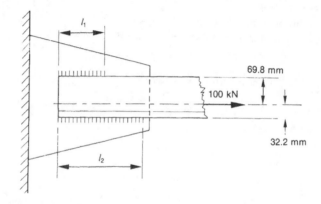

Fig. 31.5

Solution
Strength of weld per millimetre is:

$$\frac{F}{l} = 0.707fs$$
$$= 0.707 \times 136.7 \times 8$$
$$= 773.2 \text{ N/mm}$$

The required resistance at the toe of the angle is found by taking moments about the heel of the angle:

$$F_1 = \frac{100\,000 \text{ N} \times 32.2 \text{ mm}}{102 \text{ mm}}$$
$$= 31\,570 \text{ N}$$

The resistance of the weld at the heel is:

$$F_2 = 100\,000 \text{ N} - 31\,570 \text{ N}$$
$$= 68\,430 \text{ N}$$

The corresponding lengths of weld required are:

$$l_1 = \frac{31\,570 \text{ N}}{773.2 \text{ N/mm}}$$
$$= 40.8 \text{ mm}$$
$$l_2 = \frac{68\,430 \text{ N}}{773.2 \text{ N/mm}}$$
$$= 88.5 \text{ mm}$$

 Problems

31.11 For the following nominal sizes of fillet welds, determine:
(a) the throat thicknesses (mm);
(b) the safe load (N/mm) based on weld material with ultimate tensile strength equal to 410 MPa.

Nominal size *(mm)*	4	6	8	10	12	16	20	24
Throat thickness (mm)								
Safe load (N/mm)								

Note: Use the results in the above table to solve the following problems.

31.12 A 100 mm × 10 mm steel strap is to be welded to a heavy steel plate as shown in Figure 31.6. Determine the length *l* of 8 mm fillet weld to withstand a tensile force of 150 kN.

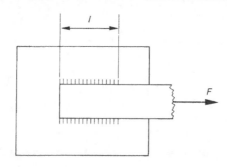

Fig. 31.6

31.13 A double-fillet lap joint is made using 12 mm welds (Fig. 31.7). Determine the load allowed on 500 mm width of plate.

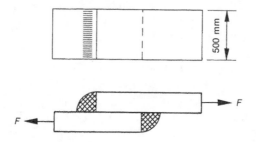

Fig. 31.7

31.14 Determine the minimum size of weld that can be used for the connection in problem 31.12 if the length of weld is not to exceed 100 mm on each side and the load is 115 kN.

31.15 A steel strap 12 mm thick by 60 mm wide is welded to a steel plate by means of 10 mm fillet welds, as shown in Figure 31.6. Determine the length l required to match the full tensile strength of the strap if the allowable stress for the strap in tension is 120 MPa.

31.16 A steel bracket is to be welded as shown in Figure 31.8. Determine the maximum mass that can be supported from this bracket.

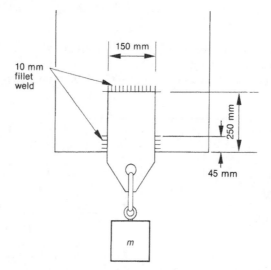

Fig. 31.8

31.17 A structural steel angle is to be welded to a gusset plate as shown in Figure 31.9. Determine the lengths of weld, l_1 and l_2, required to withstand a load of 250 kN.

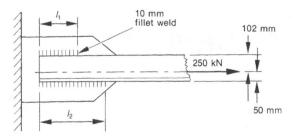

Fig. 31.9

Review questions

1. Briefly describe bolted connections and welded connections.
2. What is the difference between a lap joint and a butt joint?
3. List and explain the three most common types of failure in bolted joints.
4. Define *joint efficiency.*
5. Distinguish between fillet welds and butt welds.
6. State the formula for calculating the strength of a fillet weld.
7. Explain the effect of load eccentricity on weld design.

CHAPTER 32

Pressure vessels

Pressure vessels are cylindrical or, less frequently, spherical containers used to hold fluids, i.e. gases or liquids, under pressure. The most common form of pressure vessel, as used in steam boilers and compressed-air receivers, is the cylinder of welded construction with curved, but not quite hemispherical, ends. Spherical vessels are sometimes used, particularly for the containment of liquefied gases at high pressure and low temperature.

A pressure vessel containing a gas is subject to uniform internal pressure normal to its walls. The material of the wall, as well as the welded seams, must be sufficiently strong to resist stresses set up within the material due to the gas pressure.

The pressure responsible for stress in the walls of a vessel is the difference between internal pressure and atmospheric pressure outside the vessel. It is in fact the pressure as measured directly by a pressure gauge fitted to a pressure vessel, and is known as **gauge pressure**.

Expected learning outcomes

After carefully studying the material presented in this chapter, working through all numerical examples, and successfully completing all practice problems, students should be able to:
1. calculate stresses in the walls of cylindrical and spherical containers subjected to internal pressure;
2. calculate the safe level of internal pressure for a given container;
3. calculate the required wall thickness for a pressure vessel, given a   specified pressure, safety factor and joint efficiency.

32.1 *STRESSES IN CYLINDRICAL SHELLS*

In a cylindrical pressure vessel of welded construction, there are two types of seams that need to be considered for possible stress. These are the **longitudinal seam** along the length of the cylinder, and the **circumferential seam** (Fig. 32.1). It should be noted at the outset that stresses induced in the material of the shell are found throughout the material of the shell, whether there is an actual welded joint at a particular point or not. However, it is often useful to imagine two halves of the shell as if they were connected by some form of joint, as will be seen from the following discussion.

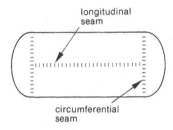

Fig. 32.1   *Seams in a cylindrical pressure vessel*

Stress on longitudinal seam: hoop stress

Let us consider one-half of a cylindrical shell separated from the other along its longitudinal seams,* as shown in Figure 32.2(a). For a given length l, and wall thickness t, the area subjected to stress is twice $l \times t$:

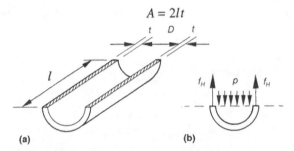

Fig. 32.2   **(a)** *Longitudinal section of cylindrical shell*   **(b)** *Hoop stress perpendicular to the cross-section of the shell material*

For the purposes of this chapter, pressure in a fluid can be regarded as similar to stress in a solid material, defined as force per unit area and measured in pascals, or its derivatives, kilopascals and megapascals.

The pressure p acts in the radial direction upon all elements of the exposed internal surface. However, it can be shown that after summation of all components of pressure perpendicular to the plane of the section, the total force F due to pressure is equal to the product of the pressure and the projected area $l \times D$, where D is the shell's internal diameter.

$$F = plD$$

Therefore, the stress in the material of the shell can be found as force per unit area:

$$f = \frac{F}{A} = \frac{plD}{2lt} = \frac{pD}{2t}$$

---

* We use the term 'seam' to mean an imaginary line of separation between two halves of a shell, forming a cross-section considered for the purpose of stress analysis. The presence, or otherwise, of an actual welded joint along that line is immaterial to our discussion, provided that the joint efficiency is 100 per cent.

This stress, known as **hoop stress,**[*] is the tensile stress in the material of the shell set up in the tangential direction, all the way along length *l* of the longitudinal seam. One should note very carefully that the direction of this stress is perpendicular to the cross-section of the shell material made by the imaginary plane of separation between the two halves (Fig. 32.2(b)). Therefore, hoop stress is:

$$f_H = \frac{pD}{2t}$$

Example 32.1

Determine the hoop stress in the material of a cylindrical air receiver 1.2 m long and 350 mm in diameter, with a wall thickness of 6 mm, subjected to a pressure of 1 MPa.

Solution

Keeping in mind that, in this context, pressure is similar to stress, we can use pressure in megapascals and dimensions in millimetres for convenience.

$$\text{Hoop stress } f_H = \frac{pD}{2t}$$
$$= \frac{1 \text{ MPa} \times 350 \text{ mm}}{2 \times 6 \text{ mm}}$$
$$= 29.2 \text{ MPa}$$

Stress on circumferential seam: axial stress

Axial stress in a shell of a pressure vessel is tensile stress in the direction of the principal axis of the cylinder. To establish the magnitude of this stress, it is necessary to consider a section along the plane of a circumferential seam as shown in Figure 32.3.

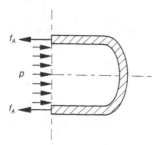

Fig. 32.3   *Axial stress in a shell of a pressure vessel*

---

[*] The term originates from a classic problem of stress analysis in a thin circular ring, or 'hoop', subjected to uniformly distributed radial forces, which produce uniform enlargement of the ring. Hoop stress is sometimes referred to as 'circumferential stress' due to its tangential direction. However, this often creates confusion in the learner's mind between circumferential stress, i.e. hoop stress, and stress on a circumferential seam which, to make things worse, is sometimes called 'longitudinal stress'. To avoid ambiguity and confusion, we shall only use the terms 'hoop stress' and 'axial stress' in this chapter.

This time, the pressure p can be regarded as acting in the axial direction, i.e. perpendicular to the projected area of the cylinder end, and equal to $\pi D^2/4$. This gives rise to a total axial force:

$$F = p\,\frac{\pi D^2}{4}$$

The area of the shell material resisting this force can be estimated approximately as the product of the circumference, based on the nominal diameter of the shell, and the wall thickness:

$$A = \pi Dt$$

The accuracy of this approximation is quite acceptable provided that the cylinder is thin, i.e. its diameter is at least ten times greater than the wall thickness, and the nominal diameter is somewhere between the inside and outside diameters of the shell.

The stress resulting from the force acting on the cross-section of the circumferential seam can be found from:

$$f = \frac{F}{A} = \frac{p\,\dfrac{\pi D^2}{4}}{\pi Dt} = \frac{pD}{4t}$$

This is axial stress:

$$\boxed{f_A = \frac{pD}{4t}}$$

Example 32.2

Calculate the axial stress in the material of the air receiver in the previous example.

Solution

$$
\begin{aligned}
f_A &= \frac{pD}{4t} \\
&= \frac{1\ \text{MPa} \times 350\ \text{mm}}{4 \times 6\ \text{mm}} \\
&= 14.6\ \text{MPa}
\end{aligned}
$$

The first step in the design of a pressure vessel is to determine the thickness of plate required. A simple comparison shows that a given pressure in a cylindrical pressure vessel causes hoop stress which is twice as high as the axial stress. This means that a longitudinal seam is more vulnerable to rupture. Therefore, if other conditions are equal, the strength of a longitudinal seam is the limiting factor in the design of cylindrical vessels, and the hoop stress formula should be used for calculating required plate thickness.

Example 32.3

Determine the minimum plate thickness required for a steam boiler drum, with diameter 1.2 m, if the maximum allowable stress in the material is 75 MPa and the pressure is 1.5 MPa.

Solution

Hoop stress $f_H = \dfrac{pD}{2t}$ is the limiting factor,

from which:

$$t = \frac{pD}{2f_H}$$

Substitute:

$$t = \frac{1.5 \text{ MPa} \times 1200 \text{ mm}}{2 \times 75 \text{ MPa}}$$
$$= 12 \text{ mm}$$

Effect of joint efficiency

For general purposes, it may be assumed that good-quality butt-welded joints in mild steel have the same strength as the plates being joined, i.e. a joint efficiency of 100 per cent is assumed. However, under certain conditions of workmanship or service, it would be safer to assume an efficiency of the joint of 90 per cent or less.[*]

It is not our purpose here to discuss various boiler and pressure vessel codes and specifications. It is sufficient, at this stage, to understand that if a joint has only a certain percentage of the strength that the solid plate has, then the thickness of the plate, to allow for the weakness of the joint, must be increased accordingly.

This in effect means that, to allow for joint efficiency, the required plate thickness must be equal to the calculated minimum thickness, t, divided by the joint efficiency.

Example 32.4

Determine the actual plate thickness that would be required in the previous example if an allowance for joint efficiency of 80 per cent is made.

Solution

Calculated thickness is $t = 12$ mm. Required thickness, to allow for joint efficiency of 80 per cent, is:

$$\frac{12 \text{ mm}}{0.8} = 15 \text{ mm}$$

32.2 *STRESSES IN SPHERICAL SHELLS*

A pressure vessel of spherical shape is symmetrical in all directions, suggesting that stresses in the material of its wall induced by internal pressure are the same at all points and in all directions.

---

[*] Another, more obvious, example is when the joint is a riveted joint. Riveted joints are never as strong as the solid plate. Consequently, an allowance must be made for the weakness of the joint. Depending on the actual type of the riveted connection used and the arrangement and size of the rivets, the efficiency of such joints may be as low as 50 per cent.

If we consider any diametral section of the sphere, as in Figure 32.4, the similarity with stresses on circumferential seams can easily be seen.

Fig. 32.4   *Diametral section of spherical pressure vessel*

The force due to pressure on a circular area $\pi D^2/4$ is equal to $p\pi D^2/4$. This force is distributed over the cross-sectional area of the material, which is equal to πDt, again assuming 'thin-wall' approximation.

Therefore, stress in the wall material is:

$$f = \frac{F}{A} = \frac{p\,\dfrac{\pi D^2}{4}}{\pi Dt} = \frac{pD}{4t}$$

Thus, for the same p, D and t, stress in a spherical shell is equivalent to axial stress on a circumferential seam in a cylindrical vessel, given by:

$$\boxed{f = \frac{pD}{4t}}$$

It should be understood, however, that a cylindrical shell of equal diameter and wall thickness, subjected to the same pressure, will also have hoop stress in its longitudinal seams of twice the magnitude of the axial stress. It follows, therefore, that since a spherical shell does not have longitudinal seams, it is not subjected to hoop stress, and can be said to be twice as strong as a corresponding cylindrical container. The following example illustrates this point.

Example 32.5

Given the ultimate strength of steel plate of 380 MPa and using a factor of safety of 5, compare the maximum allowable pressure in **(a)** spherical, and **(b)** cylindrical, pressure vessels of diameter 1 m and wall thickness 10 mm, assuming 100 per cent joint efficiency in a fully welded construction.

Solution

Allowable stress:

$$f = \frac{380 \text{ MPa}}{5}$$
$$= 76 \text{ MPa}$$

(a) *Spherical shell*

Stress in a spherical shell is given by $f = \dfrac{pD}{4t}$, from which maximum allowable pressure in a spherical pressure vessel is:

$$p = \frac{4tf}{D}$$
$$= \frac{4 \times 10 \text{ mm} \times 76 \text{ MPa}}{1000 \text{ mm}}$$
$$= 3.04 \text{ MPa}$$

(b) *Cylindrical shell*

Critical stress in a cylindrical shell is hoop stress, $f = \dfrac{pD}{2t}$, from which maximum allowable pressure in a cylindrical pressure vessel is:

$$p = \frac{2tf}{D}$$
$$= \frac{2 \times 10 \text{ mm} \times 76 \text{ MPa}}{1000 \text{ mm}}$$
$$= 1.52 \text{ MPa}$$

It is apparent that within the specified parameters, the spherical vessel can withstand twice as large a pressure as would be allowed in the cylindrical vessel.

Although theoretically quite valid, this comparison should not be overemphasised. Other comparisons on the basis of equal volume or equal cost may be more useful. Practical considerations, such as ease of manufacture and resulting costs may dictate a search for other alternatives. For example, it could be easier and cheaper to manufacture a cylindrical vessel. The resultant saving may allow the use of heavier plate for extra strength.

 Problems

32.1 Determine the stress in the material of a spherical container, diameter 600 mm and wall thickness 5 mm, if the pressure is 1 MPa.

32.2 If the maximum allowable stress in the material is 75 MPa, calculate the minimum allowable wall thickness for the container in the previous problem.

32.3 Determine the hoop stress and the axial stress in the wall of a cylindrical air receiver of diameter 300 mm and wall thickness 3 mm, when the working pressure is 720 kPa.

32.4 Find the minimum thickness of the boiler plate, with an ultimate tensile strength of 375 MPa, required for the construction of a 600 mm diameter cylindrical drum for a working pressure of 1 MPa, assuming a safety factor of 5. Compare with the answer to problem 32.2.

32.5 A cylindrical air receiver, 800 mm in diameter, is to be designed to withstand safely a working pressure of 900 kPa. Boiler plate with an allowable stress of 80 MPa will be used. Efficiency of all welded joints is assumed to be 90%. Determine the required wall thickness.

32.6 Calculate the safe working pressure for a 2 m diameter spherical container, with wall thickness 15 mm, if the ultimate tensile strength of its material is 420 MPa, the factor of safety is 6 and the efficiency of welded joints is 90%.

32.7 What is the maximum volume of a spherical container that can be made from 8 mm thick plate, having an allowable stress of 80 MPa and a joint efficiency of 100%, designed to withstand a pressure of 1.8 MPa?

32.8 Determine the factor of safety when a 1 m diameter cylindrical pressure vessel operates at an internal pressure of 950 kPa. Wall thickness is 10 mm and the ultimate tensile strength is 380 MPa.

32.9 A spherical tank, 12 m in diameter, is used to contain gas under pressure. The wall thickness is 10 mm and the joint efficiency is 90%. If the allowable stress in the wall material is 76 MPa, determine the maximum safe pressure.

32.10 A boiler drum, 1800 mm in diameter, is to have a longitudinal joint with an efficiency of 75%. The operating pressure is expected to be 1.35 MPa.

Calculate the required thickness of boiler plate with a maximum allowable tensile stress of 67.5 MPa.

Review questions

1. What is the most common shape of a pressure vessel?
2. Sketch a cylindrical pressure vessel showing circumferential and longitudinal seams.
3. What is the stress on a longitudinal seam called?
4. What is the stress on a circumferential seam called?
5. State the expressions for calculating stresses in cylindrical shells.
6. In a cylindrical shell, which is higher: hoop stress or axial stress?
7. State the expression for calculating stress in a spherical shell.
8. Explain how joint efficiency affects the wall thickness required to withstand a specified pressure.

CHAPTER 33

Power transmission by shafts

In this short chapter, which deals with power transmission by shafts, we bring together some concepts developed earlier in three different parts of this book. These include the relation between torque, rotational speed and power (Ch. 16), some general understanding of mechanical drives (Ch. 24), and the relation between torque and torsional stress and strain (Ch. 27). You may need to briefly revise these earlier topics before proceeding with the exercises presented in this chapter.

Expected learning outcomes

After carefully studying the material presented in this chapter, working through all numerical examples, and successfully completing all practice problems, students should be able to:
1. identify torque and power transmitted through different parts of a shaft;
2. calculate the torque and power that can be transmitted by a shaft within specified parameters of size and allowable stress;
3. calculate the shaft diameter required to transmit specified torque and power, without exceeding allowable stress or angle of twist.

33.1 *POWER TRANSMISSION BY SHAFTS*

A shaft can be described as a rotating machine component, subjected to torque and used for the purpose of transmitting mechanical power.

There are three important relations we should recognise. The first is the relation between the applied torque, the speed at which a shaft is rotating, and the power transmitted along the axis of the shaft from the power input point (pulley, gear or sprocket) to the power output point (another pulley, gear or sprocket). From Chapters 16 and 24, we recall that this relation is:

$$P = T\omega$$

Since rotational speed of mechanical components is usually measured and described in revolutions per minute, it is convenient for future use to incorporate the necessary conversion factors into this formula. Hence, power can also be found from:

$$P = \frac{2\pi TN}{60}$$

where N is the rotational speed in revolutions per minute.

The second important relation is the formula for torsional shear stress which is due to an applied torque:

$$f_{ts} = \frac{Tr}{J}$$

where r is the radius of the shaft

J is its polar moment of inertia

For a solid shaft:

$$r = \frac{D}{2} \text{ and } J = \frac{\pi D^4}{32}$$

For a hollow shaft, with outside diameter D_o and inside diameter D_i:

$$r = \frac{D_o}{2} \text{ and } J = \frac{\pi(D_o{}^4 - D_i{}^4)}{32}$$

The third useful relation is that between torque, the angle of twist, and the modulus of rigidity of the material from which a shaft is made:

$$\theta = \frac{TL}{JG}$$

Let us see how these relations can be put to some use.

Example 33.1

A 35 mm diameter solid shaft rotates at 1440 rpm and transmits 82.5 kW of power. Determine the stress in the shaft.

Solution

Power transmitted:

$$P = 82.5 \text{ kW}$$
$$= 82\,500 \text{ W}$$

Substitute into:

$$P = \frac{2\pi TN}{60}$$

$$82\,500 = \frac{2\pi \times T \times 1440}{60}$$

Hence torque transmitted is found to be:

$$T = 547.1 \text{ N.m}$$
$$= 547\,100 \text{ N.mm}$$

For a 35 mm diameter shaft, $r = 17.5$ mm, and:

$$J = \frac{\pi \times 35^4}{32}$$
$$= 147.3 \times 10^3 \text{ mm}^4$$

Stress in shaft:

$$f_{ts} = \frac{Tr}{J}$$
$$= \frac{547\,100 \text{ N.mm} \times 17.5 \text{ mm}}{147.3 \times 10^3 \text{ mm}^4}$$
$$= 65 \text{ MPa}$$

Example 33.2

If the material of the shaft in the previous example is steel, with modulus of rigidity $G = 80\,000$ MPa, and the shaft is 600 mm long, what is the angle of twist?

Solution

The angle of twist is found by substitution into the appropriate formula:

$$\theta = \frac{TL}{JG}$$
$$= \frac{547\,100 \text{ N.mm} \times 600 \text{ mm}}{147.3 \times 10^3 \text{ mm}^4 \times 80\,000 \text{ MPa}}$$
$$= 0.027\,86 \text{ rad}$$

Conversion to degrees yields $\theta = 1.6°$.

33.2 SHAFTS WITH MULTIPLE POWER-OUTPUT POINTS

It is not uncommon for a single shaft to have more than one power-output point, through several gears, pulleys or sprockets spaced along the length of the shaft. This means that different portions of the shaft, while all rotating at the same speed as a whole, transmit different amounts of mechanical power and experience different magnitudes of torque.

It is best to approach this problem from the point of view of energy conservation. We can use the analogy between the flow of power through different portions of the shaft and the flow of water inside pipes with several take-off taps. The principle is the same for both: what flows in must flow out. Once the flow of power has been analysed, torque which is related to corresponding amounts of power in different portions of the shaft can easily be determined.

Example 33.3

A line shaft *ABCD* rotates at 960 rpm and has 120 kW of power input through pulley *B*, with three power-output pulleys as follows:

Pulley *A*:   45 kW
Pulley *C*:   25 kW
Pulley *D*:   50 kW

Determine the torque in each portion of the shaft.

Solution

The flow of power along the shaft is shown in Figure 33.1.

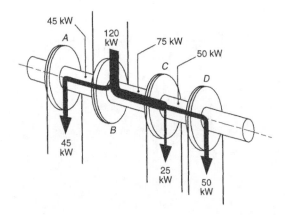

Fig. 33.1

Note that the amount of power carried by each portion of the shaft is cumulative if viewed in the reverse direction, i.e. from the extreme output ends towards the input point. Therefore, power carried by portion *BC* is the amount equal to the sum of the power outputs at *C* and *D*, while portion *CD* only needs to carry the amount of power leaving through pulley *D*.

Therefore:

$$P_{AB} = 45 \text{ kW}$$
$$P_{BC} = 75 \text{ kW}$$
$$P_{CD} = 50 \text{ kW}$$

The corresponding torque values can now be calculated using:

$$P = \frac{2\pi TN}{60}$$

Transposition gives:

$$T = \frac{60P}{2\pi N}$$

Hence:

$$T_{AB} = \frac{60 \times 45\,000}{2\pi \times 960} = 448 \text{ N.m}$$

$$T_{BC} = \frac{60 \times 75\,000}{2\pi \times 960} = 746 \text{ N.m}$$

$$T_{CD} = \frac{60 \times 50\,000}{2\pi \times 960} = 497 \text{ N.m}$$

These results can be represented in the form of a **torque distribution diagram** (Fig. 33.2), similar to shear force diagrams for beams.

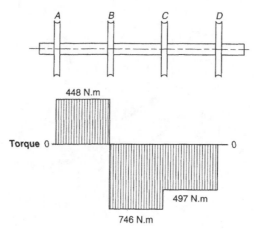

Fig. 33.2 *Torque distribution diagram*

Example 33.4

It has been decided to make a stepped shaft to suit the conditions specified in Example 33.3. If the allowable stress is 40 MPa, calculate the required diameters.

Solution

Start with the equation:

$$f_{ts} = \frac{Tr}{J}$$

and substitute $r = \frac{D}{2}$ and $J = \frac{\pi D^4}{32}$:

$$f_{ts} = \frac{Tr}{J}$$

$$= \frac{TD \times 32}{2 \times \pi D^4}$$

$$= \frac{16T}{\pi D^3}$$

Now make the unknown diameter the subject of the expression:

$$D = \sqrt[3]{\frac{16T}{\pi f_{ts}}}$$

This is a useful design-oriented formula for solid shafts. Now substitute numerical values and solve:

$$D_{AB} = \sqrt[3]{\frac{16 \times 448\,000}{\pi \times 40}}$$
$$= 38.5 \text{ mm}$$

Similarly, $D_{BC} = 45.6$ mm and $D_{CD} = 39.8$ mm.

 Problems

33.1 A shaft is to transmit 300 kW of power when running at 500 rpm. What torque will the shaft transmit?

33.2 A 100 mm diameter solid shaft transmits 740 kW of power at 1200 rpm. What is the stress in the shaft material?

33.3 A hollow shaft, with outside diameter 70 mm and inside diameter 50 mm, is to run at 670 rpm. If the allowable stress in the shaft material is not to exceed 65 MPa, how much power can it transmit?

33.4 What should be the diameter of a solid shaft that can transmit the same amount of power as the hollow shaft in problem 33.3, at the same speed and with the same allowable stress in its material?

33.5 A 1.35 m long aluminium shaft ($G = 28\,000$ MPa) must transmit 6 kW when running at 1800 rpm. If the angle of twist is to be limited to 5.6°, what should be the minimum diameter of the shaft?

33.6 A shaft *ABCDE* rotates at 1440 rpm and has power input through pulley *D*, with four power-output pulleys as follows:

Pulley *A*:   80 kW
Pulley *B*: 100 kW
Pulley *C*:   40 kW
Pulley *E*: 130 kW

Determine the magnitude of torque in each portion of the shaft.

33.7 A stepped shaft *ABC* is to be driven at 750 rpm through an input gear at *A*, with 29 kW taken off at *B* and 11 kW taken off at *C*. If the allowable stress is 51 MPa, determine to the nearest millimetre the required diameters for portions *AB* and *BC*.

Review questions

1. What are the three important relations used in the analysis of power-transmitting shafts?
2. Write the formula for each of the above.
3. How can a shaft have multiple power-output points?
4. What is the basis of the analogy between transmission of power by shafts and water flowing in pipes?
5. State the formula for calculating the required diameter of a solid shaft when torque and allowable stress are given.

APPENDIXES

≡

Nothing gives such weight and dignity to a book as an Appendix.

Herodotus of Halicarnassus
(*circa* 450 BC)

APPENDIX A

Summary of the metric system of units

This appendix is an abridged summary of the **SI units**[*] used in engineering and of the rules governing their use based on the Metric Conversion Board publication *Australia's Metric System.*

The metric system of measurement for Australia has been defined as measurement in terms of:

1. the units comprised in the International System of Units;
2. units decimally related to those units;
3. other units, declared from time to time to be within the metric system.

Parts 1 and 2 of the definition mean that the Australian metric system includes all International System (SI) units and all the decimal multiples and submultiples of each such unit that can be formed by the attachment of SI prefixes to it, provided such units are within the approval by the International Conference on Weights and Measures.

Part 3 provides for the extension of the system by introducing other units, which may owe their derivation to units outside the SI (e.g. hour, degree Celsius) or are specially named multiples or submultiples of SI units (e.g. litre, tonne).

The rules governing the formation and use of units and symbols, and the correct spellings of unit names and abbreviations, must always be followed if ambiguity and incorrect statements of measurements are to be avoided.

---

[*] This appendix contains some references to physical quantities and units not used in this book. Students will find them useful when studying other subjects, such as thermodynamics and fluid mechanics.

A.1   *THE INTERNATIONAL SYSTEM OF UNITS (SI)*

The International System of Units is founded on seven base units and two supplementary units, of which only the units tabulated below are relevant to mechanical engineering at the level treated in this book.

All physical quantities may be measured in terms of SI base and/or supplementary units taken singly or in mathematical combinations (by multiplication and/or division). Where a unit of a physical quantity is defined in terms of such a combination of SI units without the use of a numerical factor or an arbitrary constant, the unit is called an **SI derived unit**.

Table A.1   *SI base and supplementary units*

Physical quantity	SI unit	
	Name	*Symbol*
Base units		
length	metre	m
mass	kilogram	kg
time	second	s
temperature	kelvin	K
Supplementary unit		
plane angle	radian	rad

Some derived units have been allocated single-word names and special symbols. These are listed in Table A.2, together with their definitions and derivations.

Table A.2   *SI derived units with special names*

Physical quantity	SI unit			
	Name	*Symbol*	*Definition*	*Derivation*
force	newton	N	$kg.m/s^2$	
pressure, stress	pascal	Pa	N/m^2	$kg/m.s^2$
energy, work, quantity of heat	joule	J	$N.m$	$kg.m^2/s^2$
power, rate of heat flow	watt	W	J/s	$kg.m^2/s^3$

There are also many derived units with compound names, i.e. without any special name or symbol, the most common of which are listed in Table A.3, together with their symbols, which reflect their derivation.

Table A.3   *Examples of SI derived units with compound names*

Physical quantity	SI unit	
	Name	*Symbol*
area	square metre	m^2
volume	cubic metre	m^3
volumetric flow rate	cubic metre per second	m^3/s
mass flow rate	kilogram per second	kg/s
speed, velocity (linear)	metre per second	m/s
angular velocity	radian per second	rad/s
acceleration (linear)	metre per second squared	m/s^2
angular acceleration	radian per second squared	rad/s^2
density (mass density)	kilogram per cubic metre	kg/m^3
specific volume	cubic metre per kilogram	m^3/kg
torque, moment of force	newton metre	$N.m$
momentum	kilogram metre per second	$kg.m/s$
second moment of area[a]	metre to the fourth power	m^4
mass moment of inertia	kilogram square metre	$kg.m^2$
specific energy[b]	joule per kilogram	J/kg
specific thermal capacity, characteristic gas constant	joule per kilogram kelvin	$J/kg.K$
local gravitational constant	newton per kilogram	N/kg

[a] Otherwise known as moment of inertia of an area.
[b] This may include any form of energy expressed per unit mass, such as latent heat, internal energy, calorific value etc.

A.2   DECIMAL PREFIXES

Measurements of much larger or smaller magnitude than the SI units are frequently made. To avoid the inconvenience of using very large or very small numbers, the system allows for decimal multiples or submultiples of the SI units to be used.

Rather than introducing new names for the units which are larger or smaller than the SI units themselves, the system provides:

1. that the multiples and submultiples shall be decimally related to the 'parent' unit;
2. that the decimal relationships shall be indicated by the attachment of prefixes to the parent unit;
3. that each prefix shall have a standard value regardless of the unit to which it is attached.

SI prefixes are shown in Table A.4, together with the symbols for use with parent unit symbols.

One should be aware that the fact that a prefixed unit is part of the accepted Australian metric system does not necessarily mean its use is recommended. In general, prefixes for multiples and submultiples involving powers of 1000 (10^3) are preferred. This means that 'hecto', 'deka', 'deci' and 'centi' should be avoided.

Table A.4   *SI prefixes*

Factor	Prefix	Symbol
10^9	giga	G
10^6	mega	M
10^3	kilo	k
10^2	hecto	h
10	deka	da
10^{-1}	deci	d
10^{-2}	centi	c
10^{-3}	milli	m
10^{-6}	micro	μ

In particular, students of engineering must remember that linear measurements in the mechanical engineering industry should be in metres or millimetres—the centimetre is not to be used in technical applications.[*]

Special attention must be drawn to the SI unit of mass, the kilogram, which, although it contains the prefix 'kilo' for historical reasons, is nevertheless an SI unit and not a multiple. However, to avoid the use of two adjacent prefixes, the SI prefixes having their standard values, are attached to the stem word 'gram' when forming units of mass larger or smaller than the kilogram. Thus the gram itself is a submultiple of the kilogram.

A.3   *OTHER UNITS WITHIN AUSTRALIA'S METRIC SYSTEM*

Units outside the SI, but which have been declared to be within the Australian metric system, are divided into two classes:

1. units which may be used without restriction for the measurement of the particular physical quantity (Table A.5);
2. units which may be used only for particular, specified purposes (Table A.6).

---

[*] It must be noted that while it is not a technical unit, the centimetre is, nevertheless, a legal part of the metric system which is widely used in non-engineering areas of education, for general household purposes including body and clothing sizes, and for establishing the units of area and volume, such as the millilitre, which is equal to one cubic centimetre.

Table A.5   *Declared units having general application*

Description of units			
Name of unit	Symbol	Definition in terms of other metric units	Physical quantity
degree	°	$\dfrac{\pi}{180}$ rad	plane angle
minute	′	$\dfrac{\pi}{10.8} \times 10^{-3}$ rad	
second	″	$\dfrac{\pi}{648} \times 10^{-3}$ rad	
litre	L	10^{-3} m^3	volume
millilitre	mL	10^{-6} m^3	
day	d	86.4×10^3 s	time interval
hour	h	3.6×10^3 s	
minute	min	60 s	
kilometre per hour	km.h^{-1} *or* km/h	1/3.6 m.s^{-1}	velocity, speed
tonne	t	10^3 kg	mass
tonne per cubic metre	t.m^{-3} *or* t/m^3	10^3 kg.m^{-3}	density
kilogram per litre	kg.L^{-1} *or* kg/L	10^3 kg.m^{-3}	
degree Celsius	°C	the Celsius temperature, given by the relationship $t_{°C} = t_K - 273.15$	temperature
degree Celsius	°C	1 K	temperature interval

Table A.6 *Declared units having restricted application*

Name of unit	Symbol	Definition in terms of other metric units	Physical quantity	Purpose or purposes
international nautical mile	n mile	1852 m	length	marine navigation, aerial navigation, meteorology
knot	kn	$\dfrac{1852}{3600}$ m.s^{-1}	speed	marine navigation, aerial navigation, meteorology
kilowatt hour	kW.h	3.6×10^6 J	energy	measurement of electric energy

A.4 UNITS SPECIFICALLY EXCLUDED FROM AUSTRALIA'S METRIC SYSTEM

Other metric systems (e.g. the CGS and technical metric systems) contain many units which are not in Australia's metric system. Some of the units specifically excluded from the Australian metric system, together with the physical quantities they measure, are given in Table A.7. The appropriate units for use in their stead are also shown.

Table A.7 *Excluded units*

Excluded unit and symbol	Physical quantity	Australian metric unit and symbol
atmosphere, standard (atm)	pressure, stress	kilopascal (kPa)
bar (b)	pressure, stress	kilopascal (kPa)
millibar (mb)	pressure	hectopascal (hPa)[a]
calorie (cal)	energy	joule (J)
centimetre of mercury (cm Hg)	pressure	kilopascal (kPa)
kilogram force (kgf)	force	newton (N)
kilogram force per square centimetre (kgf/cm^2)	pressure, stress	kilopascal (kPa)
micron (μ)	length	micrometre (μm)[b]
torr (torr)	pressure	pascal (Pa)

[a] 1 hPa = 100 Pa, now used in meteorology to measure variations in atmospheric pressure.
[b] The term 'micron' was an earlier name for the micrometre.

A.5 *COHERENT UNITS*

SI units constitute a coherent system in which the product or quotient of any two unit quantities in the system is the unit of the resultant quantity. It follows that in a coherent system, derived units are formed without the use of numerical constants.

The coherence of SI units is probably the major benefit of the SI as it removes the need for introducing numerical factors into calculations. Therefore, if all data are expressed in terms of SI units only, results will be in terms of SI units. This leads to a series of logical steps in carrying out any calculation:

1. Express each measurement as a product of a number and an SI unit.
2. Perform the calculation, resulting in an answer in the form of a product of an SI unit and a number.
3. If necessary, select an appropriate multiple or submultiple of the SI unit to reduce the number to a convenient order of magnitude.

Example A.1

Find the power necessary to pull a load at a uniform speed of 4.2 km/hour when the load exerts a tension of 6 kN in the pulling rope.

Solution

1. Convert data to SI units:

$$4.2 \text{ km} = 4200 \text{ m}$$
$$1 \text{ h} = 3600 \text{ s}$$
$$6 \text{ kN} = 6000 \text{ N}$$

2. Perform calculation:

$$\text{Power} = \frac{\text{force} \times \text{distance}}{\text{time}}$$
$$= \frac{6000 \times 4200}{3600} \text{ N.m/s}$$
$$= 7000 \text{ J/s}$$
$$= 7000 \text{ W}$$

3. Select appropriate unit for the result:

$$7000 \text{ W} = 7 \text{ kW}$$
$$\therefore \text{ Power} = 7 \text{ kW}$$

A special mention should be made with respect to the unit of mass, the kilogram, which in spite of the inclusion of the prefix 'kilo' is a base unit and not a multiple. The kilogram should, therefore, be used in all calculations to take advantage of the coherent property.

This can be illustrated by reference to Newton's second law of motion which, when paraphrased, states that the physical quantity 'force' is related to the product of the two physical quantities 'mass' and 'acceleration'. The product of the SI units of mass and acceleration (kg and m/s^2) is the SI unit of force ($kg.m/s^2$), which is given the special name newton (symbol N). Thus the kilogram, and not the gram, is used for the formation of the newton.

The coherent property of SI units does not *generally* extend to the multiples and submultiples of SI units, nor does it apply to declared units such as the hour, used in the above example, necessitating appropriate conversion to SI units.

However, in some special cases, where repetitive calculations of the same kind are performed involving very large or very small quantities, it may be convenient to remember a *particular* coherent relationship between suitable prefixed multiple and submultiple units. For example, in Chapter 25, it was suggested that if force is expressed in newtons and dimensions in millimetres, stress can be expressed in megapascals without any conversions. All three units are usually appropriate for the quantities being measured and happen to have a coherent relationship between themselves.

Example A.2
Determine the tensile stress in a metal bar, 20 mm × 15 mm in cross-section, subjected to an axial pull of 7.5 kN.

Solution

$$\text{Stress} = \frac{\text{force}}{\text{area}}$$
$$= \frac{7500 \text{ N}}{20 \text{ mm} \times 15 \text{ mm}}$$
$$= 25 \text{ MPa}$$

The advantage of the above calculation is that an appropriate unit for the result is obtained automatically with relatively simple calculations. This advantage is somewhat outweighed by the necessity to memorise which prefixed units enter into a particular coherent relationship.

The choice of a suitable method for solving specific categories of problems is left to the discretion of the student. However, the golden rule is, if in doubt, convert all data to SI units and then perform the calculations, as in Example A.1.

A.6   GENERAL RULES

The following is a brief summary of the rules recommended for the consistent and uniform use of metric units. If followed, these rules help to achieve clarity of expression and to avoid ambiguity.

Names

1. When written in full, names of units, including prefixed units, are written in lower case (small) letters except at the beginning of a sentence. The sole exception is 'degree Celsius'.
2. Unit names take a plural 's' when associated with numbers greater than unity, e.g. 1.5 newtons.

Symbols

1. Symbols are internationally recognised mathematical representations of units and are not abbreviations of the unit names. Names and symbols must not be mixed within the same unit expression.
2. As with any other symbols, unit symbols are translated into the names when spoken.

3. Unit symbols are written in lower case letters except the symbols for units named after people, when the first letter of the symbol is a capital, e.g. pascal, Pa.*
4. Unit symbols do not take a full stop to indicate an abbreviation, nor do they take the plural form.
5. In writing, symbols should be separated from any associated numerical value by a space.

Prefixes

1. When written, the prefix is not separated from the parent unit, forming with it a new name or symbol.
2. All prefixes, except those representing a million and more, e.g. mega (M) and giga (G), have lower case symbols.
3. When a prefix is attached to a unit, it becomes an integral part of the new unit thus created and is therefore subjected to the same mathematical processes as the parent unit; for example:

$$mm^2 = mm \times mm, \text{ i.e. } 0.001 \text{ m} \times 0.001 \text{ m, not } 0.001 \text{ m}^2$$
$$= (10^{-3} \text{ m})^2$$
$$= 10^{-6} \text{ m}^2, \text{ not } 10^{-3} \text{ m}^2$$

4. No more than one prefix should be included in any unit, e.g. a tonne is a megagram and not a kilokilogram.
5. As a general rule, where a unit is expressed in the form of a product or a quotient, the prefixed unit (if any) should preferably be the first occurring unit.†

Powers

1. The raising of a unit to a power is indicated by placing the words 'squared', 'cubed', 'to the fourth power' and so on as appropriate after the unit so raised.‡
2. Powers of any unit may be shown by the appropriate indices (superscripts) to the unit symbol.

Products

1. Where a unit is derived by multiplication of two or more different units, the resulting compound name is formed by stating the names of the constituent units in succession. When writing in full, a space is left between each of the successive names.
2. Products of symbols are indicated by a full stop or a point at mid-height between the symbols.

Quotients

1. The quotient of two units is indicated by the word 'per' immediately in front of the unit(s) forming the denominator. The word 'per' should not occur more than once in any unit name.

---

* The only exception to this rule is the use of capital L for the litre in order to avoid ambiguity—see Chapter 3.
† There are, however, exceptions to this rule. In particular, because the kilogram is the SI base unit of mass, when it appears in a derived unit it is not considered to be prefixed. It is therefore preferable to write, for example, kJ/kg rather than J/g.
‡ By custom, the word 'square' or 'cubic' may be placed in front of a unit of length raised to the second or third power, e.g. square metre, cubic metre, kilogram per cubic metre.

2. Quotients are indicated by a solidus (oblique stroke /), or by a horizontal line. The solidus and the horizontal line are directly equivalent to the word 'per' used in unit names, so that no more than one solidus or horizontal line may be used in a unit symbol.

3. Alternatively, negative indices may be used to indicate those units which form the denominator of the quotient.

Appropriate units

1. When stating the value of any measurement, the appropriate unit must be chosen. No more than one unit name (or symbol) may be included in such a statement.

2. The 'appropriate unit' referred to above should generally be so chosen that the numerical value of the statement of measurement lies between 0.1 and 1000; for example:

500 kPa or 0.5 MPa, not 500 000 Pa

Pronunciation, spelling etc.

1. **gram**   The spellings 'gram' and 'gramme' are both allowed, but the shorter spelling is preferable.

2. **joule**   The correct pronunciation rhymes with 'pool'.

3. **kilo**   When used as a prefix with any unit, the pronunciation should be 'kill-o', with the accent on the first syllable and 'o' pronounced as 'oh'.

The word 'kilo' (pronounced 'kee-low'), when used as an abbreviation for kilogram, has no legal standing and its use other than in casual speech should be avoided.

4. **litre**   The spelling 'liter' is not acceptable legally.

5. **metre**   The American spelling 'meter' is not acceptable legally.

6. **tonne**   The correct pronunciation is 'tonn', with 'o' as in 'Tom'.

Symbols and formulae

B.1 *LIST OF SYMBOLS*

A	area, cross-sectional area
$\quad A_s$	area in shear
a	acceleration
$\quad a_g$	acceleration due to gravity
a, b, c	constants, side lengths of plane figures
b	base length
D	diameter
d	moment arm, perpendicular distance
E	Young's modulus of elasticity
e	direct axial strain
$\quad e_s$	shear strain
F	force
$\quad F_c$	centripetal or centrifugal force
$\quad F_f, F_n$	frictional and normal forces
$\quad F_g$	gravitational attraction
$\quad F_H, F_V$	horizontal and vertical forces
$\quad F_L, F_E$	load, effort
$\quad F_F, F_{Th}$	friction effort and theoretical effort
$\quad F_s, F_t$	slack-side and tight-side belt tensions
$\quad F_s$	shear force
$\quad F_w$	weight of a body
$\quad F_x, F_y$	components of a force
FS	factor of safety
f	stress
$\quad f_A, f_H$	axial stress and hoop stress in pressure vessels
$\quad f_c$	compressive stress
$\quad f_b$	bending stress
$\quad f_s$	shear stress
$\quad f_t$	tensile stress
$\quad f_{ts}$	torsional shear stress
G	universal gravitational constant
G	shear modulus of rigidity
g	local gravitational constant
h	hypotenuse
h	height, elevation above datum level
I	second moment of area, and mass moment of inertia
J	polar moment of inertia

KE   kinetic energy

k spring modulus

L, l length, linear dimension

M moment of force, bending moment

MA   mechanical advantage of a machine

m mass

N speed in revolutions per minute

P power

        P_{in} power input

        P_{out} power output

PE   potential energy

p pressure

R radius of curvature of beam in bending

r radius, radial distance

S linear displacement, distance travelled

        S_E distance moved by effort

        S_L distance moved by load

SE   strain energy

s nominal size of weld

T torque

t time

t thickness, weld throat thickness

t practical (Celsius) temperature

US   ultimate strength

        UCS   ultimate compressive strength

        USS   ultimate shear strength

        UTS   ultimate tensile strength

V volume

VR   velocity ratio of a machine

v linear velocity

W mechanical work

x elongation, axial deformation

        x_s deformation due to shear stress

$\bar{x}, \bar{y}$ position of centroid

y distance from neutral axis to extreme fibre of a beam

α angular acceleration

β one-half of wedge angle of V-belt

ε coefficient of restitution

η efficiency

θ angular displacement

θ angle of inclination, angle of contact, angle between two forces, angle of twist

μ coefficient of sliding friction

ν Poisson's ratio

ρ density

ϕ angle of friction

ω angular velocity

B.2   *LIST OF FORMULAE*

Chapter reference

Solution of quadratic equations

2.2

$$x = \frac{-b \pm \sqrt{b^2 - 4ac}}{2a}$$

Areas of plane figures

2.3

Square: $A = a^2$
Rectangle: $A = ab$
Triangle: $A = \frac{1}{2}bh$

Circle: $A = \dfrac{\pi D^2}{4}$

Volumes of solids

2.3

Cube: $V = a^3$
Rectangular prism: $V = abc$

Cylinder: $V = \dfrac{\pi l D^2}{4}$

Sphere: $V = \dfrac{\pi D^3}{6}$

Pythagoras' theorem

2.3

$$h^2 = a^2 + b^2$$

The sine rule

2.4

$$\frac{a}{\sin A} = \frac{b}{\sin B} = \frac{c}{\sin C}$$

The cosine rule

2.4

$$c^2 = a^2 + b^2 - 2ab \cos C$$

Rectangular components of a force

4.5

$$F_x = F \cos \theta$$
$$F_y = F \sin \theta$$

Addition of rectangular components   4.5

$$F = \sqrt{F_x^2 + F_y^2}$$
$$\tan \theta = \frac{F_y}{F_x}$$

Universal gravitation   4.8

$$F_g = G \frac{m_1 m_2}{d^2}$$
$$G = 66.7 \times 10^{-12} \text{ N.m}^2/\text{kg}^2$$

Weight of a body on Earth   4.9

$$F_w = mg$$
$$g = 9.81 \text{ N/kg}$$

Equilibrium of concurrent forces   5.1

$$\Sigma F_x = 0$$
$$\Sigma F_y = 0$$

Moment of a force   6.1

$$M = Fd$$

Equilibrium of moments   6.3

$$\Sigma M = 0$$

Equilibrium of non-concurrent forces   6.4

$$\Sigma F_x = 0$$
$$\Sigma F_y = 0$$
$$\Sigma M = 0$$

Coefficient of dry sliding friction   10.2

$$\mu = \frac{F_f}{F_n}$$

Angle of friction   10.3

$$\tan \phi = \mu$$

Helix angle of screw thread   12.1

$$\tan \theta = \frac{L}{\pi D}$$

Moment applied to a screw   12.1

$$M = \frac{F_L D}{2} \tan (\phi + \theta)$$

Reverse moment to loosen a screw   12.2

$$M' = \frac{F_L D}{2} \tan (\phi - \theta)$$

Effort required to lift a mass by a screw jack   12.3

$$F_E = \frac{mgD}{2d} \tan (\phi + \theta)$$

Equations of linear motion   13.2

$$S = t \left(\frac{v_0 + v}{2} \right)$$
$$v = v_0 + at$$
$$S = v_0 t + \frac{at^2}{2}$$
$$2aS = v^2 - v_0^2$$

Gravitational acceleration   13.3

$$a_g = 9.81 \text{ m/s}^2$$

Newton's second law   13.4

$$F = ma$$

Relationship between angular units   14.1

$$1 \text{ revolution} = 360° = 2\pi \text{ rad}$$

Equations of rotational motion   14.2

$$\theta = t\left(\frac{\omega_0 + \omega}{2}\right)$$
$$\omega = \omega_0 + \alpha t$$
$$\theta = \omega_0 t + \frac{\alpha t^2}{2}$$
$$2\alpha\theta = \omega^2 - \omega_0^2$$

Torque and rotational motion   14.4

$$T = I\alpha$$

Moment of inertia of a disc   14.4

$$I = \frac{mr^2}{2}$$

Relation between rotational and linear terms   15.1

$$S = r\theta$$
$$v = r\omega$$
$$a = r\alpha$$

Centripetal acceleration   15.2

$$a_c = \frac{v^2}{r}$$

Centripetal force   15.3

$$F_c = \frac{mv^2}{r}$$

Centripetal force and angular velocity   15.4

$$F_c = m\omega^2 r$$

Maximum velocity (km/h) before skidding   15.5

$$v = 3.6\sqrt{\mu g r}$$

Maximum velocity (km/h) before overturning   15.5

$$v = 3.6\sqrt{\frac{wgr}{2h}}$$

Angle of banking for expected speed of travel   15.6

$$\tan \theta = \frac{v^2}{gr}$$

Correct speed for a banked road   15.6

$$v = \sqrt{gr \tan \theta}$$

Mechanical work   16.1

$$W = FS \text{ (linear motion)}$$
$$W = T\theta \text{ (rotation)}$$

Power   16.2

$$P = \frac{W}{t}$$
$$P = Fv \text{ (linear motion)}$$
$$P = T\omega \text{ (rotation)}$$

Work and change of velocity   16.3

$$W = \frac{m}{2} (v^2 - v_0^2) \text{ (linear motion)}$$
$$W = \frac{I}{2} (\omega^2 - \omega_0^2) \text{ (rotation)}$$

Work done to compress a coil spring   16.4

$$W = \frac{kx^2}{2}$$

Potential energy   17.2

$$PE = mgh$$

Kinetic energy   17.3

$$KE = \frac{mv^2}{2} \text{ (linear motion)}$$

$$KE = \frac{I\omega^2}{2} \text{ (rotation)}$$

Strain energy   17.4

$$SE = \frac{kx^2}{2}$$

Transformation of potential and kinetic energy   18.1

$$PE_1 + KE_1 = PE_2 + KE_2$$

Transformation of energy involving springs   18.2

$$PE_1 + KE_1 + SE_1 = PE_2 + KE_2 + SE_2$$

Work–energy method   20.1

$$PE_1 + KE_1 \pm W = PE_2 + KE_2$$

Work–energy method involving springs   20.2

$$PE_1 + KE_1 + SE_1 \pm W = PE_2 + KE_2 + SE_2$$

Momentum   21.1

$$Momentum = mv$$

Impulse   21.2

$$Ft = mv - mv_0$$

Conservation of momentum during impact   21.3

$$m_A v_{0A} + m_B v_{0B} = m_A v_A + m_B v_B$$

Coefficient of restitution   21.3

$$\varepsilon(v_{0A} - v_{0B}) = (v_B - v_A)$$

Mechanical advantage   22.1

$$MA = \frac{F_L}{F_E}$$

Velocity ratio   22.1

$$VR = \frac{S_E}{S_L}$$

Efficiency of a simple machine   22.2

$$\eta = \frac{MA}{VR}$$

Friction effort   22.3

$$F_F = F_E - F_{Th}$$

The law of a machine   22.4

$$F_E = aF_L + b$$

Limiting efficiency   22.5

$$\eta = \frac{1}{aVR}$$

Velocity ratio of a Weston differential chain block   23.2

$$VR = \frac{2D_1}{D_1 - D_2}$$

Power transmitted by a rotating component   24.1

$$P = T\omega = \frac{2\pi NT}{60}$$

Drive efficiency   24.1

$$\eta = \frac{P_{out}}{P_{in}}$$

Torque transmitted by a belt drive   24.4

$$T = r(F_t - F_s)$$

Ratio of belt tensions

$$\frac{F_t}{F_s} = e^{\mu\theta} \text{ (flat belt)} \qquad\qquad 24.4$$

$$\frac{F_t}{F_s} = e^{\mu\theta/\sin\beta} \text{ (V-belt)} \qquad\qquad 24.5$$

$$e = 2.718$$

Ultimate strength   25.1

$$US = \frac{\text{force causing failure}}{\text{initial area}}$$

Direct axial stress   25.2

$$f = \frac{F}{A}$$

Factor of safety   25.3

$$FS = \frac{\text{ultimate strength}}{\text{working stress}}$$

Direct axial strain   25.4

$$e = \frac{x}{l}$$

Hooke's law and Young's modulus   25.5

$$E = \frac{f}{e} = \frac{Fl}{Ax}$$

Density   26.2

$$\rho = \frac{m}{V}$$

Poisson's ratio   26.4

$$v = \frac{\text{lateral strain}}{\text{axial strain}}$$

Thermal linear expansion   26.5

$$\Delta L = \alpha L_0 \, \Delta t$$

Stress due to completely prevented thermal expansion   26.6

$$f = E\alpha\Delta t$$

Shear stress   27.2

$$f_s = \frac{F_s}{A_s}$$

Shear strain   27.3

$$e_s = \frac{x_s}{l_s}$$

Modulus of rigidity   27.3

$$G = \frac{f_s}{e_s}$$
$$G = 0.4E$$

Modulus of rigidity and Poisson's ratio   27.3

$$G = \frac{E}{2(1 + v)}$$

Torsional stress   27.4

$$f_{ts} = \frac{Tr}{J}$$

Polar moment of inertia   27.4

$$J = \frac{\pi D^4}{32} \text{ (solid shaft)}$$

$$J = \frac{\pi}{32} (D_o{}^4 - D_i{}^4) \text{ (hollow shaft)}$$

Angle of twist   27.5

$$\theta = \frac{TL}{JG}$$

Centroid of plane area   29.1

$$\bar{y} = \frac{\Sigma(Ay)}{\Sigma(A)}$$

Bending stress in a beam   29.3

$$f_b = \frac{My}{I}$$

Radius of curvature of a beam   30.1

$$R = \frac{EI}{M}$$

Efficiency of bolted joints   31.2

$$\text{Joint efficiency} = \frac{\text{strength of joint}}{\text{strength of unpunched plate}}$$

Strength of fillet weld   31.3

$$F = flt = 0.707 \, fls$$

Stresses in cylindrical shells   32.1

$$f_H = \frac{pD}{2t} \text{ (hoop stress)}$$

$$f_A = \frac{pD}{4t} \text{ (axial stress)}$$

Stress in a spherical shell   32.2

$$f = \frac{pD}{4t}$$

Shaft diameter   33.2

$$D = \sqrt[3]{\frac{16T}{\pi f_{ts}}}$$

Glossary of selected terms*

Figures after each definition refer to the section or chapter in the book where the major reference to the concept may be found.

acceleration   a measure of the time rate of change in the velocity of a moving body (linear acceleration) or a rotating body (angular acceleration) [13.1 and 14.1]

angle   a measure of the inclination of one straight line to another expressed in radians, degrees or related units [3.4]

angle of friction   the angle between the resultant of the frictional and normal forces and the direction normal to the surfaces in contact at the moment of impending sliding motion [10.3]

angle of repose   the greatest angle to the horizontal which can be made by an inclined plane before an object resting on it would start sliding down the slope [10.4]

angle of twist   the angle through which one cross-section of a shaft rotates relative to another cross-section when a torque is applied [27.5]

area   a measure of the extent of a surface expressed in square metres or related units [3.5]

bending   a type of loading condition which tends to distort a component, such as a beam, from a straight into a curved shape [Chs 28–30]

bending moment   the amount of bending tendency at a cross-section of a beam measured by the summation of moments about the cross-section of all external forces on one side of the cross-section [28.4]

Bow's notation   a conventional method of identifying forces, used for systematic construction and interpretation of complex force diagrams, such as Maxwell's diagram [9.1]

Celsius scale   a practical scale of temperature measurement on which the temperature of melting ice is chosen to be zero degrees and the temperature of water boiling at atmospheric pressure to be 100 degrees, with equal unit divisions between, above and below these reference temperatures [3.4]

centrifugal force   the force acting in the radial direction away from the centre of curvature of the curved path of a body in circular motion [15.7]

centripetal acceleration   the radial component of the rate of change in the velocity of a body moving in a curved path [15.2]

centripetal force   the force acting on a moving body, directed towards the centre of its curved path, which is the cause of centripetal acceleration [15.3]

---

* The reader may find the selection of terms included in this glossary to be somewhat arbitrary, which is inevitable in a book of this size. Relative importance of a concept and frequency of its use in engineering science have been the major criteria for selection.

centroid   the point which is the geometrical centre of the area distribution of a plane area [29.1]

circular motion   motion of a body or a particle along a circular path [Ch. 15]

coefficient of friction   the ratio of the force of friction and the normal force between two surfaces in contact when sliding motion is about to occur [10.2]

coefficient of restitution   a measure of the ability of two bodies to regain their original shape after impact, defined in terms of relative velocities before and after the impact [21.3]

component of a force   the resolved part of a force in any particular direction. Any given force can be replaced by its two components in any two directions. Two mutually perpendicular components are the most frequently used [4.5]

compression   a type of loading condition which tends to shorten a component by pressing together the particles of its material, usually by axially directed pushing forces [Ch. 26]

concurrent forces   forces which intersect at a common point called the point of concurrence [Ch. 5]

conservation of energy   the law which states that in any isolated system, the total amount of energy is constant. It means that energy can be converted from one form into another, but cannot be created or destroyed [Ch. 18]

couple   a pair of forces having the same magnitude, parallel lines of action, and opposite sense. A couple produces a turning effect measured in newton metres, but its resultant force is zero [6.6]

degree   a unit of angular measure equal to 1/90 part of a right angle [3.4]

degree Celsius   a practical unit of temperature measurement, equal in magnitude to one kelvin, but used in conjunction with the Celsius scale of measurement with the origin corresponding to the melting temperature of ice [3.4]

density   the mass per unit volume of a substance, expressed in kilograms per cubic metre or related units [26.2]

displacement   a measure of the change in the position of a moving body (linear displacement) or the orientation of a rotating body (angular displacement) with respect to fixed coordinates [13.1 and 14.1]

dynamics   the branch of the mechanics of solids which deals with bodies in motion and with forces required to produce a given motion. Dynamics is divided into kinematics and kinetics [1.3]

efficiency   a measure of performance of a machine defined as the ratio of the energy output to the energy input [22.2]

elasticity   the ability of a material to return to its original size or shape after having been stretched, compressed or deformed [25.5]

energy   a physical quantity associated with a body or a substance which is a measure of its ability to do work or to release heat. Energy may be of several kinds: potential energy, kinetic energy, pressure energy, internal energy, chemical energy [17.1]

equilibrant   the force which when added to a system of forces that is not balanced will produce equilibrium. The equilibrant is a force equal and opposite to the resultant of the given system of forces [5.2]

equilibrium   the state of a body which is subjected to a system of forces that balance each other out. If a body is in equilibrium, the resultant force and the sum of the moments acting on the body are zero [5.1 and 6.4]

factor of safety   the ratio, allowed for in design, between the ultimate strength of a material and the safe working stress in a component or structure made from it [25.3]

force   any action, usually described as 'push' or 'pull' which tends to maintain the position of a body, to alter the position of a body, or to produce deformation in the size or shape of a body [4.1]

free-body diagram   a diagram showing a body subjected to a system of forces as a single object, often represented as a point, in isolation from other objects, and without unnecessary pictorial details [5.3]

friction   the resistance to sliding motion between two solid surfaces in contact with each other [Ch. 10]

friction effort   the part of total effort, applied to a machine, which is wasted in overcoming frictional resistance [22.3]

gravity   the force of mutual attraction between any two masses, proportional to the two masses and inversely proportional to the square of the distance between them [4.8]

Hooke's law   the law which states that in an elastic material, within its limit of proportionality, strain is proportional to stress [25.5]

hoop stress   the tensile stress in a longitudinal joint of a pipe or cylindrical container subjected to internal pressure [32.1]

impact   a collision between two bodies that occurs in a very short interval of time and involves relatively large forces which the two bodies exert on each other [21.3]

impulse   the product of a force and the time during which it acts [21.2]

inertia   reluctance of a body to change its state of rest or of uniform motion, which is the subject of Newton's first law of motion [13.4]

joint efficiency   a measure of effectiveness of a bolted or welded joint defined as the ratio of the strength of the joint to the strength of unpunched plate [31.2]

joule   the unit of energy, work or heat in SI, defined as the equivalent of work done by a force of one newton over a distance of one metre [16.1]

kelvin   the unit of absolute or thermodynamic temperature in SI, presently defined as equal to the fraction $1/273.16$ of the temperature of the triple point of water (the triple point of water corresponds approximately to the temperature of melting ice) [3.4]

kilogram   the unit of mass in SI, defined as the mass of the International Prototype Kilogram, i.e. the mass of a cylinder 39 mm in diameter and 39 mm high, made from an alloy containing 90 per cent platinum and 10 per cent iridium, kept at the International Bureau of Weights and Measures in Sèvres, France [3.4]

kilowatt hour   a unit of electrical energy equivalent to the work done in one hour by a device working at a constant rate of one kilowatt, i.e. at the rate of one kilojoule per second; 1 kW.h = 3.6 MJ [Appendix A]

kinematics   the part of dynamics which is concerned with the study of motion without reference to forces causing the motion [Ch. 13]

kinetic energy   a form of mechanical energy possessed by a solid body or by a quantity of fluid by virtue of its motion [17.3]

kinetics   the part of dynamics which is concerned with the relation between force and motion [Ch. 13]

law of a machine   a mathematical expression which represents the relation between load and effort applied to a simple machine [22.4]

length   a measure of distance between two points along a single straight or curved line, expressed in metres or related units [3.4]

linear motion   motion of a body or a particle along a linear path. Rectilinear motion is motion along a straight line and curvilinear motion is motion along a curve [Ch. 13]

litre   a unit of volume equal to one-thousandth part of a cubic metre [3.4]

mass   a measure of the quantity of matter in a body as evidenced by its inertia, expressed in kilograms, tonnes or related units [3.4]

mass moment of inertia   a geometrical property of mass distribution in a rigid body defined as the sum of all elementary products of mass elements and the squares of their respective distances from an axis [14.4 and 14.6]

Maxwell's diagram   a composite diagram, used for graphical analysis of internal forces in members of a truss, which combines separate force polygons for individual joints [9.4]

mechanical advantage   the ratio of the load to the effort in a simple machine [22.1]

mechanics   the study of the action of forces and of the conditions of rest or motion they produce. It is divided into mechanics of solids, mechanics of fluids and mechanics of machines [1.3]

mechanics of machines   the study of the relative motion between the parts of a machine and of the forces which act on those parts [Ch. 22]

metre   the unit of length in SI, presently defined as equal to exactly 1 650 763.73 wavelengths of the orange line in the spectrum of the krypton-86 atom in an electrical discharge [3.4]

modulus of elasticity   a measure of stiffness of a material, also known as Young's modulus of elasticity (*see* Young's modulus) [25.5]

modulus of rigidity   a measure of the ability of a material to resist deformation in shape, defined as the ratio of shear stress to shear strain, usually expressed in megapascals or related units [27.3]

moment of force   the product of the force and the perpendicular distance of its line of action from the reference point, expressed in newton metres or related units [6.1]

moment of inertia of an area   a geometrical property of area distribution, also called second moment of area, defined as the sum of all elementary products of area elements and the squares of their respective distances from the centroidal axis [29.2]

momentum   the product of the mass of a body and its linear velocity, sometimes described as the quantity of motion [21.1]

neutral plane   in a beam subjected to bending, the plane passing through the centroid of a cross-section at which stress due to bending is zero [29.3]

newton   the unit of force in SI, defined as the force which imparts an acceleration of one metre per second squared when applied to a body having a mass of one kilogram [4.2]

Newton's laws of motion   three fundamental laws of kinetics concerned with:
1. inertia of a body
2. relation between force, mass and acceleration
3. action and reaction forces [13.4]

pascal   the unit of stress and pressure in SI, defined as a force of one newton uniformly distributed over an area of one square metre [25.2]

Poisson's ratio   one of the elastic constants of a material, defined as the ratio of the lateral strain to the axial strain under direct tension or compression [26.4]

polygon of forces   a method of graphical addition of forces by drawing a polygon with sides parallel to the directions of the forces, with a head-to-tail sequence of directional arrows, and the lengths of the sides of the polygon representing the magnitude of the forces to some suitable scale [4.6]

potential energy   a form of mechanical energy possessed by a solid body or by a quantity of fluid due to its position in the gravitational field, i.e. due to its elevation above some datum level [17.2]

power   the time rate of doing work [16.2]

pressure   a measure of the intensity of normal distributed forces exerted on a surface, defined as force per unit area and expressed in pascals or related units [32.1]

radian   the unit of angular measure in SI, defined as the angle between two radii of a circle which cut off on the circumference an arc equal in length to the radius [3.4]

reactions   forces which exist between a structure, such as a beam or a truss, and its supporting surfaces. The magnitudes and directions of support reactions depend on the magnitudes and distribution of external loads and on the type of supports [5.4]

resultant force   the single force which can replace two or more forces acting on a body without changing the effect produced on the body. The resultant is a vector sum of the given forces [4.4]

rotation   the type of motion during which a body turns around a fixed axis in such a way that every particle of the body except the axis travels along a circular path [Ch. 14]

second   the unit of time in SI, presently defined as the time interval occupied by 9 192 631 770 cycles of a specified energy change in the caesium atom, as measured by the caesium atomic clock [3.4]

second moment of area   a geometrical property of area distribution, often called moment of inertia of the area, defined as the sum of all elementary products of area elements and the squares of their respective distances from the centroidal axis [29.2]

shear   a type of loading condition which tends to produce deformation in the shape of a component, in which the layers of its material move in parallel planes [27.1]

shear force   the amount of shearing tendency at a cross-section of a beam, measured by the summation of all external transverse forces on one side of the reference cross-section [28.1]

simple machines   elementary mechanical devices for overcoming a resistance at one point by the application of a force at some other point [Ch. 23]

speed   a measure of the distance travelled by a moving body per unit time without reference to the direction of the motion [13.1]

statics   the branch of the mechanics of solids which deals with bodies and structures at rest under the action of external forces in equilibrium [1.3 and 4.1]

strain   a relative measure of deformation in a material under load, defined in terms of change in dimensions (in tension or compression) or shape (in shear) of a component compared with its original unloaded condition [25.4, 26.3 and 27.3]

strain energy   energy stored in an elastic component, such as a coiled spring, when it is stretched or compressed from its free length [17.4]

strength of materials   the study of solid materials, structures and machine components with particular reference to internal forces and deformations produced by external loads [1.3]

stress a measure of the intensity of force distribution within the material of a structure or machine component, defined as force per unit area and expressed in pascals. Depending on conditions of loading, stresses may be classified as tensile, compressive, shear or bending stress [25.2, 26.2, 27.2 and 29.3]

superelevation the difference in height between the outer and inner rails on a banked railway curve [15.6]

Système International d'Unités the International System of Units, with the abbreviation SI, comprising coherent, decimally related units of measurement, adopted in 1970 as the basis of Australia's metric system [3.3]

temperature a measure of the degree of hotness or coldness of a substance with respect to a fixed scale, expressed in kelvins or in degrees Celsius [3.4]

tension a type of loading condition which tends to produce stretching in the material of a component, usually by the application of axially directed pulling forces [Ch. 25]

thermal stress stress in the material of a component caused by fully or partially prevented thermal expansion [26.6]

three-force principle a useful theorem in statics, which states that in the case of equilibrium under the action of three non-parallel forces, the lines of action of the three forces must intersect at a common point [5.5]

time a measure of the sequence of events taking place in the physical world, expressed in seconds or related units, such as minutes or hours [3.4]

tonne a unit of mass equal to one thousand kilograms [3.4]

torque the turning effort exerted by mechanical components, such as shafts, during continuous rotation, expressed in newton metres [14.3]

torsion a type of loading produced by the twisting action of torque, which tends to cause parallel cross-sections of material to rotate relative to each other [27.4]

tractive effort in railway engineering, the force exerted by a locomotive at the draw-bar. In general, the force required for the propulsion of a vehicle along a roadway or a track [13.5]

tractive resistance the sum of all frictional resistances, such as bearing friction and air resistance, expressed as a single force opposing the motion of a vehicle, often stated in units of force per unit mass of the vehicle [13.5]

triangle of forces a particular case of the polygon of forces, drawn for three forces in equilibrium at a point [4.6]

ultimate strength the highest load that can be applied to a material before fracture occurs, expressed as force per unit area. Depending on the conditions of loading, materials have different ultimate strengths in tension, compression and shear [25.1, 26.1 and 27.1]

unit of measurement an agreed-on part of a physical quantity, defined by reference to some arbitrary material standard or natural phenomenon and used as a standard of comparison in the process of measurement [3.2]

velocity a measure of the time rate of change in the position of a moving body (linear velocity) or in the orientation of a rotating body (angular velocity) [13.1 and 14.1]

velocity ratio the ratio of the distance moved by the effort on the input side of a simple machine to the distance moved by the load on the output side [22.1]

volume a measure of the amount of space occupied by an object or substance, expressed in cubic metres, litres or related units [3.4]

watt   the unit of power in SI, defined as the work done at the rate of one joule per second of time [16.2]

weight   the force of gravitational attraction exerted by the Earth on an object. On or near the surface of the Earth, the weight of any object is approximately equal to 9.81 newtons for every kilogram of its mass [4.9]

work   a form of energy transfer which occurs as a combined result of motion and effort. In linear motion, it is the product of the force applied to a body and the distance moved by the force. In rotational motion, it is the product of torque and angular displacement [16.1]

Young's modulus   a measure of the ability of a material to resist stretching, defined as the ratio of tensile stress in the material to corresponding strain, usually expressed in megapascals or related units [25.5]

Answers

Chapter 2

2.1 –33

2.2 11.5

2.3 $F = \dfrac{6EIy}{a^2(3L - a)}$

2.4 2.5

2.5 $x_1 = 5$, $x_2 = -7$

2.6 $x = 3$, $y = 6$

2.7 28.3m^3

2.8 91 mm

2.9 (a) 0.454 (b) 0.342 (c) 1.60

2.10 5.20°

2.11 2.55 m

2.12 4.41 m

Chapter 3

3.1 (a) 400 m (b) 0.053 kg (c) 0.0753 m
(d) 0.045 s (e) 0.357 kg (f) 80 kg
(g) 0.734×10^{-3} m (h) 0.54×10^9 s

3.2 (a) 12.3 km (b) 7.5 t (c) 79 ms
(d) 4.7 mm (e) 30 g (f) 85 t (g) 3 μm
(h) 4.7 mg

3.3 (a) 5×10^6 mm^2
(b) 0.75×10^{-3} m^2
(c) 0.663×10^{-3} L
(d) 0.5×10^{-3} m^3
(e) 1350 L
(f) 0.75×10^9 mm^3
(g) 0.632 m^3
(h) 47×10^6 mm^3

3.4 (a) 35.25° (b) 2.62 rad (c) 90°
(d) 0.048 rad (e) 143°14′ (f) 0.451 rad

3.5 (a) 4620 min (b) 2700 s
(c) 1 h 23 min 20 s (d) 402 s

3.6 0.263 m^2

3.7 354 mm × 354 mm

3.8 1.14 m^2, 81.7 L

3.9 35 rev/s

3.10 5.08 L/h

3.11 5 h

3.12 $33

3.13 2.4 kg

3.14 7160 L/h

Chapter 4

4.1 (a) 10 N
(b) 10 N
(c) 5 N, 5 N
(d) 5 N, 5 N
(e) 10 N, 10 N
(f) 10 N, 10 N
(g) 5 N, 5 N and 10 N
(h) 10 N, 5 N and 5 N

4.2 (a) 7 N →
(b) 1 N →
(c) 5 N ∠ 36.9°
(d) 6.08 N ∠ 25.3°
(e) 2.83 N ∠ 48.5°

4.3 96.6 kN

4.4 260 N and 150 N

4.5 *A:* 4 kN and 6.93 kN
B: –2.5 kN and 4.33 kN
C: 8.66 kN and 5 kN
D: 7.07 kN and –7.07 kN

4.6 470 N and 171 N

4.7 (a) 7.07 kN and 7.07 kN
(b) 9.66 kN and 2.59 kN

4.8 400 N and 693 N

4.9 (a) 5.39 kN ∠ 21.8°
(b) 330 N ∠ 33.3°

(c) 20.9 kN ⬈ 71.6°
(d) 0
(e) 11 kN ⬂ 17°
4.10 As for 4.9
4.11 3.52×10^{22} N
4.12 (a) 9.81 mN (b) 9.81 N (c) 9.81 kN
(d) 39.2 N (e) 0.49 N
4.13 22.6 kN
4.14 (a) 98.1 N (b) 98.1 N (c) 98.1 N
4.15 3.43 kN and 12.8 kN
4.16 12.8 kN at 72.4° to horizontal
4.17

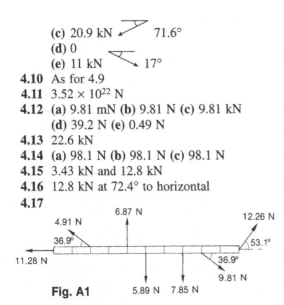

Fig. A1

Chapter 5

5.1 (a) 7.55 kN ⬈ 19.8°
(b) 49 kN ⬈ 52.1°
(c) 8.82 kN ⬈ 74.5°
(d) 0
5.2 13.2 N and 10.8 N
5.3 368 N
5.4 46.0 kN, 37.5 kN
5.5 2.76 kN, 4.51 kN, 2.86 kN, 4.64 kN
5.6 178 N ⬂ 65°, no
5.7 2.04 t, 0, 28.3 kN
5.8 $F_1 = 24.0$ kN, $F_2 = 20.8$ kN,
$F_3 = 20.8$ kN, $F_4 = 0$, $F_5 = 16.0$ kN,
$F_6 = 8.0$ kN, $F_7 = 16.0$ kN,
$F_8 = 8.0$ kN, $F_9 = 20.8$ kN,
$F_{10} = 8.0$ kN
5.9 192 N and 279 N ⬈ 20.6°
5.10 (a) 4.19 kN and 8.89 kN ⬈ 64.7°
(b) 16.8 kN and 15.4 kN ⬈ 10.2°
(c) 16.8 kN and 19.7 kN ⬈ 39.5°
(d) 25.2 kN and 24.5 kN ⬈ 21.5°
5.11 (a) 9.55 kN ⬈ 10.3° and 1.71 kN
(b) 8.31 kN ⬈ 22.8° and 3.21 kN

(c) 6.61 kN ⬈ 40.9° and 4.33 kN
5.12 406 N ⬂ 25° and
520 N ⬊ 45°
5.13 384 N and 348 N ⬈ 17°
5.14 123 N and 506 N ⬈ 76°
5.15 3.61 kN ⬈ 56.3° and 3 kN ↓
5.16 871 N ⬈ 73.3° and 250 N ←
5.17 649 N ⬈ 30° and
864 N ⬊ 49.5°
5.18 4.19 kN and 5.44 kN ⬊ 50.4°
5.19 59.5 N ⬂ 33.7°
5.20 19.2 kN ⬈ 66.8°, 0.8 m

Chapter 6

6.1 13.5 N.m, perpendicular to lever arm
6.2 75 N.m
6.3 (a) 3 kN.m ↻ and 3 kN.m ↻
(b) 1.5 kN.m ↻ and
4.5 kN.m ↻
(c) 3 kN.m ↻ and 3 kN.m ↻
(d) 3.5 kN.m ↻ and
2.5 kN.m ↻
(e) 2.6 kN.m ↻ and
2.6 kN.m ↻
6.4 17.3 kN.m ↻
6.5 (a) 28 kN.m ↻
(b) 1.32 kN.m ↻

6.6 2 kN

6.7 50 N

6.8 123 N

6.9 $\Sigma F_H = 0$, $\Sigma F_V = 0$, $\Sigma M = 0$

6.10 (a) 48 kN ↓, 4.13 m

(b) 550 kN ↓, 4.09 m

(c) 6.75 kN ↗ 79.5°, 3.81 m

(d) 4.53 kN ↘ 79.9°, 3.88 m

(e) 3.08 kN ↘ 80.5°, 3.68 m

6.11 848 N ↘ 79.8°, 4.47 m up along the ladder

6.12 666 N ↘ 11.9°, 73.2 mm

6.13 33.1 kN ↘ 65°, 4.05 m perpendicular distance

6.14 28 N.m

6.15 $\Sigma M = 0$

6.16 $\Sigma M = 0$

6.17 110 N

6.18 2 kN and 0.5 kN.m ↻

6.19 15 mm

6.20 1.3 kN and 1.95 kN.m ↻

6.21 49.1 N and 31.9 N.m ↻

(128 N couple through bolts)

Chapter 7

7.1 (a) $F_R = 9$ kN ↑, $F_L = 6$ kN ↑

(b) $F_R = 3$ kN ↑, $F_L = 3$ kN ↑

(c) $F_R = 3$ kN ↑, $F_L = 11$ kN ↑

(d) $F_R = 19.4$ kN ↑, $F_L = 21.6$ kN ↑

(e) $F_R = 2$ kN ↑, $F_L = 2$ kN ↓

(f) $F_R = 30$ kN ↑, $F_L = 10$ kN ↓

(g) $F_R = 5$ kN ↑, $F_L = 5$ kN ↓

(h) $F_R = 7.5$ kN ↑, $F_L = 2.5$ kN ↑

7.2 (a) $F = 2$ kN ↑, $M = 3.2$ kN.m ↻

(b) $F = 9$ kN ↑, $M = 16$ kN.m ↻

(c) $F = 3$ kN ↓, $M = 0$

(d) $F = 0$, $M = 12$ kN.m ↻

7.3 (a) $F_R = 12$ kN ↑, $F_L = 8$ kN ↑

(b) $F_R = 41$ kN ↑, $F_L = 15$ kN ↑

(c) $F = 18$ kN ↑, $M = 72$ kN.m ↻

7.4 $F_R = 5.02$ kN ↑, $F_{LV} = 3.35$ kN ↑, $F_{LH} = 3.47$ kN →

Chapter 8

8.1 A: $F_H = 1200$ N, $F_V = 0$

B: $F_H = 1200$ N, $F_V = 800$ N

C: $F_H = 0$, $F_V = 1600$ N

D: $F_H = 0$, $F_V = 1600$ N

E: $F_H = 1200$ N, $F_V = 1600$ N

8.2 A: $F_H = 57.7$ N, $F_V = 100$ N

B: $F_H = 57.7$ N, $F_V = 100$ N

C: $F_H = 57.7$ N, $F_V = 100$ N

8.3 A: $F_H = 12.0$ kN, $F_V = 14.7$ kN

B: $F_H = 12.0$ kN, $F_V = 4.9$ kN

C: $F_H = 12.0$ kN, $F_V = 14.7$ kN

8.4 A: $F_H = 1.25$ kN, $F_V = 3.13$ kN

B: $F_H = 1.25$ kN, $F_V = 1.88$ kN

C: $F_H = 1.25$ kN, $F_V = 0.21$ kN

Chapter 9

9.1 *AE* 2.5 kN compression

ED 2.0 kN tension

EF 0.5 kN tension

FC 2.0 kN tension

BF 2.5 kN compression

9.2 *AE* 10.0 kN compression

ED 14.1 kN tension

EF 10.0 kN compression

FD 10.0 kN tension

BG 10.0 kN compression

GF 0

CH 10.0 kN compression

HG 10.0 kN compression

HD 14.1 kN tension

9.3 *AD* 4.66 kN compression

DC 2.40 kN tension

AE 4.80 kN compression

ED 4.66 kN tension

EF 4.66 kN tension

FB 2.40 kN tension

AF 4.66 kN compression

9.4 *AF* 18.8 kN compression

FE 11.3 kN tension

AG 11.3 kN compression
GF 15.0 kN tension
GH 6.3 kN compression
HD 15.0 kN tension
HI 10.0 kN tension
IC 15.0 kN tension
AJ 11.3 kN compression
JI 6.3 kN compression
AK 18.8 kN compression
KJ 15.0 kN tension
KB 11.3 kN tension
9.5 *AE* 14.1 kN compression
ED 10.0 kN tension
AF 14.1 kN compression
FE 0
BG 14.1 kN compression
GF 4.0 kN tension
GH 5.66 kN compression
HD 14.0 kN tension
CH 19.8 kN compression
9.6 See 9.1 above.
9.7 See 9.2 above.
9.8 See 9.3 above.
9.9 See 9.4 above.
9.10 See 9.5 above.
9.11 F_1 = 8.0 kN compression
F_2 = 2.8 kN compression
F_3 = 10.0 kN tension
9.12 F_1 = 9.00 kN tension
F_2 = 5.83 kN compression
F_3 = 6.00 kN compression

Chapter 10

10.1 5.49 N
10.2 0.245
10.3 6.87 kN
10.4 0.5 kg
10.5 88.3 N
10.6 40.5 N
10.7 144 N.m
10.8 100 N
10.9 tip
10.10 31°
10.11 0.35, 19.3°, 106 N
10.12 124 N at 14° to normal
10.13 0.51, 17.5 N, 8.91 N
10.14 5.49 N
10.15 0.245
10.16 6.87 kN

10.17 0.5 kg
10.18 88.3 N
10.19 40.5 N

Chapter 11

11.1 878 N
11.2 8.49 kN
11.3 37.6 N
11.4 34.6 N
11.5 234 N, 340 N
11.6 225 kg
11.7 0.313
11.8 22.8 kg, 7.2 kg
11.9 1.67 kg, 0.6 kg

Chapter 12

12.1 9.4°
12.2 17.8 N.m
12.3 1.72 kN
12.4 11.1 N.m
12.5 yes (14° > 3°)
12.6 1.41 N.m
12.7 73 N
12.8 276 kg
12.9 128 N
12.10 10.7 kN
12.11 220 N
12.12 0.6
12.13 17.2°

Chapter 13

13.1 1.70 m/s
13.2 85 km/h, 23.6 m/s
13.3 75 km/h
13.4 2.22 m/s^2
13.5 450 m/s
13.6 5.63 km
13.7 1.07 m/s^2, 13 s
13.8 0.605 m/s^2, 25.7 s
13.9 no, 2.04 m/s
13.10 2 km, 58.8 km/h
13.11 11 s
13.12 22 s, 194 m
13.13 3.16 s, 31 m/s
13.14 62.4 m, 3.57 s
13.15 2.50 km, 4.54 km, 75.8 s

13.16 5 m/s^2
13.17 5 N
13.18 276 N
13.19 125 kN
13.20 420 N
13.21 126 s
13.22 145 kN
13.23 291 s, 3.23 km
13.24 24.6 kN
13.25 **(a)** 9.81 kN **(b)** 11.4 kN **(c)** 8.41 kN
(d) 8.81 kN
13.26 32.3 kN
13.27 **(a)** 23.6 N **(b)** 19.6 N **(c)** 15.6 N
13.28 20 N
13.29 4 s, 0.4 m/s
13.30 0.7 m/s^2
13.31 0.55 m/s^2, 4 s
13.32 35° > tan^{-1} 0.57, 2.17 m/s^2, 1.66 s
13.33 250 N

Chapter 14

14.1 **(a)** 490 rad **(b)** 557 rev **(c)** 68.8 rad/s
(d) 2320 rpm
14.2 144 rpm, 15.1 rad/s
14.3 Seconds hand: 1 rpm, 0.105 rad/s
Minutes hand: 1.67×10^{-2} rpm
1.75×10^{-3} rad/s
Hours hand: 1.39×10^{-3} rpm
1.45×10^{-4} rad/s
14.4 20 s, 3100 rad
14.5 6.28 rad/s^2, 288 rev
14.6 40 rad/s^2, 1310 rev
14.7 152 rad/s, 1450 rpm, 184 rev
14.8 11.9 min, 14 300 rev
14.9 63.6 rev
14.10 188 N.m
14.11 33.3 N
14.12 45 N.m
14.13 5.56 kN
14.14 5 rad/s^2
14.15 0.24 N.m
14.16 12.2 s, 153 rev
14.17 256 rad/s
14.18 187 N
14.19 36 s
14.20 **(a)** 7.44×10^{-3} kg.m^2
(b) 15.8×10^{-3} kg.m^2
14.21 1.67 kg.m^2

Chapter 15

15.1 0.349 rad, 244 mm
15.2 0.429 rad/s, 0.214 rad/s^2
15.3 25.1 m/s
15.4 7.59 m/s, 250 mm
15.5 3.69 rad/s^2, 55.4 rad/s, 415 rad
15.6 5.60 m/s^2
15.7 0.960 m/s^2
15.8 3.70 kN
15.9 1250 rpm
15.10 107 km/h
15.11 138 km/h
15.12 110 km/h
15.13 84 km/h, overturning
15.14 47°
15.15 140 km/h
15.16 3.18 kN radial in, 3.18 kN radial out
15.17 204 N
15.18 100 g
15.19 7.40 kN
15.20 7.81 kg

Chapter 16

16.1 240 MJ
16.2 294 kJ
16.3 7.85 kJ
16.4 56.5 kJ
16.5 11 kJ
16.6 65 kW
16.7 31.4 kW
16.8 921 kW
16.9 7.07 kW
16.10 115 N.m
16.11 10.3 kW
16.12 16.4 kW
16.13 200 kJ
16.14 59 N.m
16.15 28.3 kN, 236 kW
16.16 22.6 N.m, 711 W
16.17 25 N/mm
16.18 10 J, 1 kN

Chapter 17

17.1 790 kJ
17.2 226 kJ
17.3 1.35 m
17.4 55.7 MJ

17.5 −28 MJ
17.6 167 kJ, 296 kJ, 463 kJ
  (a) +130 kJ **(b)** +167 kJ
17.7 90 km/h
17.8 107 kJ
17.9 106 kg.m^2
17.10 +816 J
17.11 5 N/mm, 1.2 kN
17.12 74 J, 82 mm

Chapter 18

18.1 7.67 m/s
18.2 62.4 m
18.3 65.2 J, 4.32 m/s
18.4 48.9°
18.5 22.5 m/s
18.6 96.5 km/h
18.7 11 m/s
18.8 2.21 m/s
18.9 91 m
18.10 42.3 m/s, no
18.11 30 mm, 60 mm
18.12 4 m/s, 200 mm

Chapter 19

19.1 320 N, 160 N
19.2 0.866 m
19.3 0.377 m/s^2, 9.43 N
19.4 0.318 kg
19.5 11.2 kg
19.6 0.467 m/s^2
19.7 3.05 m/s^2
19.8 17.1 kg, 2.15 s
19.9 106 N, 0.946 m/s^2
19.10 0.765 m/s^2
19.11 1.55 kN
19.12 1.89 m/s^2, 10 m/s
19.13 1.84 kN.m
19.14 3.5 N.m

Chapter 20

20.1 0.32
20.2 81 km/h
20.3 2.3 m
20.4 859 N.m

20.5 30.4 N.m
20.6 1.76 m/s
20.7 1.1 kg
20.8 22.6 N
20.9 18.4 kg, 34.9 kg
20.10 699 N
20.11 33.3 m
20.12 1.5 m
20.13 349 mm
20.14 1.38 m
20.15 66.9 rpm

Chapter 21

21.1 23 300 kg.m/s
21.2 +7780 kg.m/s, −7780 kg.m/s
21.3 12.5 s
21.4 1.5 m/s, 250 kN
21.5 2.5 m/s, 3.33 m/s
21.6 0.0355 N
21.7 589 m/s, 31.1 kN
21.8 9.69 km/h
21.9 794 m/s
21.10 **(a)** 7 m/s **(b)** 2.27 m/s **(c)** 17.3 kN
21.11 25 m/s
21.12 0.5
21.13 6.44 m/s →
21.14 0.898
21.15 1.04 m/s →, 6.32 m/s →, 126 J
21.16 1.2 m/s ←, 7.6 m/s →, 0
21.17 4.4 m/s →, 197 J
21.18 0.849

Chapter 22

22.1 **(a)** 143 **(b)** 200 **(c)** 67.2 kJ, 94.0 kJ
  (d) 71.5% **(e)** 97.8 N
22.2 **(a)** **(i)**134, 67%
    (ii)146, 73%
  (b) 75%
22.3 $F_E = 0.0625F_L + 2$, 67.3%
22.4 **(a)** 20 **(b)** 98.1 N **(c)** 18 **(d)** 90%
  (e) 129 N
22.5 **(a)** 480 J **(b)** 408 J **(c)** 72 J **(d)** 85%
22.6 **(b)** $F_E = 0.0581F_L + 15$
  (c) 76% **(d)** 86.1%
22.7 **(a)** 64.8% **(b)** 79.5%

Chapter 23

23.1 7.5, 6, 5 N
23.2 59.3%
23.3 5.69, 81.3%, 9.36 N
23.4 5
23.5 577 N
23.6 75%
23.7 1.87 t, 40 rev
23.8 12.5, 270 N

Chapter 24

24.1 7.59 kW, 6.07 kW, 80%
24.2 4 rpm
24.3 2.43 t, 0.377 m/s
24.4 3, 5
24.5 19.2 N.m, 57.6 N.m, 640 N
24.6 4 kW
24.7 3
24.8 54.6 N.m, 6.86 kW
24.9 93.2 N.m, 11.7 kW
24.10 199 N.m, 497 N.m, 1.07 kN

Chapter 25

25.1 2.6:1, mild steel
25.2 130 N/mm^2
25.3 17 N/mm^2
25.4 0.9 mm
25.5 734 N
25.6 145 kN
25.7 12 mm
25.8 aluminium rod
25.9 235 MPa, 2
25.10 7
25.11 73.8 MPa
25.12 5 mm × 5 mm
25.13 0.3 × 10^{-3}
25.14 120 GPa
25.15 33.3 MPa, 6.67 mm
25.16 950 N
25.17 3 mm
25.18 83 MPa, 5, 2 mm
25.19 29.7 kN, 0.185 mm
25.20 20 mm × 20 mm
25.21 0.162 mm, 4.8

Chapter 26

26.1 32.8 kg, 321 N
26.2 31 N/mm^2
26.3 250 kN
26.4 5 MPa
26.5 401 kN
26.6 3 MPa, 0.75 mm
26.7 43.3 MPa, 0.618 × 10^{-3}, 0.155 mm
26.8 0.207 mm
26.9 0.32
26.10 79.5 mm
26.11 12.4%, 3.3%
26.12 $\dfrac{0.000\,024}{0.000\,012} = 2,\ \dfrac{0.000\,024}{0.000\,008} = 3$
26.13 8.1 mm
26.14 120 mm
26.15 4 mm
26.16 76.7°C
26.17 94.5 MPa, 90.9 kN
26.18 36.9 MPa, 35.5 kN
26.19 62.5°C
26.20 18 MPa
26.21 **(a)** 25 K **(b)** 31.5 K **(c)** 50 K

Chapter 27

27.1 35.3 kN
27.2 202 kN
27.3 6 mm
27.4 93.8 kN
27.5 90 MPa, 4
27.6 8
27.7 6 mm
27.8 24.5 kN
27.9 85.3 MPa
27.10 5.3 kN.m
27.11 8
27.12 30 kN, 200 MPa, 3.25
27.13 8 mm
27.14 **(a)** 28 000 MPa **(b)** 36 000 MPa **(c)** 48 000 MPa
27.15 0.41
27.16 4 mm, 0.24 MPa, 0.16
27.17 3.11 × 10^6 mm^4
27.18 13.7 × 10^3 mm^4
27.19 38 MPa
27.20 90 MPa
27.21 10°

27.22 13 mm

27.23 brass ($G = 35\,600$ MPa)

Chapter 28

28.1 Fig. A2 (shear force)
28.2 Fig. A3 (shear force)
28.3 Fig. A2 (bending moment)
28.4 Fig. A3 (bending moment)

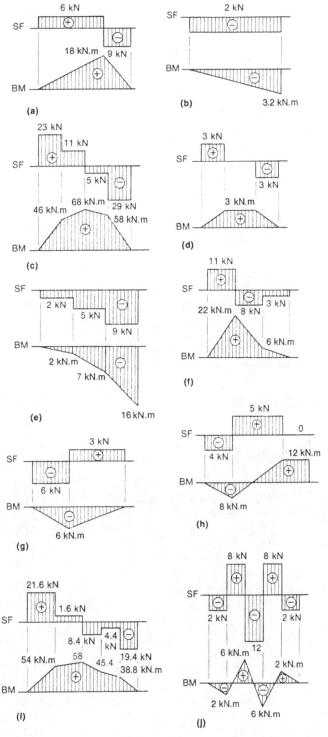

Fig. A2

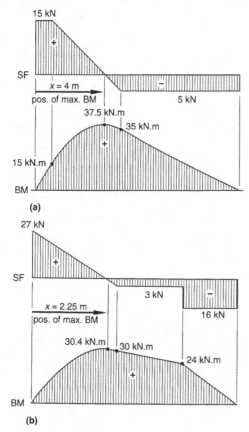

(a)

(b)

Fig. A3

Chapter 29

29.1 (a) $\bar{x} = 200$ mm, $\bar{y} = 264$ mm
(b) $\bar{x} = 15$ mm, $\bar{y} = 25$ mm
(c) $\bar{x} = 35$ mm, $\bar{y} = 21.7$ mm
(d) $\bar{x} = 11.4$ mm, $\bar{y} = 14.6$ mm
(e) $\bar{x} = 100$ mm, $\bar{y} = 216$ mm
(f) $\bar{x} = 250$ mm, $\bar{y} = 256$ mm

29.2 (a) 4.91×10^6 mm^4
(b) 1.08×10^6 mm^4
(c) 6.4×10^6 mm^4
(d) 4.17×10^6 mm^4

29.3 (a) 944×10^6 mm^4
(b) 268×10^3 mm^4
(c) 493×10^3 mm^4
(d) 85.7×10^3 mm^4
(e) 925×10^6 mm^4
(f) 2.88×10^9 mm^4

29.4 72 MPa

29.5 20.2 kN

29.6 100 mm

29.7 64 MPa

29.8 60 MPa, 80 MPa

29.9 13.3 kN.m

29.10 6.67 kN.m, 200 mm deep × 100 mm wide is better.

29.11 320 mm × 160 mm

Chapter 30

30.1 452 m

30.2 10 m

30.3 11 mm

30.4 7.77 mm

30.5 190 N/m

30.6 13 m

30.7 21 mm

30.8 19 mm

30.9 1.9 mm, 410 mm from the support nearest to the force

Chapter 31

31.1 50 MPa

31.2 80 MPa

31.3 206 MPa, no

31.4 63.6 kN, 57.8%

31.5 63.7 MPa, 31.3 MPa, 125 MPa

31.6 14.1 kN, 32.1%

31.7 (a) 6 mm (b) 2 mm (c) 18 mm

31.8 64.3%

31.9 24 mm, 16 mm, 8 mm

31.10 46.3%

31.11

4	6	8	10	12	16	20	24
2.8	4.2	5.7	7.1	8.5	11.3	14.1	17.0
387	580	773	966	1160	1550	1930	2320

31.12 97 mm

31.13 1160 kN

31.14 6 mm

31.15 45 mm

31.16 23.6 t

31.17 85 mm, 174 mm

Chapter 32

32.1 30 MPa
32.2 2 mm
32.3 36 MPa, 18 MPa
32.4 4 mm
32.5 5 mm
32.6 1.89 MPa
32.7 1.51 m^3
32.8 8
32.9 228 kPa
32.10 24 mm

Chapter 33

33.1 5.73 kN.m
33.2 30 MPa
33.3 227 kW
33.4 63.3 mm
33.5 20 mm
33.6 T_{AB} = 0.53 kN.m
T_{BC} = 1.19 kN.m
T_{CD} = 1.46 kN.m
T_{DE} = 0.86 kN.m
33.7 37 mm, 24 mm

Index

Bold figures indicate references to the Glossary of selected terms, which contains general definitions of the concepts.